INSECT TIMING:
CIRCADIAN
RHYTHMICITY TO
SEASONALITY

INSECT TIMING:
CIRCADIAN
RHYTHMICITY TO
SEASONALITY

Edited by:

D.L. Denlinger
J.M. Giebultowicz
D.S. Saunders

2001

ELSEVIER

Amsterdam · London · New York · Oxford · Paris · Shannon · Tokyo

ELSEVIER SCIENCE B.V.
Sara Burgerhartstraat 25
P.O. Box 211, 1000 AE Amsterdam, The Netherlands

First edition 2001

Library of Congress Cataloging in Publication Data
A catalog record from the Library of Congress has been applied for.

ISBN: 0 444 50608 X

♾ The paper used in this publication meets the requirements of ANSI/NISO Z39.48-1992 (Permanence of Paper).
Printed in The Netherlands.

Preface

The time of day and season of the year are among the most profound environmental factors dictating daily and seasonal patterns of insect activity. We hope that the analysis of insect timing presented in this book will prove helpful in portraying recent developments in this exciting field of research. We are grateful to the organizers of the XXI International Congress of Entomology, held in Iguassu Falls, Brazil, August 20-26, 2000, for hosting three interrelated symposia that formed the basis for this publication. The three symposia included "Circadian Clocks in Insects: Molecular and Cellular Perspectives" organized by J. M. Giebultowicz and M. D. Marques, "Photoperiodic Induction of Diapause and Seasonal Morphs" organized by D. S. Saunders and H. Numata, and "New Complexities in the Regulation of Insect Diapause and Cold Hardiness" organized by D. L. Denlinger and O. Yamashita. Chapters in the book thus range from discussions on the molecular regulation of circadian rhythms to physiological events associated with successful overwintering strategies. We appreciate the enthusiasm of the authors for this project and their timely submission of manuscripts.

We are grateful to Sandra Migchielsen and Ken Plaxton from Elsevier for their encouragement and advice, and to Irene Lenhart for her assistance in compiling the index.

D.L. Denlinger
J.M. Giebultowicz
D.S. Saunders

Foreword

James W. Truman

Department of Zoology, University of Washington, Box 351800, Seattle, WA 98195
USA

Insects, like other organisms, have evolved a diverse array of physiological and behavioral mechanisms to allow them to cope with both biotic and abiotic challenges. These mechanisms, though, are not used simply in a reactive mode that only rises to meet challenges once they are presented.   An intriguing aspect of life is that organisms continually make predictions about the future and adjust their physiology and behavior to anticipate the changes that will occur in the world around them. One class of predictive mechanisms used by insects to deal with their complex biological world involves learning processes that allow cause and effect associations to be made based on experience. To make predictions about their abiotic world, insects use mechanisms that take advantage of predictable aspects of the physical world, notably the geophysical rhythms of the daily rotation of the earth about its axis, the orbiting of the moon around the earth, and the seasonal journey of the earth around the sun. These timekeeping systems are not essential for life per se, as long as life occurs in the predictable environment of the laboratory. In the natural world, though, these timing systems fine-tune the responses of the animal to optimize its performance since physiological or behavioral mechanisms can gear-up in anticipation of expected changes in the world, rather than always being in a reactive mode.

The mechanisms that insects have used to deal with the daily and seasonal fluctuations of their world have fascinated physiologists and behaviorists through the latter half of the 20th century. Research by pioneers, such as C.S. Pittendrigh, C.M.Williams, A.D. Lees, S. Beck, J. deWilde, and A.S. Danilevskii, has illustrated the diversity and richness of phenomena relating to the adaptation of insects to daily and seasonal cycles. These researchers also showed that the study of these processes could produce fundamental insights into diverse areas of insect physiology, ecology and behavior.

This book brings together three topics that are central to the understanding of biological timing in insects: circadian rhythms, photoperiodism and diapause. The study of circadian rhythms has undergone a flowering in recent years with the molecular dissection of the components of the circadian clock. Research on diverse systems ranging from plants and fungi to insects and vertebrates shows that the molecular strategy for constructing a molecular clock is quite conserved as are some of the molecules that compose the clock. The molecular dissection of circadian clocks has open up a number of exciting new areas. One is aimed at a deeper understanding of the manner by which entraining signals interact with the molecular oscillation that comprises the workings of the clock. In addition, the issue of how the molecular oscillation is coupled with cellular outputs is a fascinating question that is just beginning to be addressed.   A second area involves a comparative approach in which the identification of clock-related genes in Drosophila has provided the molecular tools to bootstrap into the clock molecules of a diversity of insects. On the physiological side, perhaps the most exciting results come using the localization of clock proteins and their messenger RNAs to establish the cellular locations of putative clocks. It is now clear that one can no longer talk of a "master clock" that is responsible for the complete temporal organization of the animal. Organisms are more like "clock shops" with numerous cellular oscillators. In this context, though, not all clocks are created equal and we need to understand the nature and logic of the timing hierarchies. Since selected insect tissues can retain their rhythmical activities under organ culture conditions, insects are providing premier systems for studying how the hierarchical organization of clocks results in the overall temporal organization of the animal. The area of insect photoperiodism has not yet reached the level of molecular understanding that is seen for circadian rhythms. Nevertheless, photoperiodic phenomena continue to provide a fertile ground for the physiologist, ecologist and evolutionary biologist. Perhaps more than any trait, photoperiodism represents the "stamp" of the planet on a population of organisms.

In insects, the most obvious manifestation of photoperiodism is diapause. Here too, there is a theme of both unity and diversity. The spectrum of diapausing stages in insects, ranging from the egg to the adult opens up a spectrum of endocrine controls, a few of which are still being resolved. The focus of the field, though, is towards the cellular and molecular changes that occur as an insect makes the transition into and out of diapause. ·

Overall, the volume presents the rich diversity of challenges and opportunities provided by insects for the study of how life has adapted to the rhythms of the planet. It provides a good platform for the movement of the field into the new century.

TABLE OF CONTENTS

The blow fly *Calliphora vicina:* a "clock-work" insect

D. S. Saunders

Institute of Cell, Animal and Population Biology, University of Edinburgh, West Mains Road, Edinburgh EH9 3JT, Scotland, United Kingdom

The blow fly *Calliphora vicina* is used to illustrate the basic properties of the circadian system and photoperiodic induction of seasonality. In the adult female fly, a daily rhythm of locomotor activity and the (maternal) photoperiodic induction of larval diapause operate simultaneously but show features suggesting that, although both are circadian based, they are regulated by separate 'clocks' with dissimilar properties. The locomotor activity rhythm free-runs in darkness with a persistent, temperature compensated period ($\tau \sim 22.5$ hours); the circadian oscillations underlying photoperiodic time measurement, however, show a period closer to 24 hours and evidence of rapid damping. Although both clocks are brain-centred and may be entrained by extra-optic photoreceptors, they are probably located in separate sets of neurons. Some problems in circadian and photoperiodic biology are enumerated.

1. INTRODUCTION

When surveying the remarkable clock-like properties of circadian rhythms it becomes understandable why early biologists found them, at best, extraordinary. However, after more than 50 years of intensive investigation - together with a modern knowledge of genetics and evolutionary biology - we now know that endogenous circadian rhythms are the products of natural selection. Circadian rhythms were probably 'invented' by early photosynthetic procaryotes, and evolved in the face of the pervasive environmental fluctuations of light and temperature associated with the rotation of the earth around its axis. They may now be seen in organisms from cyanobacteria to 'higher' vertebrates; in eucaryotes they are probably almost universal. Other endogenous rhythms have also evolved: for example, in inter-tidal organisms, or in those exposed to the sometimes extreme annual fluctuations in day length, temperature and humidity associated with the earth's rotation around the sun. Seasonal photoperiodism, one such case, is probably one of the most important functions of the circadian system in higher organisms.

The circadian system presents a number of defining properties. These are: (1) 'daily' rhythms that persist ('free-run') in the absence of solar day signals (thereby attesting to their endogeneity). (2) a free-running rhythm (period τ hours) which is close to but rarely equal to 24 hours. (3) a well-developed homeostasis of τ when exposed to a wide range of physical, chemical and biological variables, most notably temperature (i.e. they are temperature-compensated). And (4) an ability to synchronise (entrain) to solar day variables, particularly light and temperature, so that τ becomes equal to the period of the entraining agent (or *Zeitgeber*), and adopts an adaptive phase relationship to it. Photoperiodism, itself based on the circadian system, also shows - in a covert manner - these defining properties. It provides the organism with a 'clock' which discriminates between the long days (or short nights) of summer and the short days (or long nights) of autumn and winter, thereby obtaining information on calendar time from the environment. A wide range of organisms, principally higher plants and animals living in terrestrial habitats at higher latitudes, use such 'noise-free' information to control seasonally appropriate switches in metabolism (e.g. flowering, diapause, breeding, migration), most of which have a clear functional significance or survival value.

The purpose of this chapter is to present these defining properties in an insect that presents robust circadian rhythms (of locomotor activity) *and* photoperiodic regulation (of diapause), with both clocks operating simultaneously. The insect of choice is the blow fly or bluebottle, *Calliphora vicina*. It is intended that this chapter should act as an introduction to the papers in this volume addressing current problems in circadian rhythms, photoperiodism, diapause and insect seasonality.

2. LOCOMOTOR RHYTHMICITY

2.1. Free-running locomotor rhythmicity in continuous conditions

Adult blow flies show typically robust rhythms of locomotor activity in continuous darkness (DD). The record shown in Fig. 1, and others like it, was obtained by placing a single female fly in a 9 cm Petri dish with a supply of sugar and water. The dish was then positioned in a vertical infra-red light beam in such a way that the moving fly interrupted the light. Interruptions of the light beam were recorded by computer and later assembled into the familiar 'double-plotted' actogram. The records obtained are typical for circadian activity rhythms in individual insects (Saunders, 1982) and will be described only briefly in this paper.

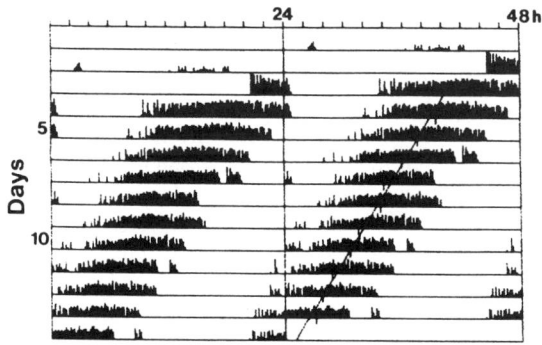

Figure 1. Locomotor activity rhythm of an adult blow fly (*Calliphora vicina*) in darkness (DD) at 20°C (double plotted actogram), showing a free-running rhythm ($\tau = 22.63$ h).

In continuous darkness at 20°C, flies showed a persistent, 'noise-free' and 'free-running' rhythm of activity and rest, with a period (τ) generally less than 24 hours. Among groups of several hundred flies, the mean value of τ was found to be about 22.5 h (Cymborowski et al., 1993; Hong and Saunders, 1998), although the range was from little over 21 h, to the longest value, just over 25 h (Fig. 2). Persistence of this rhythm for the life time of the fly (up to 7 weeks) and, of course, departure of τ from exactly 24 hours, provided cogent evidence for the endogeneity of the rhythmic mechanism underlying such activity.

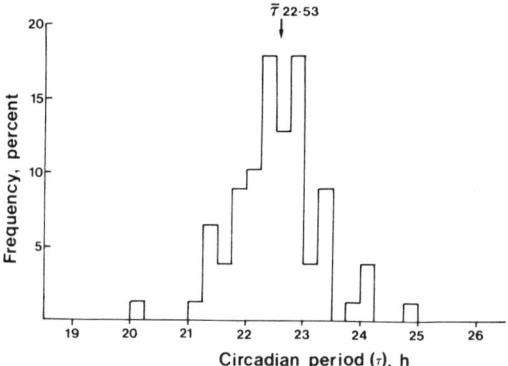

Figure 2. Distribution of free-running periods (τ values) for the locomotor activity rhythm of adult *Calliphora vicina* in DD at 20°C. Mean value of τ = 22.53 ± 0.78 h (N = 78).

Rhythms of locomotor activity also continue under continuous illumination (LL), at least at low levels of irradiance. At low intensity (up to about 0.018 Wm^{-2}) rhythmicity persisted but τ lengthened systematically, up to about 26 hours. Above about 0.033 Wm^{-2} (Fig. 3) flies

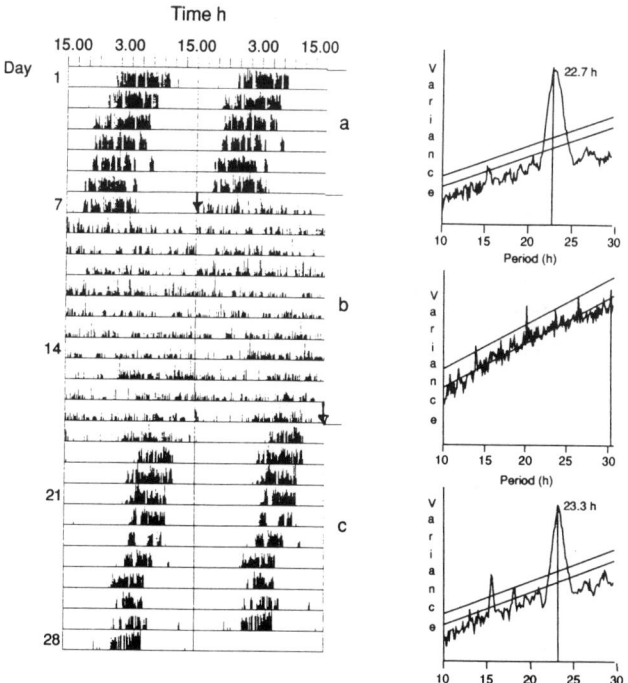

Figure 3. Locomotor activity rhythm of *C. vicina:* (a) DD, (b) behavioural arrhythmicity under LL at 0.33 Wm^{-2}, (c) DD, all at 20°C. Arrows show times of transfer from DD to LL and *vice versa.*

4

became behaviourally arrhythmic, although a free-running rhythm reappeared after the light was extinguished (Hong and Saunders, 1994). Under an intermediate irradiance (about 0.024 Wm^{-2}) flies were initially arrhythmic, but then assumed a long-period rhythm, perhaps after light adaptation of the photoreceptors.

The majority (82.9 per cent) of flies under continuous darkness showed simple (unimodal) free-running rhythms of activity; the remainder showed more complex patterns. In some (7.3 per cent), spontaneous and abrupt changes in τ occurred, either to a longer or to a shorter value. About 5 per cent were behaviourally arrhythmic, and a further 5 per cent showed complex patterns with either bimodality or internal desynchronisation into two or more 'bands' of activity, free-running with different values of τ (Kenny and Saunders, 1991; Hong and Saunders, 1998). The significance of these complex patterns, and others, will be considered below (Section 4).

2.2. Temperature compensation of the circadian period

At constant temperatures between 15 and 25°C the free-running period in DD (τ$_{DD}$) was temperature compensated, showing Q$_{10}$ values between 0.98 and 1.04 (Saunders and Hong, 2000). Single temperature steps-up (from 20 to 25°C) or steps-down (from 20 to 15°C), however, caused stable phase shifts of the activity rhythm (Fig. 4). Phase advances were dominant for steps-up and phase delays for steps-down. Phase response curves for the two types of signal were almost 'mirror images' of each other. Entrainment by temperature (and light) will be considered below.

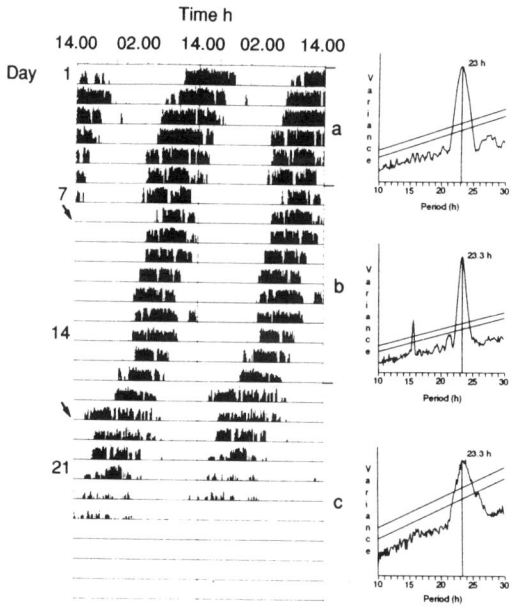

Figure 4. Locomotor activity rhythm of a female *C. vicina* under (a) 20°C, (b) 15°C and (c) 20°C, all in DD, showing phase shifts but no change in period.

2.3. Entrainment by light and temperature

Transfer of flies from continuous darkness to a light cycle leads to entrainment of the activity rhythm. The process of entrainment involves a change of τ from its free-running value in darkness (τ_{DD}) to the period of the light-dark cycle or *Zeitgeber*; in the example shown (Fig. 5) to the exact 24 hours of LD 16:8. After discontinuation of the light pulses, τ returns to a value close to τ_{DD} at the outset. When entrainment has occurred, a steady state phase relationship (ψ) to the *Zeitgeber* is achieved with ψ-values depending, among other things, on the relative periods of the endogenous rhythm and the environmental light cycle (Pittendrigh, 1981; Saunders, 1982). In the case of the blow fly, which is a diurnal insect, the clock-controlled locomotor activity becomes confined, in this example, to the 16 hours of light. The small amount of activity immediately following light-off (arrowed) is an exogenous 'masking' or 'rebound' effect caused by the abrupt ending of the light.

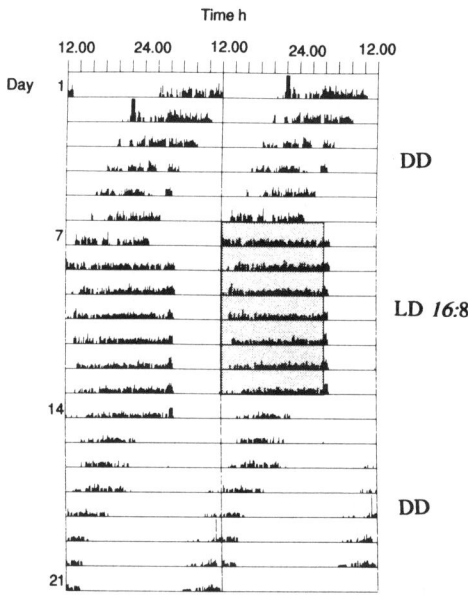

Figure 5. Entrainment of the locomotor activity rhythm of a female *C. vicina*, initially in DD at 20°C, to a light-dark cycle (LD 16:8), with a final return to DD; the free-running rhythm adopts the period of the light cycle (24 hours) during the period of illumination.

The 'mechanics' of entrainment may be explained by summing the phase shifts (advances and delays) caused by consecutive light pulses until the system achieves its steady state (Pittendrigh, 1981; Saunders, 1982). The key to this process lies in families of phase response curves (PRCs) constructed from the phase-shifting effects of single pulses of light of different duration and irradiance applied, at sequentially later circadian phases, to an otherwise free-running rhythm of activity. In short, such PRCs generally show phase delays ($-\Delta\phi$) when the light pulses fall during the early subjective night (circadian time, Ct 12 to 18), and phase advances ($+\Delta\phi$) during the late subjective night (Ct 18 to 24). On the other hand, rather small (or no) phase shifts occur during the subjective day (Ct 0 to 12), often giving rise to a so-called 'dead zone'. Light pulses of increased irradiance or duration cause larger phase shifts allowing the rhythm to entrain to a wider range of *Zeitgeber* periods, and to achieve greater values of ψ. Short or weak light pulses thus give rise to low 'amplitude' (Type 1; Winfree, 1970) PRCs with small delays and advances (see Fig. 6 for *C. vicina* exposed to 1 h pulses of light). Longer or brighter light pulses may give rise to high 'amplitude' Type 0 PRCs with delays and advances approaching one half of the circadian cycle.

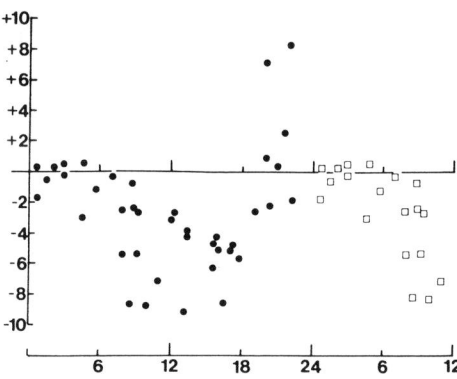

Figure 6. Phase response curve (PRC) for the locomotor activity rhythm of *C. vicina* exposed to 1 hour pulses of white light, showing phase delays (-Δϕ in hours) and phase advances (+Δϕ) as a function of circadian phase (in hours)(y axis). Data are double-plotted (● and □).

Pittendrigh's entrainment model may also be used to calculate ranges of entrainment to *Zeitgeber* cycles (T) that are multiples (i.e. T ~ 48, 72 or 96 hours) or sub-multiples (T ~ 12 hours) of τ. Such calculations have proved to be crucial to unravelling entrainment to Nanda-Hamner protocols (see below), important in the analysis of the role of circadian oscillations in photoperiodic time measurement (Saunders, 1978).

Insect circadian rhythms also entrain to temperature cycles, or thermoperiods. However, unlike light, temperature cycles differ in that the separate phase shifting effects of temperature steps-up and steps-down may be studied. This cannot be done with light because light-on automatically leads to a period of higher light intensity, which frequently causes arrhythmicity (Hong and Saunders, 1994), in which 'phase' has no clear meaning. Using protocols established by Zimmerman et al. (1968) for the rhythm of pupal eclosion in *Drosophila pseudoobscura,* Saunders and Hong (2000) used separate temperature steps-up (15 to 20°C) and down (20 to 15°C) to simulate phase shifting by a low temperature *pulse* lasting 6 hours. Computation of a theoretical PRC for 6-hour low temperature pulses, led to a prediction of steady state entrainment of the activity rhythm to a thermoperiodic *cycle* of 18 h at 20° and 6 h at 15°C. This phase relationship showed that activity was confined to the daily thermophase, as expected for a diurnally active insect.

The basic properties of the circadian system, as outlined above, have proved to be essential in understanding time measurement in the photoperiodic clock.

3. PHOTOPERIODIC TIMING

3.1. The photoperiodic response

In common with many other insects inhabiting northern latitudes, the blow fly responds to seasonal changes in day length by producing a dormant or diapausing stage as winter approaches. In *C. vicina,* the diapause stage is the third instar larva, but the stage sensitive to photoperiodic change is maternal. Thus, female flies exposed to the long days (or short nights) of summer lay eggs giving rise to larvae that develop without arrest. On the other hand, those experiencing autumnal short days (or long nights) lay eggs giving rise to larvae that enter diapause as post-feeding larvae, after burrowing into the soil, but before puparium formation (Saunders, 1987). Flies may be diverted from the diapause 'pathway', however, by temperatures in the larval habitat above about 15°C (Vaz Nunes and Saunders, 1989) or by larval over-crowding (Saunders, 1997).

Many insects entering a winter diapause acquire a range of physiological and behavioural characteristics referred to as the 'diapause syndrome'. This may include reduced metabolism, an increased storage of metabolites, especially fat, augmentation of epicuticular waxes, and an increased cold tolerance. In *C. vicina,* there is no significant increase in lipid storage (Saunders, 1997, 2000a), but they do acquire an increased cold tolerance - particularly in strains from northern latitudes - which might be an integral part of the diapause syndrome (Saunders and Hayward, 1998). The immediate cause of diapause is an alteration to the endocrine control of moulting and metamorphosis (Denlinger, 1985; Saunders, 2000b). In the case of *C. vicina* it involves a 'block' to the release of the neuropeptide prothoracicotrophic hormone (PTTH) from the brain and an associated low titre of haemolymph ecdysteroids (Richard and Saunders, 1987). It is clear, however, that diapause is not merely a cessation of activity: it is an actively induced and alternative developmental pathway regulated by the expression of a set of characteristic diapause-related genes in the brain (Denlinger et al., 1995).

Since the induction of larval diapause in *C. vicina* is of maternal origin, it is clear that the photoperiodic clock and the clock controlling the overt rhythm of locomotor activity (see above) are operating concurrently in the female fly. This raises an interesting question: are the two clocks 'the same' or 'different'? This question will be addressed in a later section.

3.2. The photoperiodic response curve

The photoperiodic responses of an insect such as *C. vicina* are best described by a photoperiodic response curve (PPRC) constructed by exposing a series of groups of the insect to stationary light cycles during their sensitive period. The PPRC then plots diapause incidence as a function of photophase. A typical curve for *C. vicina,* like most 'long day' or summer active insects, shows low diapause incidence under long days, a high incidence under short days, and a sharp discontinuity (the critical day or night length) between the two (Fig. 7). The ecologically important critical day length (CDL) separates diapause-inducing from diapause-averting photoperiods. CDL is also a function of latitude, more northerly populations having a longer CDL than populations to the south. Low temperature may affect the PPRC by raising diapause incidence under strong short days, whereas high temperatures may lower it, although different constant temperatures may have rather little affect on the value of the CDL. On the other hand, *thermoperiod* may have diapause inducing/averting affects in a similar fashion to photoperiod (Saunders, 1982).

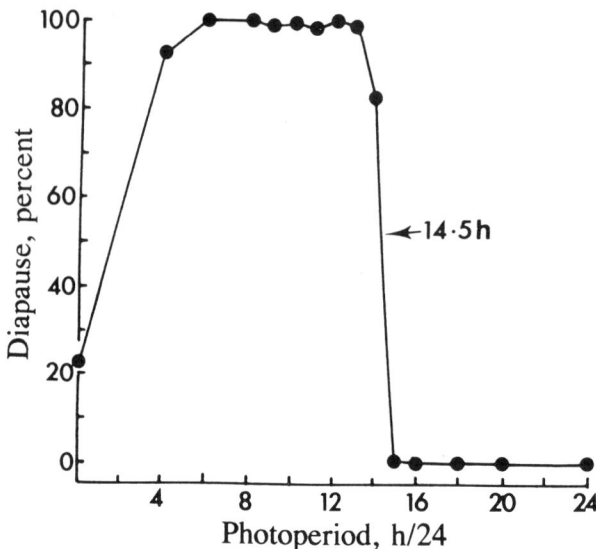

Figure 7. Photoperiodic response curve (PPRC) for the production of diapause larvae by females of *C. vicina* exposed to a range of photoperiods at 20°C. The critical day length (50% response) is at LD 14.5:8.5.

3.3. The circadian basis of photoperiodic time measurement

Erwin Bünning (1936) was the first to suggest that seasonal photoperiodic timing had its origin in the circadian system. Twenty years later, Colin Pittendrigh (1960, 1972) developed these ideas into a number of more specific models based on circadian entrainment. The literature and the evidence for this apparently causal association has been reviewed many times and will not be repeated here. For the early literature, see Saunders (1982); Vaz Nunes and Saunders (1999) provide a more recent review.

The covert operation of the circadian system in photoperiodic time measurement (PPTM) is revealed by several experimental procedures, the most widely used being the Nanda-Hamner protocol. In this type of experiment, insects are exposed to a series of non-24 hour light-dark cycles in which the light phase is held constant but darkness systematically extended through several multiples of τ (e.g. L = 12 hours; D = 8 to 72 hours). In many insects, including *C. vicina* (Fig. 8), diapause incidence rises and falls with a circadian periodicity. It is high when the overall light-dark cycle (T) is close to 24, 48 or 72 hours (i.e. when T = modulo τ) but low when T is close to 36, 60 or 84 hours (T = modulo τ + ½τ). The interval between peaks reflects the circadian period (τ). Other procedures, which will not be reviewed here, also reveal a covert rhythmicity in PPTM. These include (1) interruption of the extended night of a long cycle (e.g. LD 12:60) by systematically later short pulses of light (the Bünsow protocol), and (2) the powerful use of skeleton photoperiods in the 'zone of bistability' (Pittendrigh, 1966; Saunders, 1978; Vaz Nunes and Saunders, 1999).

These experiments demonstrate that circadian rhythmicity is involved in PPTM, but tell us little about the exact nature of that involvement. This uncertainty has given rise to a number of models for time measurement, details of which are given by Vaz Nunes and Saunders (1999).

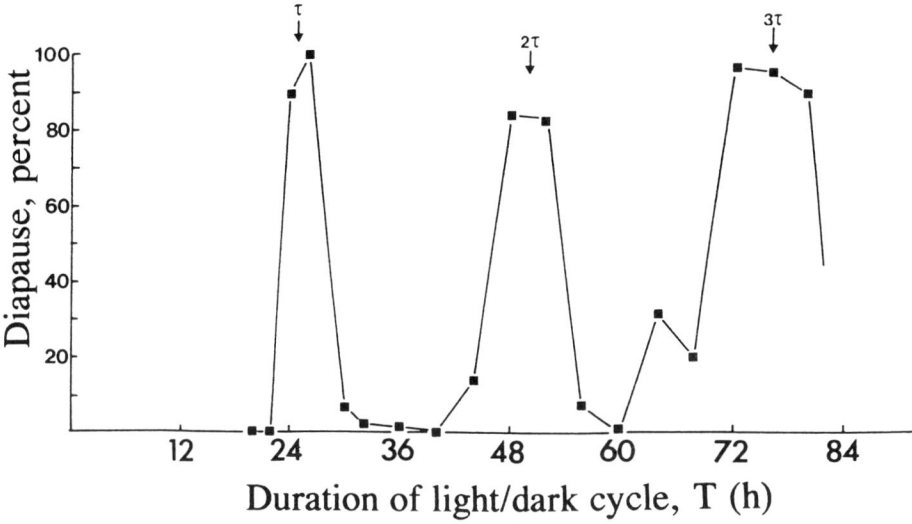

Figure 8. Nanda-Hamner experiment. Production of diapausing larvae by females of *C. vicina* exposed (at 20°C) to a range of light cycles each containing a 13 hour light component and a variable (7 to 67 hour) scotophase to give LD cycles (T) between 20 and 80 hours. Peaks of high diapause incidence occurred at about 24 hour intervals equivalent to multiples (τ, 2τ and 3τ) of the circadian period.

The 'double circadian oscillator' model (Vaz Nunes, 1998), with two oscillators separately measuring long and short nights, currently offers a good simulation of the phenomenon. 'Positive' Nanda-Hamner responses have now been recorded in about 12 species of insects and a mite. 'Negative' responses, lacking the obvious periodicity in diapause response, have however been recorded in a further 11 species. These apparently negative responses are frequently identified as evidence for a non-circadian or 'hourglass' type of clock. However, Lewis and Saunders (1987) and Vaz Nunes and Saunders (1999) conclude that hourglass-like and oscillatory timers are *both* based on the circadian system, the former showing heavy damping in extended periods of darkness and the latter more persistence. Even the 'classical' example of an hourglass type of photoperiodic clock – that in the aphid *Megoura viciae* (Lees, 1973) – has now been shown to have a circadian basis (Vaz Nunes and Hardie, 1993).

4. THE MULTIOSCILLATOR CIRCADIAN SYSTEM

Circadian rhythmicity is essentially a cellular phenomenon. Nevertheless, apparently independent or semi-independent, light entrainable circadian clocks exist at all levels of organisation from cells, through tissues to organs. Such systems have been described, *inter alia*, in nervous tissue, endocrine glands, gonads, malpighian tubules and epidermis (see Giebultowicz, 1999). Some of these oscillatory components may be truly independent; others may be part of a physiological hierarchy. There seems little doubt, however, that an insect represents a multioscillator circadian 'system'.

Even within the central nervous system that – unsurprisingly perhaps - houses circadian 'clocks' responsible for overt behavioural rhythms, pacemakers are of a multioscillator construction. Cells responsible for regulating locomotor rhythmicity in the fruit fly *Drosophila melanogaster*, for example, have been identified as the so-called lateral neurons (LNs) in the central brain (Helfrich-Förster et al., 1998). In the mutant *disconnected* most flies lack these crucial LNs and are totally arrhythmic. However, flies retaining just a single such neuron may continue to express a rhythm. Possibly homologous cells, that may also be photoreceptive, have been found in the brain of *C. vicina* (Cymborowski and Korf, 1995).

Adult females of *C. vicina*, as noted above, express locomotor rhythms and measure photoperiodic time simultaneously. Although the circadian pacemakers for both are probably located in the brain, and use a photoreceptive input that can by-pass the compound eyes and ocelli (Cymborowski et al., 1994; Saunders and Cymborowski, 1996), the two systems show quite different properties that suggest they are 'separate' parts of the multioscillator system. Locomotor rhythmicity, for example, is regulated by a self-sustaining oscillator that free-runs in darkness for life at an average period of about 22.5 hours (Kenny and Saunders, 1991). Photoperiodic induction, on the other hand, is regulated by a circadian system that expresses (through Nanda-Hamner experiments) an endogenous periodicity much closer to 24 hours, and bears all the hallmarks of a damped oscillation (Lewis and Saunders, 1987; Saunders, 1998). A similar separation of locomotor rhythmicity and photoperiodic time measurement is apparent in *D. melanogaster* where the latter is retained in behaviourally arrhythmic flies lacking the *period* gene (Saunders et al, 1989; Saunders, 1990).

5. UNRESOLVED PROBLEMS: What we know, and what we need to know

5.1. The current molecular model for circadian rhythmicity

In *D. melanogaster*, a molecular model for the generation of circadian rhythms has emerged from a welter of studies that suggest overall negative feedback loops involving the transcription and translation of so-called 'clock' genes (*period, timeless* and others). In brief, the widely accepted, or 'orthodox' view starts with the transcription of *per* and *tim*. Two other genes are involved at this stage: *Clock* and *Bmal*1. Their products are thought to form a dimer that binds to the E-boxes of *per* and *tim* to stimulate their transcription. In the cytoplasm the PER and TIM proteins form their own complex – which is variously delayed by the binding of PER with the product of another gene, *doubletime* – and this PER/TIM heterodimer later enters the nucleus during a specific 'gate' during the subjective night (Ct 19-20). The nuclear complex then binds to *Clock* and *Bmal*1 thereby removing the transcriptional activation of *per* and *tim*. Consequently, the PER/TIM dimer degenerates, transcription of *per* and *tim* restarts, and the cycle continues. According to this orthodox view, important time delays in the loop are brought about by the action of *doubletime*. Phase shifts, leading to entrainment, are caused by the action of light on TIM. Early in the subjective night, photic degeneration of TIM causes retardation of nuclear entry and hence phase delay. Later in the cycle, nuclear degeneration of TIM leads to a phase advance.

5.2. 'Clock' genes and the regulation of circadian and other periods

In *D. melanogaster*, the genes mentioned above play a clearly important role in the generation of circadian rhythmicity. But are they dedicated clock genes? Even in the fruit fly, the proposed feedback loop does not occur in all tissues: in the ovary, for example, PER remains cytoplasmic, not entering the nucleus to complete a loop. In other insects such as the house fly, *Musca domestica*, and the silkmoth, *Antheraea pernyi*, a similar lack of nuclear entry has been described, this time in brain neurons which are thought to be pacemaker cells. In *D. melanogaster*, the *per* gene is also involved in the generation of an ultradian 'love-song' rhythm which cycles with a period close to a minute. An ultradian period that short cannot be explained by a transcription-translation loop involving nuclear-cytoplasmic interactions because it is far too rapid. At the other end of the scale, the mind simply boggles at how lunar and circannual rhythms - with properties similar to circadian rhythms, but differing only in period - may be generated. It is almost as though *per* (and perhaps the other genes) had (and now have) another, more ancient, role, and have been hijacked to serve as part of a clock loop. Lastly, there may be clock loops not containing *per*. In *D. melanogaster* there are several reports of admittedly weak behavioural rhythmicity persisting in *per* null mutants (Dowse et al., 1987; Helfrich and Engelmann, 1987; McCabe and Birley, 1998). In the photoperiodic induction of diapause in *D. melanogaster,* clearly a product of the circadian system (Saunders, 1990), time measurement proceeds, albeit with an altered critical day length, in behaviourally arrhythmic per^0 flies and in flies (*per⁻*) entirely lacking the *period* locus. These studies suggest that the *period* gene is not absolutely essential, or that other cyclically transcribed genes exist which may induce some rhythmicity.

The importance of time delays to generate a near-24 hour periodicity has been recognised for a long time (e.g. Lewis and Saunders, 1987). The orthodox molecular model outlined above attributes such time delays to interactions with *doubletime* or delays in nuclear re-entry. It is difficult to see, however, how such events may be reconciled with the period-associated phenotypes of the original *period* mutants (per^S, $\tau \sim 19$ hours; per^L, $\tau \sim 29$ hours) (Konopka and Benzer, 1971). Another view is that at least some of the variation in τ arises from the multioscillator construction of the fly's circadian system, i.e. in coupling between individual cellular oscillators among the lateral neurons. Tight coupling might result in a shorter period, whereas looser coupling might give rise to a longer period. Unpublished data for *C. vicina* suggest that such a relationship does exist: in flies with a short τ the rhythm of locomotor activity is more 'precise', whereas as τ increases the rhythm 'loosens', eventually to become arrhythmic, presumably because of very weak coupling between the 'clock' cells.

5.3. Temperature compensation

Problems also exist with the generation of temperature compensation of the circadian period, one of the defining properties of a 'clock'. For *D. melanogaster,* one interesting suggestion is that period stability at different temperatures was related to the length of the threonine-glycine (T-G) encoding repeat within the 'clock' gene, *period.* Sawyer et al (1997) showed that the two major variants, T-G 17 and T-G 20, were distributed in a significant latitudinal cline in Europe and North Africa, with longer sequence flies predominating to the north. The length of the T-G repeat was related to the flies' ability to maintain their circadian period at different temperatures. The association was made plausible by the observation that per^0 flies 'rescued' with *per* lacking a T-G repeat sequence became behaviourally rhythmic but lacked temperature compensation (Ewer et al., 1990). It is difficult, however, to see a universal mechanism in this phenomenon, particularly since in some non-drosophilids the number of T-G repeats is stable (Nielsen at al., 1994). In *Lucilia cuprina,* a blow fly relative of *C. vicina,* there is no polymorphism, all flies having a T-G doublet.

5.4. The role of circadian rhythmicity in photoperiodism

If there are difficulties in providing a completely satisfactory explanation for the insect circadian clock, how much more difficult it becomes to unravel the complexities of seasonality. Photoperiodic induction may be a function of the circadian system, but this step is only one part of a long concatenation. Included in this sequence are: photoreception; measurement of night length; accumulation of successive long or short nights by a 'counter'; storage of such information; its transfer from sensitive to responsive stage; and finally, regulation of the release/retention of neurohormones controlling the onset of diapause or continuing development. The whole sequence may occupy a large part of the entire life cycle, in the case of *C.*

vicina from the adult female fly to the fully developed larva she produces. We have not seen the end of circadian research.

REFERENCES

Bǔnning, E., 1936. Die endogene Tagesrhythmik als Grundlage der Photoperiodischen Reaktion. Berichte der deutschen botanischen Gesellschaft 53, 590-607.

Cymborowski, B., Korf, H.-W., 1995. Immunocytological demonstration of S-antigen (arrestin) in the brain of the blow fly *Calliphora vicina.* Cell Tissue Research 279, 109-114.

Cymborowski, B., Gillanders, S.W., Hong, S.-F., Saunders, D.S., 1993. Phase shifts of the adult locomotor activity rhythm in *Calliphora vicina* induced by non-steroidal ecdysteroid agonist RH 5849. Journal Comparative Physiology 172, 101-108.

Cymborowski, B., Lewis, R.D., Hong, S.-F., Saunders, D.S., 1994. Circadian locomotor activity rhythms and their entrainment to light-dark cycles continue in flies (*Calliphora vicina*) surgically deprived of their optic lobes. Journal of Insect Physiology 40, 501-510.

Denlinger, D.L., 1985. Hormonal Control of Diapause. In: Kerkut, G.A., Gilbert, L.I. (Eds.) Comprehensive Insect Physiology, Biochemistry and Pharmacology. Pergamon Press Oxford, p. 353-412.

Denlinger, D.L., Joplin, K.H., Flannagan, R.D., Tammariello, S.P., Zhang, M-L., Yocum, G.D., Lee, K-Y., 1995. Diapause-specific gene expression. In: Molecular Mechanisms of Insect Metamorphosis and Diapause. Industrial Publishing & Consulting, Inc., p. 289-297.

Dowse, H., Hall, J.C., Ringo, J., 1987. Circadian and ultradian rhythms in *period* mutants of *Drosophila melanogaster.* Behavioral Genetics 17, 19-35.

Ewer, J., Hamblen-Coyle, M., Rosbash, M., Hall, J.C., 1990. Requirement for *period* gene expression in the adult and not during development for the locomotor activity rhythms of imaginal *Drosophila melanogaster.* Journal of Neurogenetics 7, 31-73.

Giebultowicz, J.M., 1999. Insect circadian clocks: is it all in their heads? Journal of Insect Physiology 45, 791-800.

Helfrich, C., Engelmann, W., 1987. Evidences for circadian rhythmicity in the *per*[O] mutant of *Drosophila melanogaster.* Zeitschrift für Naturforschung 42C, 1335-1338.

Helfrich-Foerster, C., Stengl, M., Homberg, U., 1998. Organization of the circadian system in insects. Chronobiology International 15, 567-594.

Hong, S-F., Saunders, D.S., 1994. Effects of constant light on the rhythm of adult locomotor activity in the blow fly, *Ccalliphora vicina.* Physiological Entomology 19, 319-324.

Hong, S-F., Saunders, D.S., 1998. Internal desynchronisation of the circadian locomotor rhythm of the blow fly, *Calliphora vicina,* as evidence for the involvement of a complex pacemaker. Biological Rhythm Research 29, 387-396.

Kenny, N.A.P., Saunders, D.S., 1991. Adult locomotor rhythmicity as "hands" of the maternal photoperiodic clock regulating larval diapause in the blowfly, *Calliphora vicina.* Journal of Biological Rhythms 6, 217-233.

Konopka, R., Benzer, S., 1971. Clock mutants of *Drosophila melanogaster.* Proceedings of the National Academy of Sciences, U.S.A. 68, 2112-2116.

Lees, A.D., 1973. Photoperiodic time measurement in the aphid *Megoura viciae.* Journal of Insect Physiology 19, 2279-2316.

Lewis, R.D., Saunders, D.S., 1987. A damped circadian oscillator model of an insect photo-periodic clock. I. Description of the model based on a feedback control system. Journal of theoretical Biology 128, 47-59.

McCabe, C., Birley, A., 1998. Oviposition in the *period* genotypes of *Drosophila melanogaster.* Chronobiology International 15, 119-133.

Nielsen, J., Peixoto, A.A., Piccin, A., Costa. R., Kyriacou, C.P., Chalmers D., 1994. Big flies, small repeats: The Thr-Gly repeat region on the *period* gene in Diptera. Molecular Biology and Evolution 11, 839-853.

Pittendrigh, C.S., 1960. Circadian rhythms and the circadian organization of living systems. Cold Spring Harbor Symposia Quantitative Biology 25, 159-184.

Pittendrigh, C.S., 1966. The circadian oscillation in *Drosophila pseudoobscura* pupae: a model for the photoperiodic clock. Zeitschrift für Pflanzenphysiologie 54, 275-307.

Pittendrigh, C.S., 1972. Circadian surfaces and the diversity of possible roles of circadian organization in photoperiodic induction. Proceedings of the National Academy of Sciences, U.S.A. 69, 2734-2737.

Pittendrigh, C.S., 1981. Circadian Systems: Entrainment. In: Aschoff, J. (Ed.) Handbook of Behavioral Neurobiology, vol 4 Biological Rhythms. Plenum Press, New York. 95-124.

Richard, D.S., Saunders, D.S., 1987. Prothoracic gland function in diapause and non-diapause *Sarcophaga argyrostoma* and *Calliphora vicina.* Journal of Insect Physiology 33, 385-392.

Saunders, D.S., 1978. An experimental and theoretical analysis of photoperiodic induction in the flesh-fly, *Sarcophaga argyrostoma.* Journal of Comparative Physiology 124, 75-95.

Saunders, D.S., 1982. Insect Clocks, second edition. Pergamon Press, Oxford, pp 409.

Saunders, D.S., 1987. Maternal influence on the incidence and duration of larval diapause *Calliphora vicina.* Physiological Entomology 12, 331-338.

Saunders, D.S., 1990. The circadian basis of ovarian diapause regulation in *Drosophila melanogaster:* is the *period* gene causally involved in photoperiodic time measurement? Journal of Biological Rhythms 5, 315-331.

Saunders, D.S., 1997. Under-sized larvae from short-day adults of the blow fly, *Calliphora vicina,* side-step the diapause programme. Physiological Entomology 22, 249-255.

Saunders, D.S., 1998. Insect circadian rhythms and photoperiodism. Invertebrate Neuroscience 3, 155-164.

Saunders, D.S., 2000a. Chapter 6, Arthropoda – Insecta: Diapause. In: A. Dorn (Ed) Progress in Developmental Endocrinology, Vol X Reproductive Biology of Invertebrates. (Ed. Adiyodi) Oxford & IBH Publishing Co. Pvt. Ltd., New Delhi, India.

Saunders, D.S., 2000b. Larval diapause duration and fat metabolism in three geographical strains of the blow fly, *Calliphora vicina.* Journal of Insect Physiology 46, 509-517.

Saunders, D.S., Cymborowski, B., 1996.Removal of optic lobes of adult blow flies (*Calliphora vicina*) leaves photoperiodic induction of larval diapause intact. Journal of Insect Physiology 42, 807-811.

Saunders, D.S., Henrich, V.C., Gilbert, L.I., 1989. Induction of diapause in *Drosophila melanogaster:* photoperiodic regulation and the impact of arrhythmic clock mutations on time measurement. Proceedings of the National Academy of Sciences, U.S.A. 86, 3748-3752.

Saunders, D.S., Hong, S-F., 2000. Effect of temperature and temperature-steps on circadian locomotor rhythmicity in the blow fly, *Calliphora vicina.* Journal of Insect Physiology 46, 289-295.

Saunders, D.S., Haywood, S.A.L., 1998. Geographical and diapause-related cold tolerance in the blow fly, *Calliphora vicina.* Journal of Insect Physiology 44, 541-551.

Sawyer, L.A., Hennessy, J.M., Peixoto, A.A., Rosato, E., Parkinson, H., Costa R., Kyriacou, C.P., 1997. Natural variation in a *Drosophila* clock gene and temperature compensation.

14

Science 278, 2117-2120.

Vaz Nunes, M., 1998. A double circadian oscillator model for quantitative photoperiodic time measurement in insects and mites. Journal of Theoretical Biology 194, 299-311.

Vaz Nunes, M., Hardie, J., 1993. Circadian rhythmicity is involved in photoperiodic time measurement in the aphid *Megoura viciae*. Experientia 49, 711-713.

Vaz Nunes, M., Saunders, D.S., 1989. The effect of larval temperature and photoperiod on the incidence of larval diapause in the blowfly, *Calliphora vicina*. Physiological Entomology 14, 471-474.

Vaz Nunes, M., Saunders, D.S., 1999. Photoperiodic time measurement in Insects: A review of clock models. Journal of Biological Rhythms 14, 84-104.

Winfree, A.T., 1970. Integrated view of resetting a circadian clock. Journal of Theoretical Biology 28, 327-374.

Zimmerman, W.F., Pittendrigh, C.S., Pavlidis, T., 1968. Temperature compensation of the circadian oscillation in *Drosophila pseudoobscura* and its entrainment by temperature cycles. Journal of Insect Physiology 14, 669-684.

Insect Timing: Circadian Rhythmicity to Seasonality
D.L. Denlinger, J. Giebultowicz and D.S. Saunders (Editors)
© 2001 Elsevier Science B.V. All rights reserved.

Molecular control of *Drosophila* circadian rhythms

Peter Schotland and Amita Sehgal

Department of Neuroscience, University of Pennsylvania, 232 Stemmler Hall, Philadelphia, PA 19104, USA

The ubiquity of circadian rhythms testifies to their adaptive significance. Studies in *Drosophila melanogaster* have greatly advanced our understanding of circadian rhythms at the molecular level. To date, at least seven genes in *Drosophila* have been described and shown to participate in transcription-translation based feedback loops that comprise the central oscillator controlling overt circadian behaviors. Many of these genes have functional homologues in mammals. There has also been recent progress in describing, at the molecular level, how temporal information from the environment is conveyed to the central oscillator and, how, in turn, that information is conveyed to the rest of the organism.

1. INTRODUCTION

Circadian rhythms are displayed by organisms ranging across the phyla from unicellular cyanobacteria to multicellular fungi, plants and higher order mammals, including humans (Dunlap, 1999). The universality of circadian rhythms attests to their adaptive significance. Circadian rhythms are characterized by the same three fundamental properties in different organisms, which presumably allows them to serve similar adaptive functions (Pittendrigh, 1960). 1. They can be synchronized, or entrained, by environmental stimuli called zeitgebers (German for "time givers"). The dominant zeitgebers are, not surprisingly, light and temperature as they follow closely the earth's daily rotation. 2. They persist, or freerun, under constant conditions; i.e., after removal of the entraining zeitgeber. Under these conditions the length of a complete cycle (the period) is usually between 23 and 25hr (rarely the 24hr rhythm of the zeitgeber); hence, the term "circa" (about). The freerunning period also varies somewhat between individuals. 3. The period of a circadian rhythm shows little variance with respect to (physiological) temperature, and is said to be temperature-compensated. Note that the compensation of period with temperature does not preclude temperature as a zeitgeber. Indeed, competition experiments in *Neurospora crassa* have shown temperature to be dominant over light (Liu et al., 1998). The phase of the rhythm is often temperature dependent, however.

Chronobiologists like to think heuristically in terms of a circadian system that contains a central clock, an input pathway and an output pathway. The input pathway conveys time of day information from the environment to the clock, and the output pathway conveys the internal time from the clock to other systems to allow for the temporal organization of behavior and physiology. Although input and output are often thought of as independent, linear pathways, this is probably an over-simplification of the system. It is more likely that the three components of the circadian system affect one another and, as is evident from the molecular analysis, have a significant amount of overlap. Indeed, some molecules appear to play a role in all three components of the system. There is also evidence suggesting that in some organisms there is no central oscillator, but rather several "peripheral" oscillators that can function (even entrain) independently of one another. To further complicate matters, some of the feedback mechanisms that are characteristic of the clock are sometimes also found in the input and output pathways. Nevertheless, a set of criteria was proposed that would facilitate the identification of clock components, i.e. those molecules that play a role in the time keeping mechanism (Zatz, 1992). Clock components that actually provide time cues, usually through oscillations of their abundance or activity, are called state variables.

The ubiquity of circadian rhythms provided researchers with the hope that, if the underlying mechanisms are conserved, dissection of those mechanisms in simple organisms would lead to an understanding of rhythms in higher organisms. Indeed, the molecular mechanism of the clock in most organisms studied appears to be conserved and turns out to be a transcription-translation based feedback loop comprised of cycling RNAs and proteins. While most of this characterization has been done in the bread mold, *Neurospora crassa*, and the fruit fly, *Drosophila melanogaster*, aspects of it have been described in mammals and in cyanaobacteria (Dunlap, 1999).

2. CIRCADIAN RHYTHMS IN *DROSOPHILA*

Since the pioneering work of Ron Konopka in 1971 (Konopka, 1971), more than 15 clock and clock controlled genes were identified in *Drosophila* (Dunlap, 1999). Indeed, *Drosophila* proved to be a fruitful system for the study of circadian rhythms long before the advantages of genetics came into play. Pittendrigh (Pittendrigh, 1960; Pittendrigh, 1967) demonstrated the existence of quantitative, well defined circadian behaviors in *Drosophila pseudoobscura* and developed assays still used to elucidate fundamental properties of the clock, including the discovery of the molecules discussed in this review. Using selective breeding experiments with *Drosophila pseudoobscura,* Pittendrigh demonstrated the heritability, and, hence, the genetic basis of certain clock characteristics such as period length and phase, but the focus of genetic research soon moved to *Drosophila melanogaster* because of the powerful genetic technologies being developed in that species.

Genetic studies in *Drosophila* have predominantly exploited two overt rhythms-locomotor activity and eclosion. When individual flies are placed in a glass tube with a little food, their locomotor activity can be monitored by counting the frequency of crossing of an infrared beam. When maintained in 24hr light:dark cycles, flies display a bimodal distribution of locomotor activity with peak activity occurring in late night/early morning and late day/early evening (Hamblen-Coyle, 1992). When placed in freerun, these peaks merge to a single bout of activity during the subjective day. These daily bouts of activity will persist in constant conditions, with no reference to external time, for the lifetime of the fly- as long as two months. It is extraordinary that the same holds true for many mammals, with mice and hamsters able to display circadianly gated behavior under constant conditions for years at a stretch. Eclosion, the hatching of pharate adults from their pupal casings, is regulated by the clock such that it occurs during a few hours near dawn (Pittendrigh, 1967). Eclosion and pupal development are independently regulated such that if pupal development is completed after dawn, eclosion is delayed several hours until the next dawn (Qiu, 1996). Hence, eclosion is said to be circadianly gated.

Circadian rhythmicity has been detected in other systems such as visual and olfactory sensitivity, feeding, and oviposition (Chen et al., 1992; Krishnan, 1999; McCabe & Birley, 1998). Although these clock regulated phenomena have yet to be useful in elucidating the underlying molecular clockwork, they have been shown to be regulated by the molecules discovered studying eclosion and locomotor activity.

2.1. *per, tim* and the feedback loop

The first systematic effort to isolate circadian rhythm mutants was done by Ron Konopka in 1971 (Konopka, 1971). The result, three alleles mapping to the *period* locus (*per*), is perhaps the most important in the molecular analysis of circadian rhythms. Konopka's genetic screen found pretty much every circadian mutant conceivable: a short period mutant, per^{short} (*per^s*), that reduced freerunning period to 19hr, a long period mutant, per^{long} (*per^l*), with a freerunning period of 29hr, and a mutation producing total arrhythmia, per^{01}. The *per* mutants affected both eclosion and locomotor activity implicating a role for *per* in the central clock. That single nucleotide mutations in one gene could affect a complex behavior in a variety of ways was an important discovery not only for the field of circadian rhythms, but for behavioral studies in general. For many years, the *period* locus was considered the paradigm for a behavioral gene because only behavioral defects were found in flies carrying the *per* mutations; i.e., the flies were developmentally and anatomically normal. Recently, however, mutations affecting the clock have been found in genes with roles in other processes including embryonic development (discussed below).

Several additional alleles of *per* are now known (Hamblen et al., 1998; Konopka et al., 1994). *per* remained the only known *bona fide* clock gene until the isolation of the *timeless* (*tim*) mutation in 1994 (Sehgal et al., 1994). tim^0 is a null mutation producing total arrhythmia. Since then, long and short period alleles of *timeless* have also been isolated. A 32 amino acid deletion in the N-terminal region of the *tim* protein increases freerunning period to 30-48hr (Ousley et al., 1998), the tim^{Rit} (Matsumoto et al., 1999) allele increases period to 27 hours and the $tim^{ultralong}$ allele produces 33hr rhythms (Rothenfluh et al., 2000). tim^{SL} shortens period by only 0.5 hours in a wild type background, but it can suppress the *per^l* long-period phenotype by several hours (Rutila et al., 1996). Recently, 2 short (21-22hr) and 4 long (26-28hr) period alleles of *tim* have been isolated (Rothenfluh, 2000).

Unfortunately, very little information about the biochemical function of the *per* and *tim* proteins could be inferred from their molecular sequences. The *tim* sequence is entirely unique (Myers et al., 1995), and the only protein motif found in *per* is a protein-protein interaction domain known as a PAS domain, named for the proteins it was first found in- *per, single-minded,* and the aryl hydrocarbon receptor nuclear translocator. PAS domains have since been found in other clock molecules, photoreceptors, developmental proteins and proteins involved in hypoxia and may confer on many of these proteins the ability to sense environmental signals (Crews, 1999). Such a function has, however, not yet been described for the *per* protein.

18

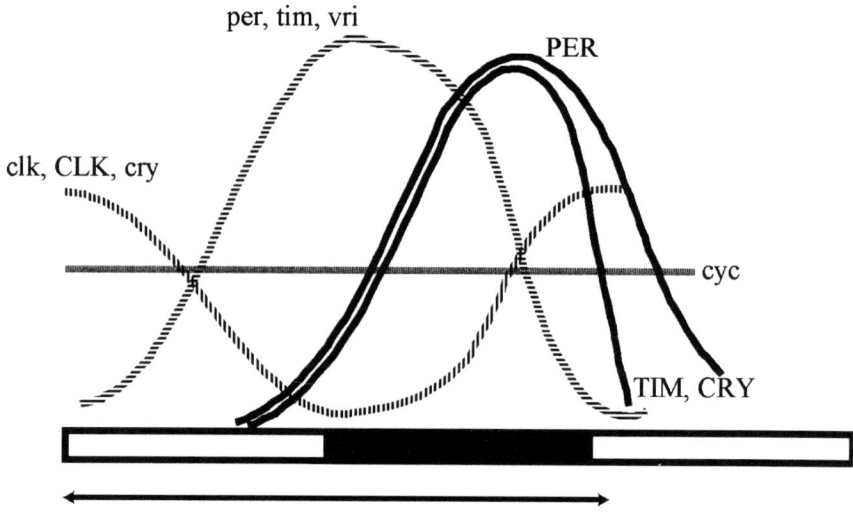

per, tim, vri

clk, CLK, cry

PER

cyc

TIM, CRY

24 hr

Fig. 1. Molecular oscillations of the clock. The abundance of various clock molecules are plotted over a 24hr, 12:12, light:dark cycle. Lower case indicates RNA, uppercase, protein. Although this plot is for a light:dark cycle, the gene products behave the same in constant darkness with two exceptions: 1. the TIM profile more closely follows PER in constant darkness as light apparently turns over TIM more rapidly than the clock at the beginning of the day. 2. CRY does not cycle in constant darkness but instead accumulates in a non-decreasing manner.

Although their biochemical properties were not obvious, analysis of the regulation of *per* and *tim* expression proved to be very informative. Measurements of the RNA and protein products of both clock genes showed that, keeping with the notion of a state variable (Zatz, 1992), oscillating gene products constitute the core of the *Drosophila* clock (see Fig. 1). Both genes encode RNAs that cycle with a circadian rhythm, such that RNA levels are high at the end of the day/beginning of the night (Hardin et al., 1990; Sehgal et al., 1994; Sehgal et al., 1995). The *per* and *tim* proteins (PER and TIM) also cycle and begin accumulating in the early evening. PER and TIM bind one another to form heterodimers that are then transported into the nucleus. Each protein is required for nuclear transport of the other; i.e., in *per* and *tim* null mutants, TIM and PER, respectively, are restricted to the cytoplasm (Hunter-Ensor et al., 1996; Myers et al., 1996; Saez & Young, 1996; Vosshall et al., 1994). The abundance of both proteins peaks in the middle of the night (Hunter-Ensor et al., 1996; Myers et al., 1996; Zeng et al., 1996).

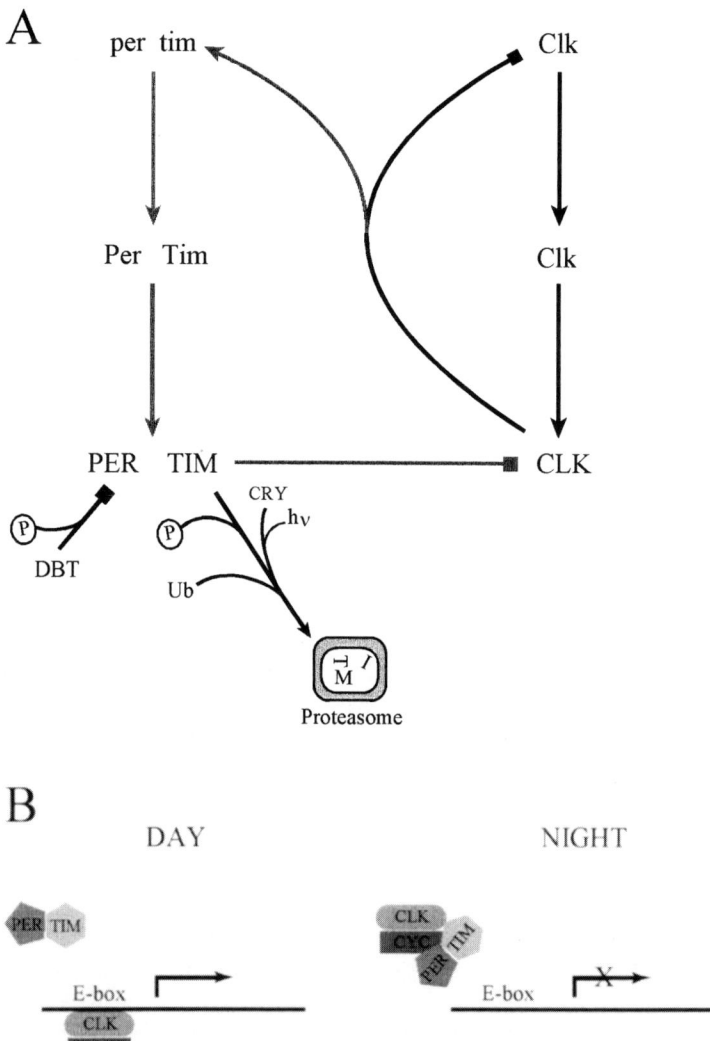

Fig. 2. The molecular feedback loop in *Drosophila*. **A** The *Drosophila* clock is comprised of interlocking feedback loops. The *per, tim* feedback loop is indicated in gray, the *dClock* loop indicated in black. Arrows indicate activation, squares inhibition. Also shown are the *dbt* dependent turnover of PER via phosphorylation, and the phosphorylation and ubiquitin dependent degradation of TIM by light. CRY likely plays a role in this process as well. **B** During the day, TIM, and, therefore, PER levels are low and insufficient to inhibit CLOCK-CYC mediated transcription. At night, TIM and PER can rise to levels sufficient to inhibit CLOCK-CYC.

Thereafter, PER levels remain high until the early morning while TIM levels drop off at the end of the night. TIM and PER are also cyclically phosphorylated, and peak phosphorylation occurs at the end of the night (Edery et al., 1994; Zeng et al., 1996). Phosphorylation probably serves many functions, one of which is to target the proteins for degradation. Phosphorylation of PER by a homologue of casein kinaseIε encoded by the *doubletime* (*dbt*) locus renders it unstable in the absence of TIM (Kloss et al., 1998; Price et al., 1998). Consistent with the notion of a state variable, the oscillations described above persist under constant conditions.

Both TIM and PER are required for cyclic expression of their mRNA's (Hardin et al., 1990; Sehgal et al., 1995). Since RNA levels are low when protein levels are high a negative feedback loop model was proposed in which PER and TIM enter the nucleus to inhibit transcription from the *per* and *tim* loci (Hardin et al., 1990) (see Fig. 2A). In support of the negative feedback hypothesis, overexpression of PER from the *rhodopsin* promoter (in photoreceptor cells) reduces levels of the endogenous *per* transcript (Zeng et al., 1994). Essential for the maintenance of such a loop is the separation of the phase of RNA synthesis from RNA inhibition (Dunlap, 1996). This separation may be achieved by the 6 hr lag between peak RNA and peak protein levels, the regulation of the time of nuclear entry of the heterodimer, or possibly temporally gated phosphorylation of PER/TIM. The rate of PER accumulation is dependent on DBT mediated destabilization of PER (Price et al., 1998).

It should be stressed, however, that this is a simplistic view and does not explain all effects of PER-TIM on their RNA levels. The negative feedback model would predict that *per* and *tim* RNA would be expressed at peak levels in flies where feedback doesn't occur because of the absence of either protein. However, per^{01} and tim^{01} flies express intermediate levels of *per* and *tim* RNA (Hardin et al., 1990; Qiu & Hardin, 1996; Sehgal et al., 1994; Sehgal et al., 1995). Transcription rates of *per* and *tim* are also intermediate in per^{01} and tim^{01} flies (So & Rosbash, 1997). This suggests that, in addition to negative feedback, the PER and TIM proteins, either directly or indirectly, also have a positive effect on the expression of the their RNAs. This positive effect derives, at least in part, from the stimulation of *Clk* gene expression by PER-TIM (see below)(Bae et al., 1998; Glossop et al., 1999). In addition, TIM increases levels of *per* RNA through a post-transcriptional mechanism (Suri et al., 1999).

Because both PER and TIM lack conventional DNA binding domains and have never been shown to associate directly with DNA, models for PER action postulated that PER associates with transcriptional activators and sequesters them, thereby preventing them from activating transcription (Huang et al., 1993). The *per* PAS domain, a protein-protein interaction domain also found in several transcription factors, was hypothesized to mediate this interaction (Huang et al., 1993). Transcriptional activators of *per* and *tim* were recently described and, as predicted, they contain PAS domains as well as bHLH motifs that allow them to bind DNA (Allada et al., 1998; Bae et al., 2000; Darlington et al., 1998; Rutila et al., 1998). These activators are encoded by the *dClock* (*dClk*) and *cycle* (*cyc*) genes, and flies mutant at either locus express very low levels of *per* and *tim* RNA (Allada et al., 1998; Rutila et al., 1998). Furthermore, as predicted by the model, cell culture studies show that PER-TIM act by inhibiting the activity of CLK/CYC (Darlington et al., 1998). The sites in the *per* and *tim* promoters that are recognized by CLK/CYC contain E-boxes (sequence = CACGTG), promoter elements that are known to be regulated by bHLH proteins. A recent study using a cell free system (Lee et al., 1999) demonstrated that CLK and CYC can bind the *per* promoter

E-box together as a heterodimer, but not individually. PER and/or TIM were capable of disrupting this binding. The molar ratio of the CLK:CYC heterodimer is unchanged by the presence of PER or TIM, indicating that the probable mechanism of transcriptional inhibition consists of blocking DNA-association of the heterodimer rather than disrupting the heterodimer itself (see Fig. 2B).

dClk RNA also cycles, antiphase to per and tim RNA (see Fig. 1) (Bae et al., 1998; Darlington et al., 1998). dClk RNA is high in the dClk null mutant compared to wild type trough levels, indicating that dClk gene products also participate in their own negative feedback loop. Additionally, dClk RNA levels are low in per^{01} and tim^{01} flies, indicating that not only does dClk positively regulate per and tim, but per and tim, directly or indirectly, are positive regulators of dClk. Thus, per, tim, and dClk are engaged in interlocked feedback loops (Glossop et al., 1999) (see Fig. 2A).

The most recently identified component of the Drosophila circadian system, and possibly of the feedback loop, is the vrille gene (Blau & Young, 1999). Expression of vri RNA is clock-controlled and cycles in phase with that of per and tim. Overexpression of vri attenuates oscillations of per/tim gene products as well as behavioral rhythms, indicating that vri can affect clock function in addition to being regulated by the clock (Blau & Young, 1999). However, the circadian phenotype produced by the null mutation is yet to be described. Homozygous loss of vri results in lethality as does loss of dbt, indicating that both these genes have other essential functions (Blau & Young, 1999; Price et al., 1998).

It should be mentioned that the phenotypes described above for the different mutants are based largely upon measurements of rest:activity, and to some extent, eclosion rhythms. Rest:activity rhythms are produced by a specific group of cells in the Drosophila brain called lateral neurons (Ewer et al., 1992; Frisch et al., 1994; Helfrich-Forster, 1998; Kaneko et al., 2000). Expression studies have focused either on these neurons or, more commonly, on assays of whole fly heads where the major source of clock RNA and protein is the photoreceptor cells of the compound eye. Photoreceptor cells are not required for activity rhythms and so clock proteins in these cells are thought to constitute an autonomous oscillator that controls an eye-specific function. Similar autonomous or semi-autonomous oscillators have been described in other parts of the fly and are the subject of another chapter in this volume (Giebultowicz, 2000).

2.2. Molecular phenotypes of period altering mutations

Thus far, these have only been described for per, tim and dbt. dbt^s and dbt^l shorten and lengthen circadian period by accelerating or delaying, respectively, the cyclic profile of PER phosphorylation (Price et al., 1998). Presumably, delayed phosphorylation increases stability of the protein so it stays around longer while accelerated phosphorylation hastens protein turnover and shortens period. Likewise, the per^s mutation shortens period by accelerating the daily decay of the protein at the end of each cycle (Marrus, 1996; Siwicki et al., 1988). The per^l mutation, on the other hand, decreases the strength of the PER-TIM interaction and delays nuclear entry (Curtin et al., 1995; Gekakis et al., 1995). As a result, period is lengthened. However, there are caveats to these simplistic models. For instance, one might expect that a decrease in stability of the protein would also delay its accumulation (given that the delay in PER appearance relative to its mRNA is due to protein turnover) and thereby lengthen period and vice versa. There is, in fact, a PER mutant that is hypophosphorylated, shows increased stability, accumulates faster and shortens period (Schotland et al., 2000). However, the short period phenotype of this mutant may be due to an additional defect in its ability to effect negative feedback. Clearly, there are many different aspects of protein regulation and most, if not all likely contribute to the determination of period.

Mutations in *tim* have similar molecular phenotypes. TIM^{UL} shows prologed nuclear localiation and protracted repression of *per/tim* transcription suggesting that its stability is increased (Rothenfluh et al., 2000). TIM^{Rit} suppresses PER cycling and reduces overall PER levels (Matsumoto et al., 1999). These effects on PER are apparently mediated through post-transcriptional effects on *per* RNA. Finally, the *tim*^{SL} mutation changes the profile of TIM phosphorylation to somehow rescue the delayed nuclear entry and longer period conferred by the *per*^{l} mutation (Rutila et al., 1996). Since *tim*^{SL} does not affect the PER^{L}-TIM interaction, as measured in yeast, it is thought to be a bypass suppressor (Rutila et al., 1996).

2.3. Mechanisms that entrain the clock to light

If the *per* and *tim* gene products are state variables whose levels provide time-of-day cues to the organism, it follows that resetting of the clock in response to any stimulus would be accompanied by changes in the levels of these components. The question is which clock component changes first and mediates the resetting response? In *Drosophila*, the TIM protein shows an acute response to light such that it is degraded within 30-90 minutes of light treatment (based upon the cell type being studied) (Hunter-Ensor et al., 1996; Myers et al., 1996; Zeng et al., 1996). We now know that TIM is first phosphorylated on tyrosine residues in response to light, ubiquitinated and then degraded by the proteasome (Naidoo et al., 1999). A number of lines of evidence support the idea that the TIM response mediates behavioral resetting: (1) Dose response studies show that light pulses of different intensity produce corresponding changes in the TIM response and the behavioral response to light (Suri et al., 1998; Yang et al., 1998). (2) The action spectrum for the TIM response to light matches that of the behavioral response- both are most sensitive in the blue light region of the spectrum (Suri et al., 1998). (3) Mutants that affect the TIM response to light also affect behavioral resetting (Yang et al., 1998). (4) TIM interacts with the circadian photoreceptor, cryptochrome (CRY), in a light-dependent manner (Ceriani et al., 1999).

Cryptochromes are flavin-binding molecules with homology to DNA photolyases. They were first identified in plants where they also turn out to have a role in circadian photoreception (Cashmore et al., 1999). In *Drosophila*, the only known CRY is essential for some circadian responses to light such as appropriate behavioral resetting in response to light pulses (nonparametric entrainment) and arrhythmic locomotor activity in the presence of constant light (Emery et al., 2000; Stanewsky et al., 1998). However, entrainment to light:dark cycles can also be mediated, either directly or indirectly, by the visual system (Stanewsky et al., 1998). It is probably beneficial to the organism to have some redundancy built into the circadian entrainment system. The role of CRY in TIM degradation has not yet been reported, but the light-dependent TIM-CRY interaction in cultured cells rapidly blocks negative feedback by PER-TIM (Ceriani et al., 1999). Presumably this is followed by TIM degradation which then effects a permanent change in the phase of the clock.

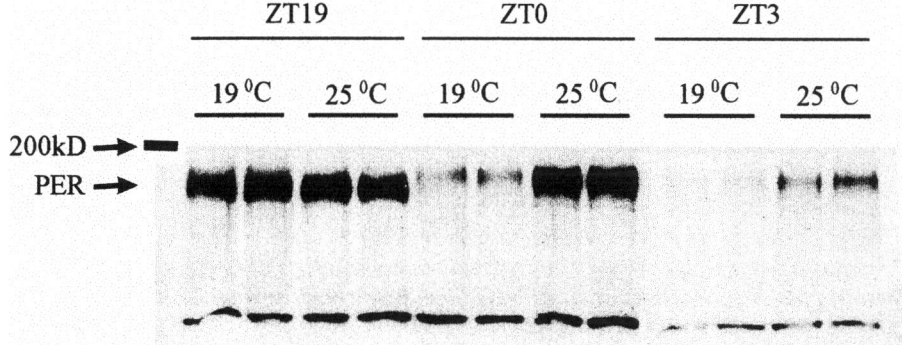

Fig. 3. PER abundance is temperature dependent. To examine the effect of temperature on PER abundance *Drosophila* were maintained at either 19°C or 25°C on a 12:12, light:dark cycle and collected at hours 19, 0 and 3. Total protein from heads was extracted, western blotted and stained with an anti-PER antibody. Samples were run in duplicate. PER displays an increase in temperature at all three temperatures tested.

2.4. Effects of temperature on the clock

Temperature affects the clock in multiple ways. First, temperature sets the limits of rhythmicity- organisms tend to lose rhythmicity at very high or very low temperatures. Within this temperature range the period of the rhythm is compensated such that it does not change. However, the phase of the rhythm can be reset by temperature. As one might imagine, it is difficult to tease apart the mechanisms involved in these different responses.

Levels of clock components change with the ambient temperature. As demonstrated by Majerak et al (Majercak et al., 1999), levels of PER are higher at all times at 29°C as compared to 19°C. We have seen the same result at 19°C vs 25°C (see Fig. 3). Thus, the protein oscillates around a higher level at the higher temperature. In addition, the daily protein profile is altered such that the protein accumulates earlier in a light:dark cycle. This is apparently mediated by a temperature-dependent splicing event in the *per* RNA (Majercak et al., 1999). The behavioral consequences of the altered protein profile are that locomotor activity tends to be concentrated more in the daytime hours at low temperatures and distributed in the morning and evening hours at higher temperatures (Majercak et al., 1999).

Normally, higher levels of PER result in a shorter period (Baylies et al., 1987; Cooper et al., 1994; Cote & Brody, 1986; Smith & Konopka, 1981). However, this does not happen at higher temperatures due to compensation mechanisms that are largely unknown. Although a number of mutants that affect temperature compensation have been identified and some mechanisms have been proposed (Hamblen et al., 1998; Huang et al., 1995; Leloup & Goldbeter, 1997; Price, 1997; Sawyer et al., 1997), there is no consensus on the process involved. The mutations that affect temperature compensation map to various regions of the PER protein, suggesting that anything that changes the conformation/folding of the protein can produce defects in temperature compensation. Finally, the mechanisms underlying temperature entrainment in Drosophila are not yet understood either.

2.5. Output mechanisms

The output pathway remains the least understood aspect of the circadian system. While inroads have been made into our understanding of the clock itself and entrainment pathways in *Drosophila* (at least in response to light, if not temperature), little is known about how the clock transmits time-of-day cues and produces overt rhythms. This paucity of information at the output end is true of all organisms where clocks have been studied at a molecular level. As peripheral oscillators are discussed in the chapter by Giebultowicz et al, here we will focus on the control of the rest:activity output.

The lateral neurons release a neuropeptide, pigment-dispersing factor (PDF), which is required for rest:activity rhythms (Renn et al., 1999). It may also be required for eclosion rhythms as these rhythms are disrupted when PDF is overexpressed in specific locations (Helfrich-Forster et al., 2000). Using antibodies to PDF, arborizations of the axons extended by lateral neurons have been traced (Helfrich-Forster & Homberg, 1993). They innervate large regions of the Drosophila optic lobes and also extend to the lateral neurons on the other side of the brain. However, the nature of the cells innervated is not known nor, for the most part, are the molecules that act downstream of the clock. Thus, the molecular links between the clock proteins and PDF and between PDF and the activity rhythm have not been identified. Although protein kinase A (Levine et al., 1994; Majercak et al., 1997) is known to be an output molecule required for rest:activity rhythms, its location in the output pathway remains a mystery. Interestingly, CREB (Belvin et al., 1999) which sometimes, but not always, is downstream of PKA, affects PER protein cycling, but PKA itself does not.

3. CONCLUSION

Along with the recent explosion in our understanding of the molecular underpinnings of the clock, it has been a delight to see that the work described here is applicable to other systems. *per, tim, dClk, cyc, dbt,* and *cry* all have homologues in the mammalian system, many of them functional counterparts (see the review by Dunlap) (Dunlap, 1999). There are 3 mammalian *per* homologues that cycle with a circadian rhythm and function as (putative) negative elements in a feedback loop. Mammalian *Clock* and *cyc* (*bmal1/mop3*) can heterodimerize and activate transcription from E-boxes (Gekakis et al., 1998; Rutila et al., 1998; Shearman et al., 2000), and *tau* (mammalian *dbt*) has been shown to phosphorylate the *mper* proteins *in vitro* (Lowrey et al., 2000).

Some important differences between the mammalian and *Drosophila* systems should be mentioned. The two known mammalian *cry* homologues have not been shown to mediate circadian photoreception but, surprisingly, appear instead to have a role as negative elements in the feedback loop (Griffin et al., 1999; Kume et al., 1999; Shearman et al., 2000; Vitaterna et al., 1999) and are required for behavioral rhythmicity (van der Horst et al., 1999; Vitaterna et al., 1999). *tim*'s role in the mammalian system is still unknown although, in contrast with *Drosophila*, *tim* is essential for embryonic development (Gotter et al., 2000). Finally, light sensitivity does not appear to be mediated by a light sensitive protein such as TIM, but rather via photic induction of the *mper* RNAs (reviewed by Dunlap, 1999).

In addition to the molecular conservation seen in animals, the mechanism of the transcription-translation based negative feedback loop is found across the phyla in cyanobacteria, fungi, and plants.

REFERENCES

Allada, R., White, N.E., So, W.V., Hall, J.C. and Rosbash, M. 1998. A mutant Drosophila homolog of mammalian Clock disrupts circadian rhythms and transcription of period and timeless. Cell 93, 791-804.

Bae, K., Lee, C., Hardin, P.E. and Edery, I. 2000. dCLOCK is present in limiting amounts and likely mediates daily interactions between the dCLOCK-CYC transcription factor and the PER-TIM complex. Journal of Neuroscience 20, 1746-53.

Bae, K., Lee, C., Sidote, D., Chuang, K.Y. and Edery, I. 1998. Circadian regulation of a Drosophila homolog of the mammalian Clock gene: PER and TIM function as positive regulators. Molecular & Cellular Biology 18, 6142-51.

Baylies, M.K., Bargiello, T.A., Jackson, F.R. and Young, M.W. 1987. Changes in abundance or structure of the per gene product can alter periodicity of the Drosophila clock. Nature 326, 390-2.

Belvin, M.P., Zhou, H. and Yin, J.C. 1999. The Drosophila dCREB2 gene affects the circadian clock. Neuron 22, 777-87.

Blau, J. and Young, M.W. 1999. Cycling vrille expression is required for a functional Drosophila clock. Cell 99, 661-71.

Cashmore, A.R., Jarillo, J.A., Wu, Y.J. and Liu, D. 1999. Cryptochromes: blue light receptors for plants and animals. Science 284, 760-5.

Ceriani, M.F., Darlington, T.K., Staknis, D., Mas, P., Petti, A.A., Weitz, C.J. and Kay, S.A. 1999. Light-dependent sequestration of TIMELESS by CRYPTOCHROME [see comments]. Science 285, 553-6.

Chen, D.M., Christianson, J.S., Sapp, R.J. and Stark, W.S. 1992. Visual receptor cycle in normal and period mutant Drosophila: microspectrophotometry, electrophysiology, and ultrastructural morphometry. Visual Neuroscience 9, 125-35.

Cooper, M.K., Hamblen-Coyle, M.J., Liu, X., Rutila, J.E. and Hall, J.C. 1994. Dosage compensation of the period gene in Drosophila melanogaster. Genetics 138, 721-32.

Cote, G.G. and Brody, S. 1986. Circadian rhythms in Drosophila melanogaster: analysis of period as a function of gene dosage at the per (period) locus. Journal of Theoretical Biology 121, 487-503.

Crews, S.F., CM. 1999. Remembrance of things PAS: regulation of development by bHLH-PAS proteins. Current Opinion in Genetics and Development 9, 580-587.

Curtin, K.D., Huang, Z.J. and Rosbash, M. 1995. Temporally regulated nuclear entry of the Drosophila period protein contributes to the circadian clock. Neuron 14, 365-72.

Darlington, T.K., Wager-Smith, K., Ceriani, M.F., Staknis, D., Gekakis, N., Steeves, T.D.L., Weitz, C.J., Takahashi, J.S. and Kay, S.A. 1998. Closing the circadian loop: CLOCK-induced transcription of its own inhibitors per and tim [see comments]. Science 280, 1599-603.

Dunlap, J.C. 1996. Genetics and molecular analysis of circadian rhythms. Annual Review of Genetics 30, 579-601.

Dunlap, J.C. 1999. Molecular bases for circadian clocks. Cell 96, 271-90.

Edery, I., Zwiebel, L.J., Dembinska, M.E. and Rosbash, M. 1994. Temporal phosphorylation of the Drosophila period protein. Proceedings of the National Academy of Sciences of the United States of America 91, 2260-4.

Emery, P., Stanewsky, R., Hall, J.C. and Rosbash, M. 2000. A unique circadian-rhythm photoreceptor. Nature 404, 456-7.

Ewer, J., Frisch, B., Hamblen-Coyle, M.J., Rosbash, M. and Hall, J.C. 1992. Expression of the period clock gene within different cell types in the brain of Drosophila adults and mosaic analysis of these cells' influence on circadian behavioral rhythms. Journal of Neuroscience 12, 3321-49.

Frisch, B., Hardin, P.E., Hamblen-Coyle, M.J., Rosbash, M. and Hall, J.C. 1994. A promoterless period gene mediates behavioral rhythmicity and cyclical per expression in a restricted subset of the Drosophila nervous system. Neuron 12, 555-70.

Gekakis, N., Saez, L., Delahaye-Brown, A.M., Myers, M.P., Sehgal, A., Young, M.W. and Weitz, C.J. 1995. Isolation of timeless by PER protein interaction: defective interaction between timeless protein and long-period mutant PERL [see comments]. Science 270, 811-5.

Gekakis, N., Staknis, D., Nguyen, H.B., Davis, F.C., Wilsbacher, L.D., King, D.P., Takahashi, J.S. and Weitz, C.J. 1998. Role of the CLOCK protein in the mammalian circadian mechanism [see comments]. Science 280, 1564-9.

Giebultowicz, J.M. 2000. Molecular mechanism and cellular distribution of insect circadian clocks. Annual Review of Entomology 45, 769-93.

Glossop, N.R., Lyons, L.C. and Hardin, P.E. 1999. Interlocked feedback loops within the Drosophila circadian oscillator. Science 286, 766-8.

Gotter, A.L., Manganaro, T., Weaver, D.R., Kolakowski, L.F., Jr., Possidente, B., Sriram, S., MacLaughlin, D.T. and Reppert, S.M. 2000. A time-less function for mouse timeless. Nature Neuroscience 3, 755-6.

Griffin, E.A., Jr., Staknis, D. and Weitz, C.J. 1999. Light-independent role of CRY1 and CRY2 in the mammalian circadian clock. Science 286, 768-71.

Hamblen, M.J., White, N.E., Emery, P.T., Kaiser, K. and Hall, J.C. 1998. Molecular and behavioral analysis of four period mutants in Drosophila melanogaster encompassing extreme short, novel long, and unorthodox arrhythmic types. Genetics 149, 165-78.

Hamblen-Coyle MJ, W.D., Rutila JE, Rosbash M, Hall JC. 1992. Behavior of period-altered circadian rhythm mutants of Drosophila in light:dark cycles. Journal of Insect Behavior 5, 417-445.

Hardin, P.E., Hall, J.C. and Rosbash, M. 1990. Feedback of the Drosophila period gene product on circadian cycling of its messenger RNA levels. Nature 343, 536-40.

Helfrich-Forster, C. 1998. Robust circadian rhythmicity of Drosophila melanogaster requires the presence of lateral neurons: a brain-behavioral study of disconnected mutants. Journal of Comparative Physiology A-Sensory Neural & Behavioral Physiology 182, 435-53.

Helfrich-Forster, C. and Homberg, U. 1993. Pigment-dispersing hormone-immunoreactive neurons in the nervous system of wild-type Drosophila melanogaster and of several mutants with altered circadian rhythmicity. Journal of Comparative Neurology 337, 177-90.

Helfrich-Forster, C., Tauber, M., Park, J.H., Muhlig-Versen, M., Schneuwly, S. and Hofbauer, A. 2000. Ectopic expression of the neuropeptide pigment-dispersing factor alters behavioral rhythms in Drosophila melanogaster. Journal of Neuroscience 20, 3339-53.

Huang, Z.J., Curtin, K.D. and Rosbash, M. 1995. PER protein interactions and temperature compensation of a circadian clock in Drosophila [see comments]. Science 267, 1169-72.

Huang, Z.J., Edery, I. and Rosbash, M. 1993. PAS is a dimerization domain common to Drosophila period and several transcription factors. Nature 364, 259-62.

Hunter-Ensor, M., Ousley, A. and Sehgal, A. 1996. Regulation of the Drosophila protein timeless suggests a mechanism for resetting the circadian clock by light. Cell 84, 677-85.

Kaneko, M., Park, J.H., Cheng, Y., Hardin, P.E. and Hall, J.C. 2000. Disruption of synaptic transmission or clock-gene-product oscillations in circadian pacemaker cells of Drosophila cause abnormal behavioral rhythms. Journal of Neurobiology 43, 207-33.

Kloss, B., Price, J.L., Saez, L., Blau, J., Rothenfluh, A., Wesley, C.S. and Young, M.W. 1998. The Drosophila clock gene double-time encodes a protein closely related to human casein kinase Iepsilon. Cell 94, 97-107.

Konopka, R.J. and Benzer, S. 1971. Clock mutants of Drosophila melanogaster. Proceedings of the National Academy of Sciences of the United States of America 68, 2112-6.

Konopka, R.J., Hamblen-Coyle, M.J., Jamison, C.F. and Hall, J.C. 1994. An ultrashort clock mutation at the period locus of Drosophila melanogaster that reveals some new features of the fly's circadian system. Journal of Biological Rhythms 9, 189-216.

Krishnan B, D.S., Hardin PE. 1999. Circadian rhythms in olfactory responses of *Drosophila melanogaster*. Nature 400, 375-378.

Kume, K., Zylka, M.J., Sriram, S., Shearman, L.P., Weaver, D.R., Jin, X., Maywood, E.S., Hastings, M.H. and Reppert, S.M. 1999. mCRY1 and mCRY2 are essential components of the negative limb of the circadian clock feedback loop. Cell 98, 193-205.

Lee, C., Bae, K. and Edery, I. 1999. PER and TIM inhibit the DNA binding activity of a Drosophila CLOCK-CYC/dBMAL1 heterodimer without disrupting formation of the heterodimer: a basis for circadian transcription. Molecular & Cellular Biology 19, 5316-25.

Leloup, J.C. and Goldbeter, A. 1997. Temperature compensation of circadian rhythms: control of the period in a model for circadian oscillations of the per protein in Drosophila. Chronobiology International 14, 511-20.

Levine, J.D., Casey, C.I., Kalderon, D.D. and Jackson, F.R. 1994. Altered circadian pacemaker functions and cyclic AMP rhythms in the Drosophila learning mutant dunce. Neuron 13, 967-74.

Liu, Y., Merrow, M., Loros, J.J. and Dunlap, J.C. 1998. How temperature changes reset a circadian oscillator. Science 281, 825-9.

Lowrey, P.L., Shimomura, K., Antoch, M.P., Yamazaki, S., Zemenides, P.D., Ralph, M.R., Menaker, M. and Takahashi, J.S. 2000. Positional syntenic cloning and functional characterization of the mammalian circadian mutation tau. Science 288, 483-92.

Majercak, J., Kalderon, D. and Edery, I. 1997. Drosophila melanogaster deficient in protein kinase A manifests behavior-specific arrhythmia but normal clock function. Molecular & Cellular Biology 17, 5915-22.

Majercak, J., Sidote, D., Hardin, P.E. and Edery, I. 1999. How a circadian clock adapts to seasonal decreases in temperature and day length [see comments]. Neuron 24, 219-30.

Rutila, J.E., Zeng, H., Le, M., Curtin, K.D., Hall, J.C. and Rosbash, M. 1996. The timSL mutant of the Drosophila rhythm gene timeless manifests allele-specific interactions with period gene mutants. Neuron 17, 921-9.

Saez, L. and Young, M.W. 1996. Regulation of nuclear entry of the Drosophila clock proteins period and timeless. Neuron 17, 911-20.

Sawyer, L.A., Hennessy, J.M., Peixoto, A.A., Rosato, E., Parkinson, H., Costa, R. and Kyriacou, C.P. 1997. Natural variation in a Drosophila clock gene and temperature compensation. Science 278, 2117-20.

Schotland, P., Hunter-Ensor, M., Lawrence, T. and Sehgal, A. 2000. Altered entrainment and feedback loop function effected by a mutant period protein. Journal of Neuroscience 20, 958-68.

Sehgal, A., Price, J.L., Man, B. and Young, M.W. 1994. Loss of circadian behavioral rhythms and per RNA oscillations in the Drosophila mutant timeless [see comments]. Science 263, 1603-6.

Sehgal, A., Rothenfluh-Hilfiker, A., Hunter-Ensor, M., Chen, Y., Myers, M.P. and Young, M.W. 1995. Rhythmic expression of timeless: a basis for promoting circadian cycles in period gene autoregulation [see comments]. Science 270, 808-10.

Shearman, L.P., Sriram, S., Weaver, D.R., Maywood, E.S., Chaves, I., Zheng, B., Kume, K., Lee, C.C., van der Horst, G.T., Hastings, M.H. and Reppert, S.M. 2000. Interacting molecular loops in the mammalian circadian clock [see comments]. Science 288, 1013-9.

Siwicki, K.K., Eastman, C., Petersen, G., Rosbash, M. and Hall, J.C. 1988. Antibodies to the period gene product of Drosophila reveal diverse tissue distribution and rhythmic changes in the visual system. Neuron 1, 141-50.

Smith, R.F. and Konopka, R.J. 1981. Circadian clock phenotypes of chromosome aberrations with a breakpoint at the per locus. Molecular & General Genetics 183, 243-51.

So, W.V. and Rosbash, M. 1997. Post-transcriptional regulation contributes to Drosophila clock gene mRNA cycling. EMBO Journal 16, 7146-55.

Stanewsky, R., Kaneko, M., Emery, P., Beretta, B., Wager-Smith, K., Kay, S.A., Rosbash, M. and Hall, J.C. 1998. The cryb mutation identifies cryptochrome as a circadian photoreceptor in Drosophila. Cell 95, 681-92.

Suri, V., Lanjuin, A. and Rosbash, M. 1999. TIMELESS-dependent positive and negative autoregulation in the Drosophila circadian clock. EMBO Journal 18, 675-86.

Suri, V., Qian, Z., Hall, J.C. and Rosbash, M. 1998. Evidence that the TIM light response is relevant to light-induced phase shifts in Drosophila melanogaster. Neuron 21, 225-34.

van der Horst, G.T., Muijtjens, M., Kobayashi, K., Takano, R., Kanno, S., Takao, M., de Wit, J., Verkerk, A., Eker, A.P., van Leenen, D., Buijs, R., Bootsma, D., Hoeijmakers, J.H. and Yasui, A. 1999. Mammalian Cry1 and Cry2 are essential for maintenance of circadian rhythms [see comments]. Nature 398, 627-30.

Vitaterna, M.H., Selby, C.P., Todo, T., Niwa, H., Thompson, C., Fruechte, E.M., Hitomi, K., Thresher, R.J., Ishikawa, T., Miyazaki, J., Takahashi, J.S. and Sancar, A. 1999. Differential regulation of mammalian period genes and circadian rhythmicity by cryptochromes 1 and 2. Proceedings of the National Academy of Sciences of the United States of America 96, 12114-9.

Marrus SB, Z.H., Rosbash M. 1996. Effect of constant light and circadian entrainment of *perS* flies: evidence for light-mediated delay of the negative feedback loop in *Drosophila*. EMBO J 15, 6877-6886.

Matsumoto, A., Tomioka, K., Chiba, Y. and Tanimura, T. 1999. timrit Lengthens circadian period in a temperature-dependent manner through suppression of PERIOD protein cycling and nuclear localization. Molecular & Cellular Biology 19, 4343-54.

McCabe, C. and Birley, A. 1998. Oviposition in the period genotypes of Drosophila melanogaster. Chronobiology International 15, 119-33.

Myers, M.P., Wager-Smith, K., Rothenfluh-Hilfiker, A. and Young, M.W. 1996. Light-induced degradation of TIMELESS and entrainment of the Drosophila circadian clock [see comments]. Science 271, 1736-40.

Myers, M.P., Wager-Smith, K., Wesley, C.S., Young, M.W. and Sehgal, A. 1995. Positional cloning and sequence analysis of the Drosophila clock gene, timeless [see comments]. Science 270, 805-8.

Naidoo, N., Song, W., Hunter-Ensor, M. and Sehgal, A. 1999. A role for the proteasome in the light response of the timeless clock protein. Science 285, 1737-41.

Ousley, A., Zafarullah, K., Chen, Y., Emerson, M., Hickman, L. and Sehgal, A. 1998. Conserved regions of the timeless (tim) clock gene in Drosophila analyzed through phylogenetic and functional studies. Genetics 148, 815-25.

Pittendrigh, C. 1960. Circadian rhythms and the circadian organization of living things. Cold Spring Harbor Symposium on Quantitative Biology 25, 159-184.

Pittendrigh, C. 1967. Circadian Systems I. The driving oscillation and its assay in *Drosophila pseudoobscura*. Proceedings of the. National. Academy of Sciences. USA 58.

Price, J.L. 1997. Insights into the molecular mechanisms of temperature compensation from the Drosophila period and timeless mutants. Chronobiology International 14, 455-68.

Price, J.L., Blau, J., Rothenfluh, A., Abodeely, M., Kloss, B. and Young, M.W. 1998. double-time is a novel Drosophila clock gene that regulates PERIOD protein accumulation. Cell 94, 83-95.

Qiu, J. and Hardin, P.E. 1996. per mRNA cycling is locked to lights-off under photoperiodic conditions that support circadian feedback loop function. Molecular & Cellular Biology 16, 4182-8.

Qiu J, H.P. 1996. Developmental state and the circadian clock interact to influence the timing of eclosion in *Drosophila melanogaster*. Journal of Biological Rhythms 11.

Renn, S.C., Park, J.H., Rosbash, M., Hall, J.C. and Taghert, P.H. 1999. A pdf neuropeptide gene mutation and ablation of PDF neurons each cause severe abnormalities of behavioral circadian rhythms in Drosophila [published erratum appears in Cell 2000 Mar 31;101(1):following 113]. Cell 99, 791-802.

Rothenfluh, A., Young, M.W. and Saez, L. 2000. A TIMELESS-independent function for PERIOD proteins in the Drosophila clock. Neuron 26, 505-14.

Rothenfluh A, A.M., Price JL, Young MW. 2000. Isolation and Analysis of Six timeless Alleles That Cause Short- or Long-Period Circadian Rhythms in Drosophila. Genetics 156, 665-675.

Rutila, J.E., Suri, V., Le, M., So, W.V., Rosbash, M. and Hall, J.C. 1998. CYCLE is a second bHLH-PAS clock protein essential for circadian rhythmicity and transcription of Drosophila period and timeless. Cell 93, 805-14.

Vosshall, L.B., Price, J.L., Sehgal, A., Saez, L. and Young, M.W. 1994. Block in nuclear localization of period protein by a second clock mutation, timeless [see comments]. Science 263, 1606-9.

Yang, Z., Emerson, M., Su, H.S. and Sehgal, A. 1998. Response of the timeless protein to light correlates with behavioral entrainment and suggests a nonvisual pathway for circadian photoreception. Neuron 21, 215-23.

Zatz, M. 1992. Perturbing the pacemaker in the chick pineal. Discoveries in Neuroscience 8, 67-73.

Zeng, H., Hardin, P.E. and Rosbash, M. 1994. Constitutive overexpression of the Drosophila period protein inhibits period mRNA cycling. EMBO Journal 13, 3590-8.

Zeng, H., Qian, Z., Myers, M.P. and Rosbash, M. 1996. A light-entrainment mechanism for the Drosophila circadian clock. Nature 380, 129-35.

Insect Timing: Circadian Rhythmicity to Seasonality
D.L. Denlinger, J. Giebultowicz and D.S. Saunders (Editors)
© 2001 Elsevier Science B.V.   All rights reserved.

Organization of the insect circadian system: spatial and developmental expression of clock genes in peripheral tissues of *Drosophila melanogaster*

J. M. Giebultowicz, M. Ivanchenko and T. Vollintine

Oregon State University, Department of Entomology, 2046 Cordley Hall, Corvallis, OR 97331, USA

Insects display circadian rhythms in all aspects of their lives. The term circadian clock is used to describe the mechanism that generates daily rhythms in insects and other organisms. Genetic and molecular studies on *Drosophila melanogaster* revealed that the clock mechanism comprises a suite of genes regulated via feedback loops which results in their cycling with a circa 24 h period. Two key clock components, genes *period (per)* and *timeless (tim)*, cycle with a peak in the late day followed by a peak in their protein products PERIOD (PER) and TIMELESS (TIM) during the night. Oscillations of these clock molecules are found within and outside of the central nervous system. In a systematic study of fly internal organs, we identified new tissues, such as alimentary tract and fat body, in which PER and TIM cycle in light/dark and constant conditions implying the existence of multi-cellular local oscillators. We then used transgenic flies carrying *per*-lacZ or *tim*-GFP reporter constructs to determine the activity of *per* and *tim* in peripheral organs at different stages of metamorphic development. We found that the two clock genes are turned on at different metamorphic stages in different tissues. Thus, the timing of clock gene activation in the periphery is tissue-specific and, therefore, cannot be accounted for by development-dependent hormonal fluctuations in the hemolymph. Moreover, at least some peripheral oscillators in adults also appear independent of the fly's hormonal milieu with regard to their phases. These oscillators are self-sustained and directly photoreceptive in vitro. Collectively, these observations imply that the fly circadian system is not organized hierarchically, but rather, includes independently operating oscillators, synchronized by the external light/dark cycles.

1. INTRODUCTION

Insect life functions are tightly synchronized with the predictable environmental changes associated with the succession of day and night (Saunders 1982). Such synchronization is achieved by the circadian system, or clock, which generates daily rhythms at the biochemical, physiological and behavioral levels. The defining properties of circadian rhythms (see D. Saunders, this volume) include their entrainment by environmental cycles and persistence in constant environment with a free-running, temperature compensated period. The molecular mechanism of circadian timing has been studied vigorously in *Drosophila melanogaster*, as well as in other organisms, and recent advances have been summarized in several reviews (Dunlap 1999; Giebultowicz 2000; Scully and Kay 2000). Briefly, the clock comprises a set of genes controlled through autoregulatory feedback loops by their proteins, which form heterodimers and enter the cell nuclei at specific times of the day/night cycle. In *Drosophila*, proteins dCLOCK and CYCLE, encoded by *dclock* (*dclk*) and *cycle* (*cyc*) genes, form dCLK-CYC dimers that activate transcription of two other essential clock genes, *period* (*per*) and *timeless* (*tim*), during the day. At the same time, dCLK-CYC dimers inhibit transcription of the *dclk* gene. During the late evening, however, PER-TIM dimmers enter the nucleus, and bind dCLK-CYC complexes, thereby repressing the *per* and *tim* genes but releasing the repression of *dclk* (Glossop et al. 1999). Recent discovery of similar interdependent molecular loops in the mammalian (Shearman et al. 2000) and fungal (Lee et al. 2000) circadian systems suggests that such loops represent a widespread feature of the clock regulation. It is also known that the phase of the clock oscillations can be reset through a rapid change in the level of an essential clock component in response to an environmental signal. For example, in *Drosophila*, resetting of the clock by light involves degradation of TIM protein (reviewed in Young 1998).

To understand the organization of the insect circadian system, two important questions need to be addressed: Which cells in the insect body are bestowed with special abilities to keep track of time and how are the time-keeping centers synchronized with each other? One approach to identify timing centers responsible for behavioral rhythms is monitoring clock outputs after surgical manipulations of the central nervous system (CNS). Clocks controlling locomotor activity rhythms have been mapped to specific areas of the optic lobes in cockroaches and crickets (see review by K. Tomioka, this book). A series of elegant studies in *D. melanogaster* has demonstrated that the pacemaking center for the locomotor activity rhythm is a group of cells called lateral neurons (Helfrich-Forster et al. 1998; Kaneko 1998). In silkmoths, the central brain has been identified as essential for the rhythms of eclosion and locomotor activity (Truman 1972), however, the rhythm in eclosion-preparatory behavior was found to be brain-independent (Truman 1984). Some other rhythms also persist independently of the CNS. For example, the rhythms of cuticle deposition in locusts and sperm release in moths, are maintained in isolated organs in vitro, indicating that they are driven by local clocks (for review see Giebultowicz 1999). This data, along with observation of differences in formal properties of multiple output rhythms in a single organism (Saunders 1986), provided early evidence that the insect circadian system may have multi-oscillatory organization with local clocks controlling tissue-specific functions.

2. EXPRESSION OF CLOCK GENES IN PERIPHERAL TISSUES

The cloning of the *per* gene in *Drosophila melanogaster* and the generation of antibodies against PER protein, made it possible to demonstrate the broad distribution of *per* mRNA and PER protein within and outside of the CNS (reviewed by Hall 1995). Cyclic expression of PER was detected in lateral neurons controlling locomotor activity, in a few other subsets of neurons, and in large groups of glial cells in the brain, as well as in photoreceptor cells comprising the compound eyes (Siwicki et al. 1988; Zerr et al.

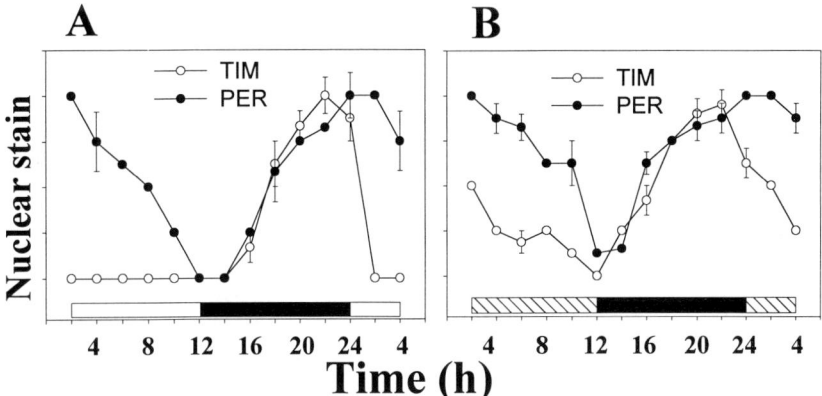

Figure 1. TIM and PER cycling in *Drosophila* Malpighian tubules, assayed by immunofluorescence. Tissues were dissected from flies kept in a normal 12:12 LD cycle (A), or from flies that were transferred to DD for 2 days (B). White bars indicate day, black bars indicate night, and hatched bars indicate subjective day (adapted from Ivanchenko *et al*., submitted).

1990; Kaneko et al. 1997). Among peripheral tissues, the most complete picture of the expression patterns of PER and TIM proteins was obtained for the Malpighian tubules (Hege et al. 1997; Ivanchenko et al. submitted). There are two pairs of Malpighian tubules in *D. melanogaster* consisting of excretory epithelium, which is dominated by large principal cells, all expressing clock molecules. In 12:12 h light-dark cycles (LD), accumulation of PER and TIM in the nucleus begins simultaneously at Zeitgeber Time (ZT) 16, consistent with the fact that these proteins enter the cell nuclei as PER-TIM dimers (Fig. 1A). TIM peaks late at night, and then abruptly disappears after lights-on. PER accumulates in parallel with TIM but persists in the cell nuclei during the beginning of the light phase. In constant darkness (DD), TIM and PER are also rhythmic, however, both proteins are present in the cell nuclei for longer periods of time, compared to LD (Fig. 1B). Similar profiles of TIM and PER have been demonstrated in brain neurons and in whole heads, as determined by immunocytochemistry and Western blotting, respectively (Marrus et al. 1996; Kaneko et al. 1997).

Although Malpigian tubules are the only peripheral organ in which single cell resolution was used to demonstrate cycling in *both* PER and TIM, the presence of *per* mRNA and PER protein was reported in many other organs of *D. melanogaster* (Liu et al. 1988; Saez and Young 1988; Zerr et al. 1990; Hege et al. 1997; Plautz et al. 1997; Emery et al. 1997). Recent use of transgenic flies carrying regulatory *tim* sequences fused to the Green Fluorescent Protein (*tim*-GFP) demonstrated that *tim* is co-expressed with *per* in several tissues (Kaneko and Hall 2000). However, it was not previously demonstrated whether both genes show cyclic and free-running expression in any tissue other than Malpighian tubules. We conducted a systematic survey of peripheral organs with respect to the expression of clock proteins PER and TIM in LD and free-running conditions, determining their levels at 6 h intervals. Simultaneous staining of the same

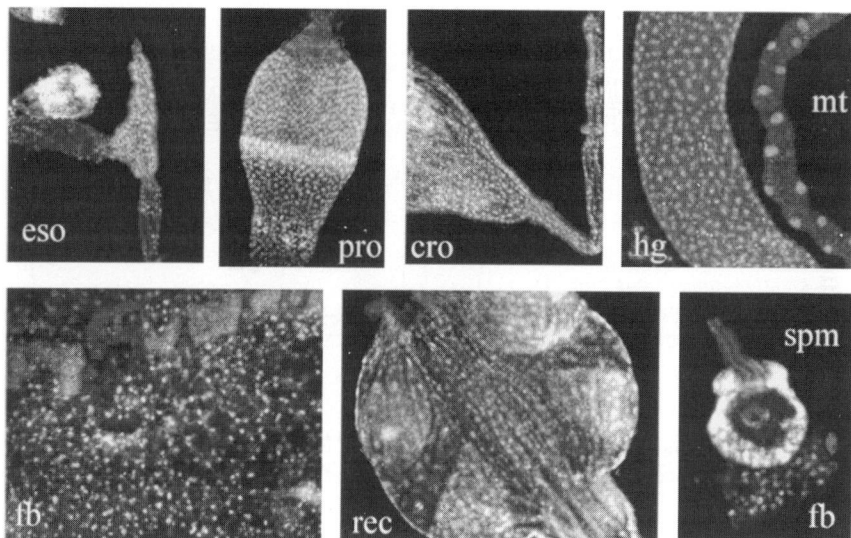

Figure 2. Internal organs of *D. melanogaster* that show cyclic co-expression of PER and TIM, as assayed by double immunofluorescense. Tissues were dissected at 6 h intervals, fixed, and reacted with a mixture of anti-PER and anti-TIM antisera at their optimal concentrations (Giebultowicz and Hege 1997). Immunoreactivity was detected with Alexa flurophores (Molecular Probes). Images represent nuclear TIM signal at the time of its peak (ZT22); PER was co-localized in the nuclei of the same cells (not shown). Six flies were used per time point. Abbreviations: eso, esophagus; pro, proventriculus; hg, hindgut; mt, Malpighian tubules; fb, fat body; rec, rectum, spm, spermatheca.

organs with antibodies against PER and TIM followed by appropriate fluorescent markers allowed us to establish temporal patterns of the two clock proteins. We observed that both PER and TIM were clearly rhythmic in several tissues in LD and DD. PER and TIM peaked in the cell nuclei late at night, at ZT 22 (Fig. 2) and were at a minimum late during the day, at ZT 8, essentially confirming the oscillatory pattern observed in Malpighian tubules (Fig. 1). Entrained and free-running rhythms of PER and TIM were detected in most segments of the alimentary canal, including esophagus, crop, proventriculus (a.k.a. cardia), hindgut, and rectum. PER and TIM were also rhythmic in all examined regions of the fat body, including the subcutaneous fat layer, and fat cells associated with the gut and the reproductive organs. More discrete staining patterns were observed in the female reproductive system. Rhythmic and nuclear expression of PER and TIM were evident in the paired spermathecae and paraovaria, but the signal was absent in the seminal receptacle, the lateral and common oviducts, and uterus. Similarly discrete distribution of clock proteins was observed in the male reproductive system: oscillations in PER and TIM were detected in the testis base, seminal vesicles, and ejaculatory ducts, but not in the main body of the testes or paragonial accessory glands (B. Gvakharia, personal communication). Finally, some tissues including the epidermis, skeletal muscles and tracheal epithelium did not show detectable levels of either PER or

TIM, demonstrating that clock proteins are not ubiquitously present in all tissues but rather are limited to specific, albeit multiple, organs or their specific segments.

A few studies performed on other insects provide additional evidence of clock genes being active in peripheral tissues. Daily oscillations of *per* mRNA and protein were detected in the larval gut of the silkworm, *Antheraea pernyi* (Sauman and Reppert 1998), and in the reproductive system of the codling moth, *Cydia pomonella* (Gvakharia et al. 2000). Insects are not an exception with regard to widespread activity of the timing genes in their bodies; a similar picture has emerged from studies of vertebrates. From fishes to mammals, mRNAs coding for *clk*, *BMAL1* (the vertebrate equivalent of *dcyc*), and *per* were detected in organs such as heart, lungs, kidney, and testis (King et al. 1997; Tei et al. 1997; Whitmore et al. 1998; Yamazaki et al. 2000).

3. AUTONOMY OF THE PERIPHERAL CLOCKS.

The cycling of clock proteins in different fly tissues shows remarkable synchrony of phase (with the exception of certain brain neurons (see, Kaneko 1998). The times of nuclear translocation of PER and TIM and the times of their maximal expression are similar in peripheral organs (Fig. 1 and 2), in the brains of adult flies (Stanewsky et al. 1997), and in the lateral neurons of larval and pupal brains (Kaneko et al. 1997; 2000; Ivanchenko, submitted). There are two possible explanations for the synchrony between peripheral oscillators and those located in the brain. First, the brain oscillator entrained by the LD cycles could coordinate peripheral oscillators via blood-borne factor(s). Second, central and peripheral oscillators could operate independently, achieving synchronization via direct entrainment to the LD cycles. Several lines of experimental evidence, discussed below, support the second possibility.

One piece of early evidence for the existence of autonomous circadian clocks in insects came from studying the rhythms of sperm release in moths. In many moth species, the release of sperm bundles from the testis to the vas deferens shows a daily rhythm (reviewed by Giebultowicz 1999). The rhythm continues in cultured testis-vas deferens complexes and can be entrained in vitro by LD cycles (Giebultowicz et al. 1989). Thus, a whole circadian system including photoreceptors, the clock pacemaker and the output rhythms, is located in this specific portion of the male moth reproductive system. Importantly, the very same tissues rhythmically express *per* mRNA and PER protein (Gvakharia et al. 2000). Another example of a brain-independent cycle is the daily pattern of cuticle deposition in the integument of some insects, which arises due to rhythmic changes in the orientation of secreted cuticular layers (Neville 1970). This rhythmic activity continues in pieces of integument cultured *in vitro*, providing indirect evidence for the existence of an autonomous circadian oscillator in epidermal cells (Weber 1995)

Convincing evidence for the existence of autonomous local oscillators was gained using transgenic *D. melanogaster* that express luciferase under *per* or *tim* regulatory sequences (*per*-luc and *tim*-luc lines); luciferase acts as a real time reporter for the activity of the respective clock genes (Brandes et al. 1996). Owing to the fact that such transgenic flies produce measurable light proportional to clock gene activities, one can determine whether those genes continue to oscillate in internal organs cultured in vitro. A group of tissues with self-sustained and light-entrainable rhythms of *per* was identified this way. One example is the prothoracic (ring) gland, which produces the insects molting hormone ecdysone. In fly pupae, *per* gene and protein are rhythmically expressed in the ring gland and this expression continues in vitro (Emery et al. 1997). Rhythms in *per*-luc activity were also found in chemosensory hairs located on the flies' antennae, proboscis, wing margin, and legs maintained in vitro; these rhythms persisted in DD and were shifted in response to a change in the LD cycle (Plautz et al. 1997). We recently demonstrated that both *per*:luc and *tim*:luc oscillate in hindgut-rectum complexes and Malpighian tubules in vitro (Giebultowicz et al. 2000). These oscillations persisted in constant darkness with a period of nearly 24 h, and oscillation amplitude increased upon return to LD, consistent with the sensitivity of these tissues to changes in the environmental LD cycles.

The assumption that different local time-keeping centers widely distributed in the fly body may be autonomous in their entrainment to the environmental LD cycles requires that these clocks are equipped with their own light-sensing devices. Some organs in *Drosophila* are not innervated, thus lacking any direct connection with the fly external photoreceptors. We have recently studied the circadian photoreception in such an organ, the Malpighian tubules. We determined that the blue light circadian photoreceptor CRYPTOCHROME (CRY), recently identified in plants and insects (reviewed by Hall 2000), is present in Malpighian tubules. This photoreceptor entrains the clock in Malpighian tubules, as it does the central clock in the brain (Stanewsky et al. 1998). CRY mediates rapid degradation of the clock protein TIM, known to entrain the fly clock in response to the environmental dark-light changes (Young 1998) both in Malpighian tubules and in the brain lateral neurons. In addition to CRY, the clock in Malpighian tubules is entrained by an unknown tissue-autonomous mechanism that does not directly affect the level of TIM (Ivanchenko et al, submitted).

The physiological and molecular evidence presented above support the hypothesis that the peripheral oscillators in insects are self-sustained and photoreceptive when cultured in vitro. One may expect that these oscillators would also have a high degree of autonomy in vivo, although phase-imposing effects of the brain clock cannot be excluded *a priori*. To address this issue we monitored the free-running cycles of TIM protein in *Drosophila* Malpighian tubules that were transplanted to host flies entrained to reverse LD cycles with respect to the donor flies. TIM in the transplanted tubules cycled 12 hours out of phase compared to host tubules, suggesting that different clocks in one organism may operate independently despite sharing the same hormonal milieu (Giebultowicz et al. 2000). It appears, based on these observations, that circadian coordination of physiological sub-systems in insects may be achieved via direct entrainment of light-sensitive autonomous oscillators by environmental signals.

4. ACTIVATION OF CLOCK GENES DURING METAMORPHOSIS.

One of the important considerations in trying to understand how the insect circadian system is organized relates to the developmental origins of the circadian clocks. It has been known that the phase of adult behavioral rhythms in fruit flies can be set by a pulse of light applied in early larval life to insects otherwise held in DD (reviewed in (Kaneko 1998). Examination of PER and TIM staining patterns in the CNS revealed that their cyclic expression persist in the lateral neurons of larval and pupal brains, providing

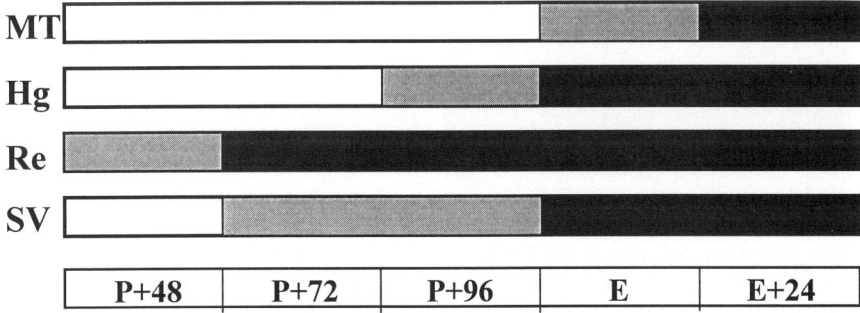

Figure 3. Activation of *tim* and *per* genes during metamorphosis in selected tissues of D. *melanogaster*, as shown by the by *per*-LacZ and *tim*-GFP reporters. The levels of expression were scored subjectively as negative (white), intermediate (gray) and strong (black). Developmental stages (see text) are indicated at the bottom. Tissues from 3 independently transformed lines for each reporter gave similar results. No fluorescence was observed in control flies carrying the GFP reporter without the *tim* regulatory regions.

the molecular basis for circadian time memory (Kaneko et al. 1997). Unlike the lateral neurons in the brain, which are preserved from larvae to adult, most adult internal organs differentiate de novo from the larval histoblast during metamorphosis. We examined the developmental activation of the two clock genes, *per* and *tim*, in selected fly organs using transgenic flies carrying *per*-lacZ (Stanewsky et al. 1997) or *tim*-GFP (Kaneko and Hall 2000) reporter constructs. Developing adults were staged according to Bainbridge and Bownes (1981) and examined with respect to expression of clock genes at approximately 24 h intervals after pupariation. Activity of lacZ was detected in fixed organs as described (Hege et al. 1997), and the intensity of GFP signal was observed in live organs using a Zeiss Axiovert microscope with a GFP-optimized filter set. In all examined organs, the onset of both *per* and *tim* gene expression occurred simultaneously; this may signify the activation of local clock function, since both genes are required for initiation of circadian cycles. However, the activation of clock genes occurred at different metamorphic stages in different organs (Fig. 3), indicating that it is developmentally regulated, but in a tissue-specific way. The earliest expression of *per* and *tim* was detected approximately 48 hours after puparium formation in the four rectal pads but not in other parts of the rectum (Fig. 4). Activation of the genes in the male seminal vesicles occurred approximately 24 hours later (P+72). Several other tissues, including the hindgut and the remainder of the rectum, showed weak expression of clock genes one day before adult eclosion (P+96) and a strong expression 24 h later, at the time of eclosion. Finally, the Malpighian tubules were the last tissue in which the timing mechanism was activated; weak expression of *per* and *tim* was detected at eclosion and maximal expression was achieved one day after eclosion.

To gain a more complete picture of the developmental pattern of clock gene expression, we also studied the activity of such genes in larvae of *D. melanogaster*. Most larval tissues die during metamorphosis and adult tissues differentiate de novo within the

tim-GFP   *per*-lacZ

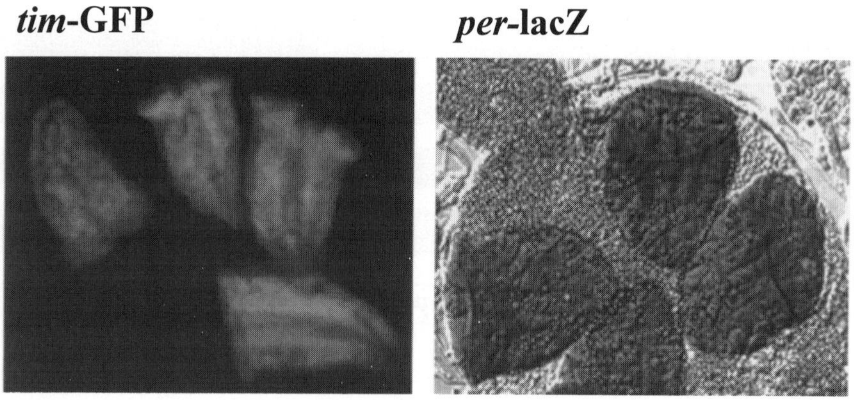

Figure 4. Expression of *tim* as reported by *tim*-GFP and *per* as reported by *per*-lacZ in the rectum of *D. melanogaster* 72 hours after pupariation. The four rectal pads are strongly expressing both genes at this time. Note the excretory material in the rectum appearing like granulated deposits on the *per*-lacZ image.

pupal case, with the exception of the Malpighian tubules, which function as an excretory organ in larvae and then survive metamorphosis to resume the same function in adults. Do larval tubules harbor a circadian oscillator? The answer appears to be negative, since immunocytochemistry failed to detect PER or TIM proteins in larval tubules and *tim*-GFP reporter gave no signal (Giebultowicz, unpublished). Thus, this tissue presents an interesting case where the timing mechanism is absent in the organ at the larval stage but becomes active in the same organ in adults. We also tested the digestive and excretory systems taken from third instar larvae carrying *tim*-GFP and did not detect *tim* activity in these tissues. The widespread presence of the timing oscillators in the alimentary tract of adults but not in larvae may be related to different lifestyles and feeding habits of the two life stages. Larvae seem to feed continually, whereas adults may feed periodically in correlation with their rest/activity cycles, although this was not yet rigorously tested.

The recruitment of timing mechanisms in specific physiological contexts is supported by the fact that the expression of clock genes is initiated at different times in different internal organs during adult development. Such tissue-by-tissue activation of clock genes suggests that peripheral clocks may be turned on by tissue-specific signals, rather than by a central mechanism. The independent onset of clock gene expression is likely to be correlated with tissue-specific needs for circadian synchronization. The very early activation of clock genes in the developing adult rectal pads may be related to the accumulation of excretory material observed in the rectum during metamorphosis (Fig 4). One may speculate that a putative oscillator in the rectal epithelium may be required for excretory cycles associated with metamorphic development.

5. FUNCTIONS OF THE INSECT PERIPHERAL OSCILLATORS.

The oscillations of clock molecules in insect peripheral organs lead to the expectation that these local oscillators impose daily rhythms on tissue-specific processes.

Unfortunately, cellular and physiological outputs are not yet known for most of the newly identified peripheral oscillators, therefore, their clock status must remain tentative. However, there are few cases in which the role of local oscillators has become apparent, providing the first evidence that these oscillators are essential components of the multi-oscillatory circadian system that seems to operate in insects.

One prominent example of a biologically relevant local oscillator is the brain-independent clock located in the vas deferens of male moths. Several circadian rhythms occurring in daily succession were identified in this complex (reviewed by Giebultowicz 1999). First, sperm is released from the testis during a circadian gate at the end of the day and stored in the vas deferens. This rhythm is correlated with the rhythmic secretion of glycoproteins from vas deferens epithelium into its lumen. Several hours later, sperm is pushed out of the vas deferens by daily increases in contraction intensity of the vas deferens wall. Thus, a locally operating clock appears to synchronize multiple rhythms involved in sperm release and maturation. Disruption of the circadian rhythms by constant light leads to male sterility, demonstrating their critical role in reproduction. A similar mechanism may operate in *D. melanogaster*. A putative circadian oscillator was identified in specific parts of the male fly reproductive system that are involved in sperm release and maturation (Gvakharia et al, in preparation). Involvement of local circadian clocks in fly reproduction is further suggested by the fact that the activation of clock genes in the male reproductive tissues (Fig. 3) precedes the first sperm release in developing adults (Giebultowicz, unpublished).

A defined output function has also been assigned to the peripheral oscillators in the chemosensory hairs on the fly antennae. These organs display rhythm in electrophysiological responses to two different classes of olfactory stimuli. Genetic tests demonstrated that olfactory rhythms are driven by the oscillations of locally expressed clock genes, rather than those active in the central brain (Krishnan et al. 1999).

Although the biological significance of the multiple insect peripheral clocks remains obscure, physiological functions may emerge from co-localization of clock genes with rhythmic output genes. A case in point is the identification of oscillatory expression of the *takeout* gene, which codes for a ligand-binding protein, in segments of the alimentary canal (Sarov-Blat et al. 2000) in which we observed cycling of PER and TIM proteins.

6. CONCLUSIONS

The existence of multi-oscillatory systems in complex animals was predicted in the past based on the dissimilar characteristics of output rhythms in single organisms (see, for example, Saunders 1986; Tossini and Menaker 1998). The use of molecular tools for mapping the activity of clock genes confirms these predictions by revealing the widespread distribution of these genes in peripheral organs. Rhythmic expression of clock genes that free-runs in constant darkness provide the initial clue as to which insect organs may posses circadian clock function. However, final proof of such function will be the identification of physiological output rhythms. This has not been achieved for most clock-positive tissues. An even bigger future task is developing an understanding of the regulatory cascades leading from the clock genes to overt rhythms via clock-controlled effector molecules.

Despite a lack of physiological data, we have made substantial progress in constructing an overall map of the fly circadian system. Several lines of evidence discussed here suggest that some of the identified putative local oscillators may have remarkable degrees of autonomy. First, they are self-sustained and photoreceptive in vitro, and, second, they can be made to oscillate out of phase with each other in vivo. Finally, we demonstrate here that initiation of clock oscillations during metamorphosis occurs independently in different tissues. All these observations lead to the tentative conclusion that circadian coordination of physiological sub-systems in insects may be achieved through independently operating clocks that are directly entrained by external cycles of light and darkness.

ACKNOWLEDGMENTS

We thank J. Hall for sharing transgenic flies and PER antiserum and M. Young for TIM antiserum. This work was supported by grants from NSF (99723227-IBN) and USDA NRI (9802598) to JMG.

REFERENCES

Bainbridge, S.P., Bownes, M., 1981. Staging the metamorphosis of *Drosophila melanogaster*. Journal of Embryology and Morphology 66, 57-80.

Brandes, C., Plautz, J., D., Stanewsky, R., Jamison, C.F., Straume, M., Wood, K.V., Kay, S.A., Hall, J.C., 1996. Novel features of *Drosophila period* transcription revealed by real-time luciferase reporting. Neuron 16, 687-692.

Dunlap, J.C., 1999. Molecular bases for circadian clocks. Cell 96, 271-290.

Emery, I.F., Noveral, J.M., Jamison, C.F., Siwicki, K.K., 1997. Rhythms of *Drosophila period* gene expression in culture. Proceedings of the National Academy of Science USA *94*, 4092-4096.

Giebultowicz, J.M., 1999. Insect circadian rhythms: Is it all in their heads? Journal of Insect Physiology 45, 791-800.

Giebultowicz, J.M., 2000. Molecular mechanism and cellular distribution of insect circadian clocks. Annual Review of Entomology 45, 767-791.

Giebultowicz, J.M., Riemann, J.G., Raina, A.K., Ridgway, R.L., 1989. Circadian system controlling release of sperm in the insect testes. Science 245, 1098- 1100.

Giebultowicz, J.M., Stanewsky, R., Hall, J.C., Hege, D.M., 2000. Transplanted *Drosophila* excretory tubules maintain circadian clock cycling out of phase with the host. Current Biology 10, 107-110.

Glossop, N.R., Lyons, L.C., Hardin, P.E., 1999. Interlocked feedback loops within the *Drosophila* circadian oscillator. Science 286, 766-768.

Gvakharia, B.O., Kilgore, J.A., Bebas, P., Giebultowicz, J.M., 2000. Temporal and spatial expression of the *period* gene in the reproductive system of the codling moth. Journal of Biological Rhythms 15, 27-35.

Hall, J.C., 1995. Tripping along the trail to the molecular mechanisms of biological clocks. Trends in Neuroscience 18, 230-240.

Hall, J.C., 2000. Cryptochromes: sensory reception, transduction, and clock functions subserving circadian systems. Current Opinion in Neurobiology 10, 456-486.

Hege, D.M., Stanewsky, R., Hall, J.C., Giebultowicz, J.M., 1997. Rhythmic expression of a PER-reporter in the Malpighian tubules of decapitated *Drosophila*: evidence for a brain-independent circadian clock. Journal of Biological Rhythms 12, 300-308.

Helfrich-Forster, C., Stengl, M., Homberg, U., 1998. Organization of the circadian system in insects. Chronobiology International 15, 567-594.

Kaneko, M., 1998. Neural substrates of *Drosophila* rhythms revealed by mutants and molecular manipulations. Current Opinion in Neurobiology 8, 652-658.

Kaneko, M., Hall, J.C., 2000. Neuroanatomy of cells expressing clock genes in *Drosophila*: transgenic manipulation of the *period* and *timeless* genes to mark the perikarya of circadian pacemaker neurons and their projections. Journal of Comparative Neurology 422, 66-94.

Kaneko, M., Hamblen, M.J., Hall, J.C., 2000. Involvement of the *period* gene in developmental time-memory: effect of the *per*Short mutation on phase shifts induced by light pulses delivered to *Drosophila* larvae. Journal of Biological Rhythms 15, 13-30.

Kaneko, M., Helfrich-Forster, C., Hall, J.C., 1997. Spatial and temporal expression of the *period* and *timeless* genes in the developing nervous system of *Drosophila*: Newly identified pacemaker candidates and novel features of clock gene product cycling. Journal of Neuroscience 17, 6745-6760.

King, D.S., Zhao, Y., Sangoram, A.M., Wilsbacher, L.D., Tanaka, M., Antoch, M.P., Steeves, T.D.L., Vitaterna, M.H., Kornhauser, J.M., Lowrey, P.L., Turek, F.W., Takahashi, J.S., 1997. Positional cloning of the mouse circadian *clock* gene. Cell 89, 644-654.

Krishnan, B., Dryer, S.E., Hardin, P.E., 1999. Circadian rhythms in olfactory responses of *Drosophila melanogaster*. Nature 400, 375-378.

Lee, K., Loros, J.J., Dunlap, J.C., 2000. Interconnected feedback loops in the *Neurospora* circadian system. Science 289, 107-110.

Liu, X., Lorenz, L., Yu, Q., Hall, J.C., Rosbash, M., 1988. Spatial and temporal expression of the *period* gene in *Drosophila melanogaster*. Genes & Development 2, 228-238.

Marrus, S.B., Zeng, H., Rosbash, M., 1996. Effect of constant light and circadian entrainment of *per*^{S} flies: evidence for light-mediated delay of the negative feedback loop in *Drosophila*. EMBO Journal 15, 6877-86.

Neville, A.C. 1970. Cuticle ultrastructure in relation to the whole insect. In: *Insect Ultrastructure*. Neville, A. C. Oxford, Blackwell Scientific. pp 17-39.

Plautz, J.D., Kaneko, M., Hall, J.C., Kay, S.A., 1997. Independent photoreceptive circadian clocks throughout *Drosophila*. Science 278, 1632-1635.

Saez, L., Young, M.W., 1988. *In situ* localization of the *per* clock protein during development of *Drosophila melanogaster*. Molecular and Cellular Biology 8, 5378-5385.

Sarov-Blat, L., So, W.V., Liu, L., Rosbash, M., 2000. The *Drosophila takeout* gene is a novel molecular link between circadian rhythms and feeding behavior. Cell 101, 647-656.

Sauman, I., Reppert, S.M., 1998. Brain control of embryonic circadian rhythms in the silkworm, *Antheraea pernyi*. Neuron 20, 741-748.

Saunders, D.S. (1982). Insect Clocks. Pergamon Press, Oxford.

Saunders, D.S., 1986. Many circadian oscillators regulate developmental and behavioral events in the flesh fly, *Sarcophaga argyrostoma*. Chronobiology International 3, 71-83.

Scully, A.L., Kay, S.A., 2000. Time flies for *Drosophila*. Cell 100, 297-300.

Shearman, L.P., Sriram, S., Reppert, S.M., 2000. Interacting molecular loops in the mammalian circadian clock. Science 288, 1013.

Siwicki, K.K., Eastman, C., Petersen, G., Rosbash, M., Hall, J.C., 1988. Antibodies to the *period* gene product of *Drosophila* reveal diverse tissue distribution and rhythmic changes in the visual system. Neuron 1, 141-150.

Stanewsky, R., Frisch, B., Brandes, C., Hamblen-Coyle, M.J., Rosbash, M., Hall, J.C., 1997. Temporal and spatial expression patterns of transgenes containing increasing amounts of the *Drosophila* clock gene *period* and a *lacZ* reporter: mapping elements of the PER protein involved in circadian cycling. Journal of Neuroscience 17, 676-696.

Stanewsky, R., Kaneko, M., Emery, P., Beretta, B., Wager-Smith, K., Kay, S.A., Rosbash, M., Hall, J.C., 1998. The cry^b mutation identifies cryptochrome as a circadian photoreceptor in *Drosophila*. Cell 95, 681-692.

Tei, H., Okamura, H., Sakaki, Y., 1997. Circadian oscillation of a mammalian homologue of the *Drosophila* period gene. Nature 389, 512-515.

Tossini, G., Menaker, M., 1998. Multioscillatory circadian organization in a vertebrate, *Iguana iguana*. Journal of Neuroscience 18, 1106-1114.

Truman, J.W., 1972. Physiology of insect rhythms II: the silkworm brain as the location of the biological clock controlling eclosion. Journal of Comparative Physiology 81, 99-114.

Truman, J.W., 1984. The preparatory behavior rhythm of the moth *Manduca sexta*: an ecdysteroid triggered circadian rhythm that is independent of the brain. Journal of Comparative Physiology A 155, 521-528.

Weber, F., 1995. Cyclic layer deposition in the cockroach (*Blaberus cranifer*) endocuticle: a circadian rhythm in leg pieces cultured *in vitro*. Journal of Insect Physiology 41, 153-161.

Whitmore, D., Foulkes, N.S., Strahle, U., Sassone-Corsi, P., 1998. Zebrafish *Clock* rhythmic expression reveals independent peripheral circadian oscillators. Nature Neuroscience 1, 701-708.

Yamazaki, S., Numano, R., Abe, M., Hida, A., Takahashi, R., Ueda, M., Block, G.D., Sakaki, Y., Menaker, M., Tei, H., 2000. Resetting central and peripheral circadian oscillators in transgenic rats. Science 288, 682-5.

Young, M.W., 1998. The molecular control of circadian behavioral rhythms and their entrainment in *Drosophila*. Annual Review of Biochemistry 67, 135-152.

Zerr, D.M., Hall, J.C., Rosbash, M., Siwicki, K.K., 1990. Circadian fluctuations of *period* protein immunoreactivity in the CNS and the visual system of *Drosophila*. Journal of Neuroscience 10, 2749-2762.

Insect Timing: Circadian Rhythmicity to Seasonality
D.L. Denlinger, J. Giebultowicz and D.S. Saunders (Editors)
© 2001 Elsevier Science B.V. All rights reserved.

The circadian clock system of hemimetabolous insects

K. Tomioka, A. S. M. Saifullah and M. Koga

Department of Physics, Biology and Informatics, Faculty of Science and Research Institute for Time Studies, Yamaguchi University, Yamaguchi 753-8512, Japan

The circadian system of hemimetabolous insects, such as crickets and cockroaches, is a unique model for the study of the circadian timekeeping mechanism. The circadian clock is located one in each optic lobe, regulates a variety of physiological functions, such as locomotor activity, sensitivity of the visual system and neuronal activity, and synchronizes to a light dark cycle perceived in most cases by the compound eye. The exact location of the clock cells is unknown, although the accessory medulla was recently proposed as a candidate locus. The oscillatory mechanism of the clock is still largely unknown, but experiments with a cytosolic translation inhibitor, cycloheximide, revealed that at least protein synthesis is a required process for the clock functioning in crickets. The bilaterally redundant clocks retain their synchrony and provide a stable temporal structure to the animal's behavior by interacting with one another through neural pathway. In the interaction, so-called medulla bilateral neurons, serotonergic neurons and pigment dispersing hormone-immunoreactive neurons are suggested to be involved. The hemimetabolous clock system may contribute to the understanding of the circadian organization that maintains the well-ordered temporal structure in physiology and behavior in multi-oscillator systems.

1. INTRODUCTION

Hemimetabolous insects, such as crickets and cockroaches, provide good animal models for the study of insect circadian system (Helfrich-Förster et al., 1998). The reasons for this are that they are rather large insects enabling easy surgical manipulation and that they show a clear circadian locomotor or stridulatory activity that is easy to record with rather simple techniques. The circadian system is composed of three major constituents, i.e., circadian clock that is a self-sustaining oscillator generating the circadian oscillation, input system which is necessary for synchronization to environmental 24 hr cycles, and output system that regulates the various physiological functions such as behavioral activity, hormonal secretion, neuronal activity and sensitivity of

This work was supported in part by grants from the Ministry of Education, Science, Sports and Culture of Japan, The Inamori Foundation and The Sumitomo Foundation.

sensory systems (Page, 1985a). To understand the mechanism that regulates the circadian rhythms, it is required to identify all these constituents in animal's central nervous system and elucidate the interaction among them as well as the oscillatory mechanism of the clock. In this review we will summarize our current understanding of hemimetabolous circadian clock.

2. SEARCH FOR THE CLOCK LOCUS

The optic lobe has been recognized as a circadian clock locus for many years since Nishiitsutsuji-Uwo and Pittendrigh (1968) reported that arrhythmic activity was induced by complete ablation of the optic lobe in the cockroach *Leucophaea maderae*. It is composed of three major neuropiles, i.e., lamina, medulla and lobula, from distal to proximal in that order (Fig. 1). A small neuropile called the accessory medulla was recently found between the medulla and lobula. A series of surgical lesioning experiments followed the report by Nishiitsutsuji-Uwo and Pittendrigh to localize the clock within the optic lobe in cockroaches. Roberts (1974) found that surgical lesioning of medulla area often resulted in the loss of the locomotor rhythm and concluded that the medulla and lobula are crucial elements for the functional clock. With electrolytic lesioning, somata in the ventral part of the inner chiasma and lobula were subsequently

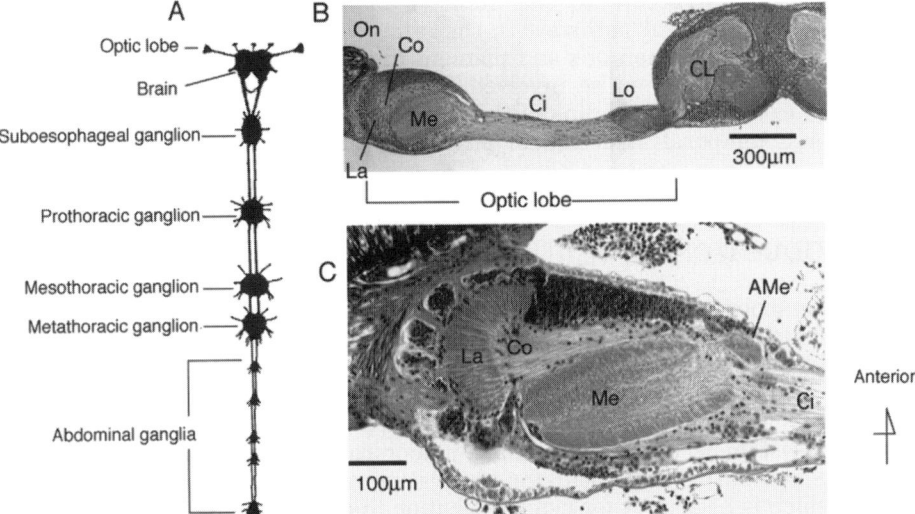

Figure 1. The schematic drawing of the central nervous system (A) and haematoxylin-eosin stained horizontal 8 μm section of the optic lobe (B,C) of the cricket *Gryllus bimaculatus*. AMe, accessory medulla; Ci, inner chiasma; CL, cerebral lobe; Co, outer chiasma; La, lamina; Lo, lobula; Me, medulla; On, optic nerve. The inner chiasma forms a long nerve tract, called in crickets the optic stalk.

suggested to be crucial (Sokolove, 1975). Page (1978) further extended this hypothesis and concluded that the cells in the ventral half of the optic lobe near the lobula are critical for maintenance of the rhythm. Experiments measuring electroretinogram (ERG) further confirmed the hypothesis (Wills et al., 1985).

Also in crickets, studies with surgical lesioning suggested the optic lobe to be the clock locus (Loher, 1972; Tomioka and Chiba, 1984; Abe et al., 1997; Shiga and Numata, 1999). In *Gryllus bimaculatus*, the fact that the ERG rhythm persisted even after the optic stalk was severed (Tomioka and Chiba, 1982) and that circadian electrical activity rhythms of optic lobe efferent neurons persisted in the isolated optic lamina-medulla complex (Tomioka and Chiba, 1986, 1992) unequivocally show that this complex is the site of the circadian clock.

Recently the accessory medulla was proposed as the clock locus both in cockroaches and crickets (Helfrich-Förster et al., 1998). The neuropile is rather peculiar both in its structure and development. Unlike other optic neuropiles, it does not have layer structure corresponding to the retinotopic organization of the visual system. In *G. bimaculatus*, it grows only up to 5 times the size of the 1st instar nymph even in adults (Tomioka and Terada, unpublished data). The most characteristic feature of this neuropile is its innervation by neurons immunoreactive to anti-pigment dispersing hormone antibody (PDH-IRNs). The PDH-IRNs are now strongly hypothesized as the circadian clock neurons by Stengl, Homberg and their colleagues. The data on which the hypothesis is based are as follows. First, in *Drosophila*, the lateral neurons, which are considered to be the circadian clock cells regulating the circadian locomotor rhythm, are also immunoreactive to anti-PDH (Helfrich-Förster, 1995); they are recently

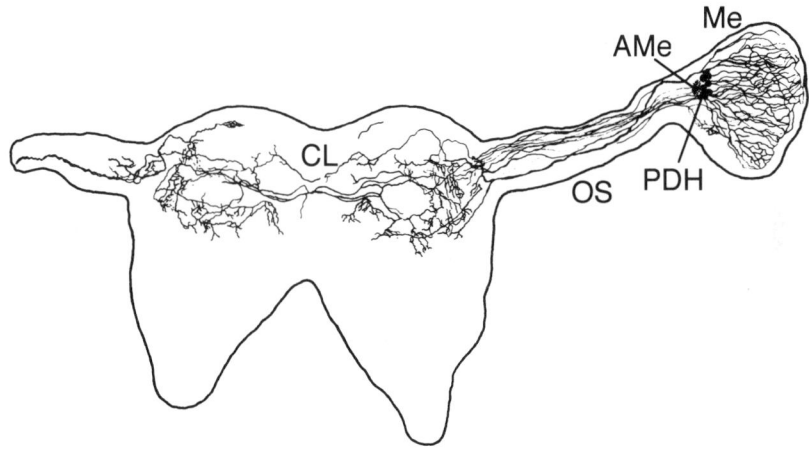

Figure 2. Camera lucida drawing of the PDH-IRNs in the brain and the optic lobe of a cricket, *Gryllus bimaculatus*, that showed arrhythmic activity after removal of lamina and 1/3 of medulla in addition to the contralateral lamina-medulla complex removal. AMe, accessory medulla; CL, cerebral lobe; Me, medulla; OS, optic stalk; PDH, cell bodies of PDH-IRNs.

confirmed to produce PDF (pigment dispersing factor) (Renn et al., 1999). Second, there is a positive correlation between the restoration of the locomotor rhythm and the regeneration of PDH-IRNs in cockroaches with optic tract severed. Circadian locomotor rhythms are often restored after several weeks of bilateral optic tract severance (Page, 1983). In cockroaches with restored rhythms, the PDH-IRNs regenerate to project to certain parts of the cerebral lobe (Stengl and Homberg, 1994). Third, crickets with bilateral optic lobe removed still showed persistent rhythmicity if the fragment of the PDH-IRNs remained undegenerated in the brain (Stengl, 1995).

These recent arguments are somewhat different from the previous hypothesis proposed on the results of mainly physiological and behavioral studies. We recently re-examined the role of optic lobes in the generation of locomotor rhythm in the cricket *G. bimaculatus*, with surgical lesioning experiments coupled with immunohistochemistry using anti-PDH antibody (Okamoto et al, 2000). One optic lobe was removed and the contralateral lobe was partially removed to various extents. The rhythm persisted if the lamina was removed, but no rhythm was exhibited at all after the removal of distal 1/3 to 1/2 of the medulla. The accessory medulla was left intact and the PDH-IRNs in the proximal medulla were with nearly intact shape (Fig. 2). The results suggest neither that PDH-IRNs in the medulla are the clock neurons nor that the accessory medulla is the clock locus. It is more likely that the distal area of the medulla or the area of the outer chiasma is important for the rhythm generation. This hypothesis is supported by the observation that a cluster of neurons immunoreactive to *Periplaneta*-PERIOD protein are located in the dorsal and ventral part of the outer chiasma in the cockroach *Periplaneta americana* (M. Takeda, personal communication).

3. MOLECULAR DISSECTION OF THE CIRCADIAN CLOCK

It is now widely accepted that the circadian clock in *Drosophila* consists of a feedback loop in which clock genes, such as *period* and *timeless*, are rhythmically expressed (Dunlap, 1999). The validity of this hypothesis has been examined in several insect species. In the cockroach *P. americana*, the protein shows neither daily change in its abundance nor translocation into the nucleus (M. Takeda, personal communication). Although a daily oscillation in mRNA and protein levels of *period* has been shown in silkmoths *Anthraea pernyi*, the PER protein stays in cytoplasm and does not translocate to the nucleus in cerebral clock cells. Therefore, it needs to be carefully examined whether the autoregulatory feedback loop found in *Drosophila* is functioning in other hemimetabolous as well as holometabolous insects.

Transcription and translation are likely involved in cricket circadian rhythms. It has been shown that RNA transcription and protein synthesis are greater during the light phase in cerebral neurosecretory cells of *Acheta domesticus*, which are reportedly involved in the regulation of the locomotor rhythm (Cymborowski and Dutkowski, 1969, 1970). We have examined the effects of cycloheximide (CHX), a cytosolic translation inhibitor, on the cricket (*G. bimaculatus*) optic lobe circadian clock *in vitro*. Continuous treatment with CHX resulted in arrhythmicity in both spontaneous and light-induced activity of the optic lobe efferent neurons (Fig. 3). The treatment did not disturb the responsiveness to the visual stimulus. Thus it is concluded that the treatment had an effect specific to the circadian clock. Short-term treatments (10^{-5} M, 6hr) resulted in phase shifts of the rhythm: the magnitude and direction of the phase shift were dependent on the time of treatment, yielding a phase response curve with delays during the late subjective day to early subjective night and phase advances during the late subjective night (Fig. 3E) (Tomioka, 2000). These data

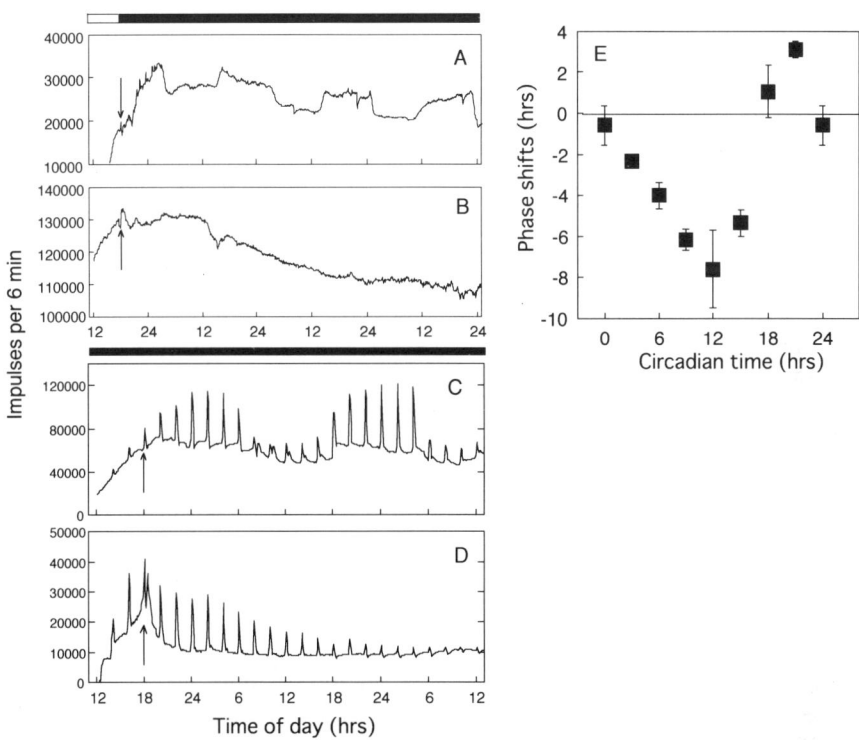

Figure 3. A-D: Effects of continuous treatment with CHX on the circadian rhythm of spontaneous (A, B) and light induced neuronal activity (C, D) in the cricket (*Gryllus bimaculatus*) optic lamina-medulla compound eye complex *in vitro*. The frequency of optic stalk impulses is plotted as a function of the time of day. The isolated complexes were transferred to constant darkness at 18:00 (A, B) or 12:00 (C, D). White and black bars indicate light (white) and dark (black). (A, C) Rhythms from control complexes in which the medium was replaced at 18:00 (arrow). (B, D) The records from complexes treated with CHX of 10^{-4}M (B) or 10^{-3} M (D). Exposure of the complex to CHX began at 18:00 (arrow). In C and D, sharp increases in neuronal activity indicate the response to 15 min light pulses given every 2 hrs. (E) Phase response curve obtained by 6 hr CHX (10^{-5} M) pulses.

indicate that protein synthesis is necessary for the movement of the optic lobe circadian clock and suggest that it occurs during the late subjective day to the late subjective night. The phase of the required protein synthesis in the cricket well corresponds to that suggested in *Drosophila* clock system. The phase of clock protein synthesis is different from mammals and mollusks (Rothman and Strumwasser, 1976; Yeung and Eskin, 1988; Wollnik et al., 1989) where the electrical activity is higher during the subjective day (Jacklet, 1969; Inouye and Kawamura, 1979). It seems to be a common rule that the phase of protein synthesis corresponds to the phase at which the electrical activity of the circadian clock tissue is higher.

We have also searched for the clock related proteins occurring in the cricket (*G. bimaculatus*) optic lobe with two dimensional gel electrophoresis. A 31.4 kDa protein with pI 6.5 was found to show a circadian fluctuation in its abundance in LD and DD, increasing during the subjective night (Ishibashi and Tomioka, 1998). Since the time course of its occurrence is in good agreement with that expected from the results of studies with CHX and since it is highly specific to the optic lobe, the 31.4 kDa protein may be a clock related protein.

4. PHOTIC ENTRAINMENT PATHWAY

Light information necessary for the photic entrainment is perceived by the compound eye in cockroaches and crickets. There seems to be no specific area in the compound eye for this entrainment in the cricket *G. bimaculatus*. Since partial reduction of the compound eye results in the reduced entrainability of the locomotor rhythm, it is likely that the photic information from the different area of the compound eye additively works in entrainment (Tomioka et al., 1990).

In two orthopteran species, extraretinal photoreception for photic entrainment has been reported. In the New Zealand weta, *Hemideina thoracica*, synchronization of the locomotor rhythm to light cycles persists even after optic nerve severance and complete occlusion of the compound eye and the ocelli (Waddel et al., 1990). The bandlegged crickets, *Dianemobius nigrofasciatus*, also showed a faint but significant entrainment to light cycles after bilateral severance of the optic nerves (Shiga and Numata, 1999). Recent studies in *Drosophila* revealed that extraretinal photoreception is attributable to the newly found blue light photoreceptor molecule, cryptochrome (CRY) (Emery et al., 1998, 2000; Stanewsky et al., 1998). It is an interesting issue whether CRY is also involved in the photoreception in the hemimetabolous insects.

The sensitivity of the circadian photoreceptor is known to be regulated by the optic lobe circadian clock. In cockroaches and crickets, the electroretinographic amplitude shows a clear circadian rhythm peaking during the subjective night (Tomioka and Chiba, 1982; Wills et al., 1985). The ERG rhythm persisted after the optic tract was severed, but was lost when the optic lobe was ablated (Tomioka and Chiba, 1982; Wills et al., 1985). This indicates that the circadian clock in the optic lobe receives the photic information that is circadianly oscillating through the regulatory pathway even in the constant illumination level. Although the significance of this circadian regulation has not been experimentally studied yet, it seems reasonable that the sensitivity is high during the subjective night when the reset of the clock is mainly performed. It is also likely that the clock feeds back its own oscillation through the photic entrainment pathway to stabilize its oscillation even in constant conditions.

5. MULTI-OSCILLATOR SYSTEM

There are often two or more circadian clocks in a single insect. Such multi-oscillator system was analyzed in detail by Pittendrigh et al. (1958) in *Drosophila pseudoobscura* to explain transient cycles during phase shifts after a single light pulse. They postulated the light sensitive A (master)-oscillator and temperature sensitive B (slave)-oscillator. The A-oscillator completes the phase shift immediately after a light pulse, while the B-oscillator regulating the overt eclosion rhythm slowly synchronizes to the phase of A-oscillator: The process of the resynchronization of B-oscillator to A is observed as transient cycles.

A similar master-slave model has been proposed in cockroaches and crickets (Rence and Loher, 1975; Page, 1985b). These insects become arrhythmic after bilateral removal of the optic lobe at constant conditions, while they show rhythmic locomotor or singing rhythms under temperature cycles within limited period lengths (Ts). In an analogical view to the *Drosophila* two oscillator model, it has been suggested that there is a temperature sensitive slave oscillator(s) outside the optic lobe which usually synchronizes to the optic lobe clock and regulates the locomotor activity.

Apart from the master-slave relationship, there is another system composed of two, bilaterally redundant clocks, one in each optic lobe in cockroaches and crickets. These clocks govern the circadian rhythms in overt behaviors and have almost same freerunning periods and output waveforms (Page et al., 1977; Okada et al., 1991). The biological significance of the redundant existence of two clocks is still unclear, but is often said that it is to secure persistence of the rhythm even when one clock is damaged. It seems more likely, however, that it is because of the bilateral structural redundancy of the insect nervous system. The bilaterally paired clocks have the necessity to keep a steady phase relationship to maintain a stable temporal structure in a single animal. This requirement has developed a mechanism to synchronize the clocks. In the cricket *G. bimaculatus*, the interaction between the clocks has been profoundly studied (Tomioka et al., 1991, 1994; Tomioka, 1993; Yukizane and Tomioka, 1995; Tomioka and Yukizane, 1997). When the optic nerve was severed on one side and the animal was placed in LD13:13, there appeared two rhythmic components in its locomotor activity: one ran in synchrony with the given light cycle, the other ran free with marked fluctuation in its freerunning period (Fig. 4A). The fluctuation in the freerunning period (tau) of the freerunning (F) component was clearly dependent on its phase relationship to the entrained (E) component: tau of F-component was lengthened when E-component occurred during its subjective night, while shortened when E occurred during its subjective day, yielding a period modulation curve (Fig. 4B). This means that the clock on the intact side dominates the blinded side and shifts the phase toward its own phase. The amplitude of the period modulation depends on the light intensity, being larger with higher intensities. Under constant dimlight conditions, the phase of the intact side was also observed to be modulated. These observations suggest that the period modulation occurs mutually and there are photic and clock information involved in the coupling mechanism. In experiments using different Ts, the mutual coupling between the clocks was revealed rather weak and was maintained only within 23 hr < T< 25 hr.

Electrophysiological studies revealed that the medulla bilateral neurons (MBNs) mediate the signals necessary for the coupling (Fig. 4C). If the tract including the MBNs was severed, the period modulation did not occur any more. Spontaneous and light-evoked responses of the MBNs showed circadian rhythms peaking in the subjective night (Tomioka et al., 1994; Tomioka and Yukizane, 1997).

Serotonin, a well-known neuroactive substance in the insect central nervous system, seems to be involved in this cricket clock coupling mechanism. Administration of exogenous serotonin induces the phase shift of the pacemaker *in vitro* in phase dependent and dose dependent manner. The PRC thus yielded

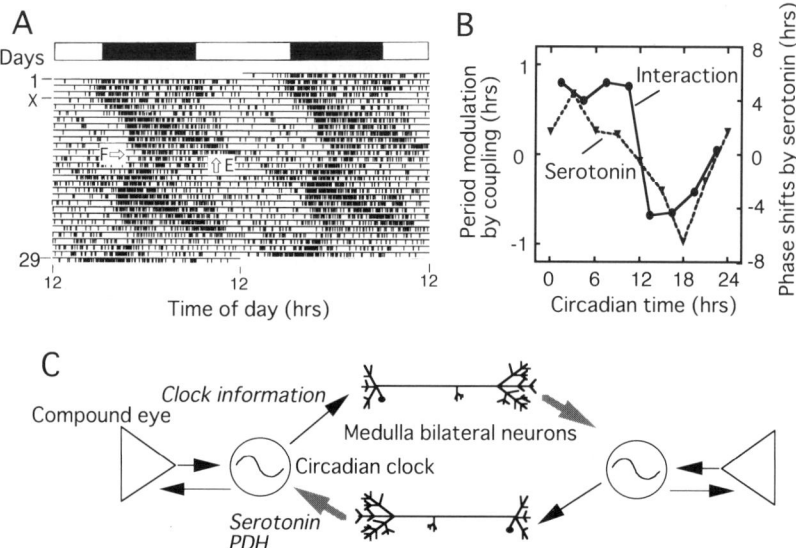

Figure 4. Mutual interaction between the bilaterally paired optic lobe circadian clocks in the cricket (*Gryllus bimaculatus*). (A) Locomotor activity record of a cricket that was placed in LD13:13 after the optic nerve was unilaterally severed (X). White and black bars indicate the light (white) and dark (black) cycle in which the cricket was exposed for the first 3 days. In LD13:13, two rhythmic components appeared. One (E-component) was entrained to the light cycle and its onset corresponded to the lights-off, the other (F-component) ran free with fluctuation in its freerunning period. (B) Period modulation of F-component by coupling force from E-component and phase response curve of optic lobe neuronal activity rhythm to 6 hr serotonin pulses. (C) Possible coupling mechanism between the optic lobe clocks. The circadian clocks communicate with their contralateral partner through medulla bilateral neurons. The information may be mediated by serotonin and PDH.

shows phase delays during the subjective night and phase advances during the subjective day (Fig. 4B). The shape of PRC is quite similar to that yielded by the mutual coupling between the bilateral circadian clocks (Tomioka, 1999). In cockroaches, serotonin only induces phase delays during the early subjective night, however (Page, 1987). When 5,7-DHT was injected into the optic lobe, the coupling between the clocks weakened (M. Koga and K. Tomioka, unpublished observation). Taken together, serotonin seems to be a likely candidate for the mediator of coupling signal. The phase shifts induced by serotonin during the subjective day were totally suppressed by CHX, suggesting that the phase shifts by serotonin require protein synthesis like in the mollusk, *Aplysia* (Eskin et al., 1984).

Serotonin suppresses the responsiveness of MBNs in a dose dependent and time dependent manner in crickets (A.S.M. Saifullah and K. Tomioka, unpublished observation). Thus it is likely that the circadian clock on one side stimulates serotonin release in the contralateral optic lobe through the coupling signal mediated by the MBNs. This causes not only the phase shifts of the contralateral clock but also suppression of the responsiveness of the contralateral MBNs to keep this particular clock's own phase stable (Fig. 4C).

PDH is hypothesized as the other candidate molecule included in mutual coupling between the bilateral optic lobe clocks in cockroaches (Petri and Stengl, 1997). This hypothesis is based on the observations that the PDH-IRNs connect bilateral optic lobes directly and that injection of synthetic PDH into the optic lobe resulted in the phase shifts of the circadian clock: the PRC thus yielded showed only delays during the late subjective day to early subjective night. The relationship between MBNs, PDH-IRNs and serotonergic neurons is required to be elucidated for further understanding the coupling mechanism between the bilaterally paired circadian clocks.

6. CONCLUSIONS

The circadian system of hemimetabolous insects has been extensively studied neuroethologically. These studies revealed that the optic lobe is the locus of the circadian clock regulating the overt rhythms through neural pathways, that the compound eye is the major photoreceptor for photic entrainment, and that the two bilaterally redundant clocks are coupled to one another to maintain a stable internal temporal structure. This system is helping to understand how the clock regulates the overt rhythm at the neuronal level, but further progress depends on the identification of the clock cells in the brain.

REFERENCES

Abe Y., Ushirogawa H., Tomioka K., 1997. Circadian locomotor rhythms in the cricket *Gryllodes sigillatus*. I. Localization of the pacemaker and the photoreceptor. Zoological Science 14, 719-727.

Cymborowski B., Dutkowski A., 1969. Circadian changes in RNA synthesis in the neurosecretory cells of the brain and suboesophageal ganglion of the house cricket. Journal of Insect Physiology 15, 1187-1197.

Cymborowski B., Dutkowski A., 1970. Circadian changes in protein synthesis in the neurosecretory cells of the central nervous system of *Acheta domesticus*. Journal of Insect Physiology 16, 341-348.

Dunlap J. C., 1999. Molecular bases for circadian biological clocks. Cell 96, 271-290.

Emery P., So W. V., Kaneko M., Hall J. C., Rosbash M., 1998. CRY, a *Drosophila* clock and light-regulated cryptochrome, is a major contributor to circadian rhythm resetting and photosensitivity. Cell 95, 669-679.

Emery P., Stanewsky R., Hall J. C., Rosbash M., 2000. *Drosophila* cryptochromes: a unique circadian-rhythm photoreceptor. Nature 404, 456-457.

Eskin A., Yeung S. J., Klass M. R., 1984. Requirement for protein synthesis in the regulation of a circadian rhythm by serotonin. Proceedings of National Academy of Science U.S.A. 81, 7637-7641.

Helfrich-Förster C., 1995. The *period* clock gene is expressed in central nervous system neurons which also produce a neuropeptide that reveals the projections of circadian pacemaker cells within the brain of *Drosophila melanogaster*. Proceedings of National Academy of Science U.S.A. 92, 612-616.

Helfrich-Förster C., Stengl M., Homberg U., 1998. Organization of the circadian system in insects. Chronobiology International 15, 567-594.

Inouye S.-I. T., Kawamura H., 1979. Persistence of circadian rhythmicity in a mammalian hypothalamic "island" containing the suprachiasmatic nucleus. Proceedings of National Academy of Science U.S.A. 76, 5962-5966.

Ishibashi H., Tomioka K., 1997. Two-dimensional gel electrophoretic analysis of proteins involved in the optic lobe circadian clock in the cricket *Gryllus bimaculatus*. Zoological Science 14Suppl.,115.

Jacklet J. W., 1969. Circadian rhythm of optic nerve impulses recorded in darkness from isolated eye of *Aplysia*. Science 164, 562-563.

Loher W., 1972. Circadian control of stridulation in the cricket *Teleogryllus commodus* Walker. Journal of Comparative Physiology 79, 173-190.

Nishiitsutsuji-Uwo J. and Pittendrigh C. S., 1968. Central nervous system control of circadian rhythmicity in cockroach. III. The optic lobes, locus of the driving oscillation? Zeitschrift für vergleihende Physiologie 58, 14-46.

Okada Y., Tomioka K., Chiba Y., 1991. Circadian phase response curves for light in nymphal and adult crickets, *Gryllus bimaculatus*. Journal of Insect Physiology 37, 583-590.

Okamoto A., Mori H., Tomioka K., 2000. The role of optic lobe in circadian locomotor rhythm generation in the cricket *Gryllus bimaculatus*, with special reference to PDH-immunoreactive neurons. Journal of Insect Physiology (in press).

Page T. L., 1978. Interaction between bilaterally paired components of the cockroach circadian system. Journal of Comparative Physiology 124, 225-236.

Page T. L., 1983. Effects of optic tract regeneration on internal coupling in the circadian system of the cockroach. Journal of Comparative Physiology 153, 231-240.

Page T. L., 1985a. Clocks and circadian rhythms. In G. A. Kerkut, L. I. Gilbert eds., Comprehensive Insect Physiology, Biochemistry and Pharmacology. Vol. 6, Nervous system: Sensory, Pergamon Press, Oxford, pp.577-652.

Page T. L., 1985b. Circadian organization in cockroaches: effects of temperature cycles on locomotor activity. Journal of Insect Physiology 31, 235-243.

Page T. L., 1987. Serotonin phase-shifts the circadian rhythm of locomotor activity in the cockroach. Journal of Biological Rhythms 2, 23-34.

Page T. L., Caldarola P. C., Pittendrigh C. S., 1977. Mutual entrainment between bilaterally distributed circadian pacemakers. Proceedings of National Academy of Science U.S.A. 74, 1277-1281.

Petri B., Stengl M., 1997. Pigment-dispersing hormone shifts the phase of the circadian pacemaker of the cockroach *Leucophaea maderae*. Journal of Neuroscience 17, 4087-4093.

Pittendrigh C. S., Bruce V. G., Kaus P., 1958. On the significance of transients in daily rhythms. Proceedings of National Academy of Science, U.S.A. 44, 965-973.

Rence B. G., Loher W., 1975. Arrhythmically singing crickets: thermoperiodic reentrainment after bilobectomy. Science 190, 385-387.

Renn S. C. P., Park J. H., Rosbash M., Hall J. C., Taghert P. H., 1999. A *pdf* neuropeptide gene mutation and ablation of PDF neurons each cause severe abnormalities of behavioral circadian rhythms in *Drosophila*. Cell 99, 791-802.

Roberts S. K., 1974. Circadian rhythms in cockroaches: effects of optic lobe lesions. Journal of Comparative Physiology 88, 21-30.

Rothman B. S., Strumwasser F., 1976. Phase shifting the circadian rhythm of neuronal activity in the isolated *Aplysia* eye with puromycin and cycloheximide. Journal of General Physiology 68, 359-384.

Shiga S., Numata H., 1999. Localization of the photoreceptor and pacemaker for the circadian activity rhythm in the band-legged ground cricket, *Dianemobius nigrofasciatus*. Zoological Science 16, 193-201.

Sokolove P. G., 1975. Localization of the cockroach optic lobe circadian pacemaker with microlesions. Brain Research 87, 13-21.

Stanewsky R., Kaneko M., Emery P., Beretta B., Wager-Smith K., Kay S. A., Rosbash M., Hall J. C., 1998. The cry^b mutation identifies cryptochrome as a circadian photoreceptor in *Drosophila*. Cell 95, 681-692.

Stengl M., 1995. Pigment-dispersing hormone-immunoreactive fibers persist in crickets which remain rhythmic after bilateral transection of the optic stalks. Journal of Comparative Physiology A 176, 217-228.

Stengl M., Homberg U., 1994. Pigment dispersing hormone-immunoreactive neurons in the cockroach *Leucophaea maderae* share properties with circadian pacemaker neurons. Journal of Comparative Physiology A 175, 203-213.

Tomioka K., 1993. Analysis of coupling between optic lobe circadian pacemakers in the cricket *Gryllus bimaculatus*. Journal of Comparative Physiology A 172, 401-408.

Tomioka K., 1999. Light and serotonin phase-shift the circadian clock in the cricket optic lobe in vitro. Journal of Comparative Physiology A 185, 437-444.

Tomioka K., 2000. Protein synthesis is a required process for the optic lobe circadian clock in the cricket *Gryllus bimaculatus*. Journal of Insect Physiology 46, 281-287.

Tomioka K., Chiba Y., 1982. Persistence of circadian ERG rhythms in the cricket with optic tract severed. Naturwissenschaften 69, 355-356.

Tomioka K. and Chiba Y., 1984. Effects of nymphal stage optic nerve severance or optic lobe removal on the circadian locomotor rhythm of the cricket, *Gryllus bimaculatus*. Zoological Science 1, 385-394.

Tomioka K., Chiba Y., 1986. Circadian rhythms in the neurally isolated lamina-medulla complex of the cricket, *Gryllus bimaculatus*. Journal of Insect Physiology 32, 747-755.

Tomioka K. and Chiba Y., 1992. Characterization of optic lobe circadian pacemaker by in situ and in vitro recording of neuronal activity in the cricket *Gryllus bimaculatus*. Journal of Comparative Physiology A 171, 1-7.

Tomioka K., Nakamichi M., Yukizane M., 1994. Optic lobe circadian pacemaker sends its information to the contralateral optic lobe in the cricket *Gryllus bimaculatus*. Journal of Comparative Physiology A 175, 381-388.

Tomioka K., Okada Y., Chiba Y., 1990. Distribution of circadian photoreceptors in the compound eye of the cricket *Gryllus bimaculatus*. Journal of Biological Rhythms 5, 131-139.

Tomioka K., Yamada K., Yokoyama S., Chiba Y., 1991. Mutual interactions between optic lobe circadian pacemakers in the cricket *Gryllus bimaculatus*. Journal of Comparative Physiology A 169, 291-298.

Tomioka K., Yukizane M., 1997. A specific area of the compound eye in the cricket *Gryllus bimaculatus* sends photic information to the circadian pacemaker in the contralateral optic lobe. Journal of Comparative Physiology A 180, 63-70.

Waddel B., Lewis R. D., Engelmann W., 1990. Localization of the circadian pacemakers of *Hemideina thoracica* (Orthoptera; Stenopelmatidae). Journal of Biological Rhythms 5, 131-139.

Wills S. A., Page T. L., Colwell C. S., 1985. Circadian rhythms in the electroretinogram of the cockroach. Journal of Biological Rhythms 1, 25-37.

Wollnik F., Turek F. W., Majewski P., Takahashi J. S., 1989. Phase shifting the circadian clock with cycloheximide: response of hamsters with an intact or a split rhythm of locomotor activity. Brain Research 496, 82-88.

Yeung S. J., Eskin A., 1988. Responses of the circadian system in the *Aplysia* eye to inhibitors of protein synthesis. Journal of Biological Rhythms 3, 225-236.

Yukizane M., Tomioka K., 1995. Neural pathways involved in mutual interactions between optic lobe circadian pacemakers in the cricket *Gryllus bimaculatus*. Journal of Comparative Physiology A 176, 601-610.

Insect Timing: Circadian Rhythmicity to Seasonality
D.L. Denlinger, J. Giebultowicz and D.S. Saunders (Editors)
© 2001 Elsevier Science B.V. All rights reserved.

Cellular circadian rhythms in the fly's visual system

Elzbieta Pyza

Zoological Museum, Institute of Zoology, Jagiellonian University,
Ingardena 6, 30-060 Kraków, Poland

The visual systems of the flies *Musca domestica, Drosophila melanogaster* and *Calliphora vicina* show circadian rhythms in the structure of their cells and in their subcellular organization. These rhythms have been detected in the first optic neuropile, the lamina, in the number of synaptic contacts and the vertical migration of screening pigment in photoreceptor terminals, and in size and shape changes in two of their interneurons, L1 and L2. These are regulated by the lamina's neurotransmitters and controlled by so-called "clock genes", such as the *per* gene. Although the changes in L1 and L2's sizes are driven by inputs from the retina and a circadian clock located in the central part of brain, the epithelial glial cells in the lamina are also involved in rhythmic changes in L1 and L2's lamina axon sizes. Even though the function of circadian rhythms in L1 and L2 is still unknown, these seem to correlate with cyclical variation in the fly's motor activity and with activity in the visual system. The mechanism of these rhythmic neuronal size changes is not known, but disrupting the cytoskeleton or blocking the proton pump, vacuolar ATPase, in cells of the optic lobe abolishes the daily rhythm of size changes in L1 and L2.

1. INTRODUCTION

Circadian rhythms in the visual system have been detected in both vertebrates (Morin, 1994) and invertebrates (Barlow et al., 1989) studied to date. They have been mostly studied in the retinal photoreceptors of animals. They include circadian rhythms in the sensitivity of the eye measured as the electroretinogram (ERG) (Wills et al., 1986; Jacklet, 1991; Remé et al., 1991), in the structure of insect ommatidia (Ferrel and Reitcheck, 1993), in disc-shedding in vertebrate photoreceptors (Remé et al., 1991), and in the migration of the screening pigment in the photoreceptor and pigment cells of arthropods (Barlow et al., 1989). Moreover, in the retina, circadian oscillations have been detected in the concentration of melatonin and dopamine in mammals (Remé et al., 1991) and of octopamine in the horseshoe crab, *Limulus polyphemus* (Barlow et al., 1989). In mammals, the retina's rhythms are regulated by a circadian input from the pacemaker located in the suprachiasmatic nucleus and by an oscillator located in the retina itself (Remé et al., 1991). As in the pineal gland, the mammalian retina exhibits circadian rhythms of melatonin synthesis and this process is controlled by the retina's own circadian oscillator, since this rhythm can be maintained in an isolated retina cultured *in vitro* (Tosini and Menaker, 1996).

In invertebrates circadian rhythms in the eye have been mainly studied in molluscs and arthropods. The eyes of *Aplysia* and *Bulla*, two species of sea snails, contain circadian clock neurons which generate spontaneous, circadian oscillations in electric activity which can be entrained by light (Jacklet and Rolerson, 1982; Block and Wallace, 1982). In arthropods, intensive studies have been carried out in the lateral compound eye of the horseshoe crab *Limulus polyphemus*. The *Limulus* eye shows an especially large number of circadian rhythms in retinal properties. During the night when *Limulus* is behaviourally active, its eyes undergo dramatic structural and physiological changes to increase their sensitivity at night. In the photoreceptors, the photosensitive membranes of the rhabdoms are compressed, the shape of the photoreceptors and pigment cells changes, the eye aperture enlarges to give a larger acceptance angle and catch more photons, and the screening pigment disperses (Barlow et al., 1989). These changes are suppressed during the day by input, mediated by octopamine, from the circadian clock located in the brain. Since *Limulus* is an ancient creature, we might surmise that circadian changes in the visual system are evolutionarily old processes.

Circadian rhythms in the compound eye of insects have also been detected in the ERG (Wills et al. 1986; Colwell and Page, 1989; Chen D.-M. et al., 1992; Chen, B. et al., 1999) and in some insect species in the structure of the rhabdomeres (Williams, 1982). In the eye of Diptera neither the ultrastructure of the retinula cells nor the size of the individual rhabdomeres shows obvious structural differences during the day and night, even though the visual pigment turnover is cyclical (Sapp et al., 1991). Moreover, the photoreceptors of the fruit fly *Drosophila melanogaster* show cyclical expression of the *period* (*per*), and *timeless* (*tim*) "clock genes" (Siwicki et al., 1988; Zerr et al., 1990; Hunter-Ensor et al., 1996). The levels of retinal PER and TIM proteins oscillate during the day and night and are higher in the middle of the night. This cycle is the same as in the lateral neurons (LNs), the so-called "clock neurons" in *Drosophila* which are essential for expression of circadian rhythms in behaviour (Ewer et al., 1992). Detailed studies show, however, that *Drosophila*'s photoreceptors have an autonomous circadian oscillator which operates in the absence of *per* expression in the brain (Cheng and Hardin, 1998). With the exception of clock gene expression and the underlying processes of membrane and visual pigment turnover, circadian processes in the *Drosophila*'s visual system has not been intensively studied because of the small size of the eye, the difficulty of studying its cells by light microscopy and the even greater difficulties of making rigorous electrophysiological recordings. For example, the circadian rhythm in the ERG of *Drosophila* was detected only after combining recordings obtained from several flies (Chen D.-M. et al., 1992). A clear circadian rhythm in the ERG was recorded for several days in single flies, but this was detected in the larger blow fly *Calliphora vicina* (Chen B. et al., 1999), which is more amenable to electrophysiology. Even the housefly *Musca domestica* fails to show an ERG rhythm (Pyza E. and Shaw S.R., unpublished data). In the housefly, however, a circadian rhythm in the eye's electric activity has been reported from the surface of the eye (Gunning and Shipp, 1976).

A new group of circadian rhythms have been found in the first optic neuropile, the lamina (Fig. 1A), of the housefly's optic lobe (Meinertzhagen and Pyza, 1996). These rhythms are a new example of structural circadian rhythms in neurons other than photoreceptor cells. Since these changes are recurrent, they can also be viewed as a new example of neuronal plasticity controlled by a circadian clock.

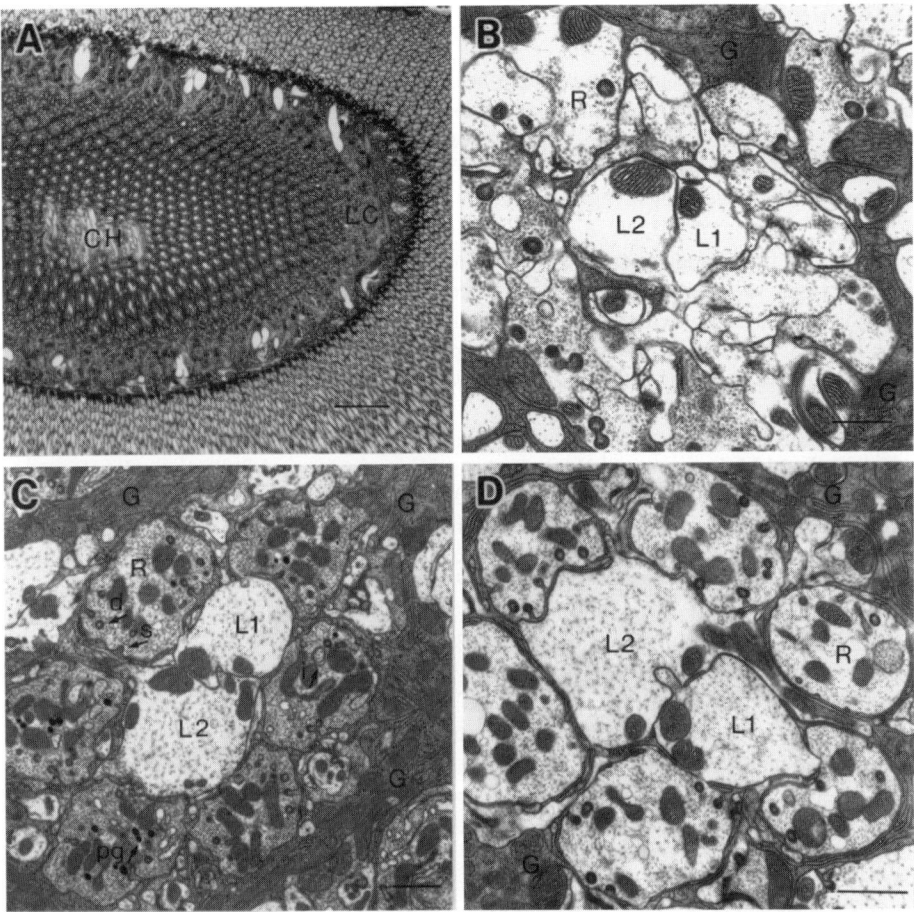

Figure 1. A. Tangential section of the lamina in the housefly, *Musca domestica*, showing cross-sectioned cartridge profiles. LC - lamina cortex; CH - chiasma, where monopolar cell axons traverse between the lamina and medulla, where they terminate. Magnification x200, scale bar = 50 μm; B-D. A single cartridge cross-section of the lamina in *Drosophila melanogaster* (B), *Musca domestica* (C), and *Calliphora vicina* (D). R: one of six photoreceptor terminals; L1 and L2: axon profiles of monopolar cells, G: glial cells, arrowhead: presynaptic element of the tetrad synapse, pg: pigment granules, s: shallow capitate projection, d: deep capitate projection, i: inter-receptor invagination. Magnifications: x11,000 (A), x6,280 (B), x7,000 (C). Scale bars = 1 μm.

2. CIRCADIAN RHYTHMS IN THE FLY'S FIRST OPTIC NEUROPILE

In the lamina, circadian rhythms have been detected in the number of synaptic contacts between the photoreceptor terminals and their first-order interneurons, monopolar cells L1 and L2 (Pyza and Meinertzhagen, 1993), as well as in changes in the axon sizes of L1 and L2 (Pyza and Meinertzhagen, 1995) and in the number of two types of organelles, pigment granules and inter-receptor invaginations into the photoreceptor terminals (Fig. 1B-D) (Pyza and Meinertzhagen, 1997a).

The lamina has a modular structure and is composed of an array of cartridges (Fig. 1A) each of which is innervated by six photoreceptor terminals, R1-R6, and comprises five monopolar cells, L1-L5, with terminals located in the deeper optic neuropiles (Trujillo-Cenóz, 1965; Braitenberg, 1967). Each lamina cartridge is surrounded by three epithelial glial cells (Fig. 1B-D). Among the lamina monopolar cells, L1 and L2 are the largest and their axons are located at the cartridge axis (Fig. 1B-D). They receive equal photic input from the retina's photoreceptors R1-R6, at so-called tetrad synapses (Fig. 1B-D) (Burkhardt and Braitenberg, 1976; Nicol and Meinertzhagen, 1982), and in addition in *Musca* L2 forms feedback synapses back upon the terminals of R1-R6 (Strausfeld and Campos-Ortega, 1977). The numbers of both types of synapses in *Musca* oscillate during the day and night, in opposite phases to each other (Pyza and Meinertzhagen, 1993). The number of tetrad synapses shows a modest increase at the beginning of the day while the L2 feedback synapses have a clear increase at the beginning of night. The rhythm in the frequency of feedback synapses is circadian because it is also observed under constant darkness (DD). In the case of the tetrad synapses, changes in their number in DD are weak and the circadian nature of this rhythm uncertain. The number of tetrad synapses seems instead to be regulated rather by direct light exposure than by a circadian mechanism (Pyza and Meinertzhagen, 1993).

The circadian rhythm in the size changes in L1 and L2, measured as the cross-sectional area of their axons at different depths in the lamina, was first found in the housefly (Pyza and Meinertzhagen, 1995). In this species, L1 and L2 enlarge the girth of their axons during the day but this then decreases during the night. Changes in L1 and L2 are maintained under constant darkness (DD) as well as in continuous light (LL), which indicates that these rhythms are circadian in origin. The rhythmic changes in L1 and L2 have also been detected in two other flies species, *Drosophila melanogaster* (Pyza and Meinertzhagen, 1999a) and *Calliphora vicina* (Pyza and Cymborowski, 1999), indicating that structural rhythms in L1 and L2 are a general phenomenon in Diptera. *Drosophila*, however, has a different shape in the lamina axons of its L1 and L2 cells, which are an inverted conical shape, while in *Musca* they taper distally, towards the retina (Fig. 2A,B). Moreover, additionally to size changes, L1 and L2 also change in the shape of their axons, being conical during the day and more cylindrical at night in *Drosophila* (Fig. 2B). These cells have not been studied in other insect species, especially those with a nocturnal pattern of activity. The structure of neurons others than photoreceptors, and their adaptations during the day or night, have been studied, however, in mammals. In the rat's retina, adaptational structural changes have been observed in terminals of rod bipolar cell axons (Behrens and Wagner, 1996) which can be thought of as functional counterparts to the insect's lamina monopolar cells. If structural day/night changes occur in such evolutionarily distant animals as insects and mammals, such changes among visual interneurons might constitute a general phenomenon in the visual system of all animals.

In the housefly's lamina, in addition to circadian changes in neuron size the epithelial glial cells surrounding the lamina cartridges also change their sizes. The glial cells swell during the night and shrink during the day, in a reciprocal pattern to the one seen in L1 and L2 (Pyza et al., 1997). These rhythmic changes in the glial cells seem to offset the changes in neurons. Unlike the changes in the monopolar cells, L1 and L2, and their epithelial glial cell envelopment, the cross-sectional areas of photoreceptor terminals do not change significantly during the day and night. In the photoreceptor terminals, however, the numbers of two types of organelles, the screening pigment granules and inter-receptor invaginations, cycle under

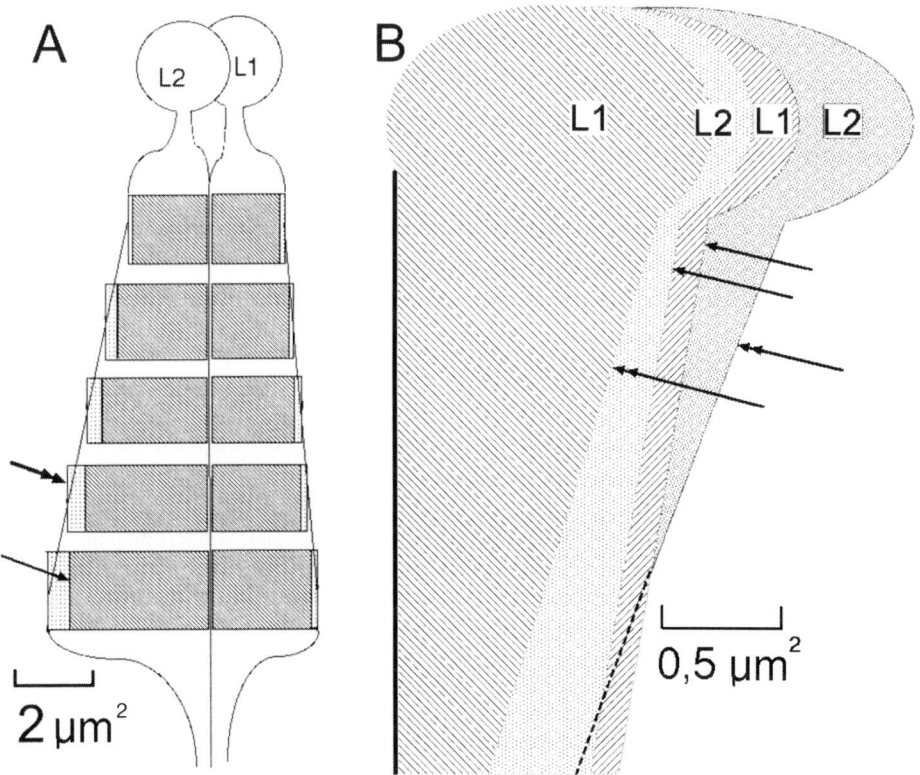

Figure 2. Example of day/night changes in the axon cross-sectional areas of L1 and L2 in *Musca*(A) and *Drosophila*(B). Double heads and one head arrows indicate cell shapes and sizes during the day and during the night, respectively. See text for further explanations.

day/night (LD) and DD conditions. During light adaptation in the retina, screening pigment granules in the photoreceptors move horizontally towards the rhabdomeres and this process is well known as a pupil mechanism (Ro and Nilsson, 1994). In *Musca* the screening pigment granules also move in a vertical direction, however, because at the end of day there are more pigment granules at the proximal part of the lamina than in the distal lamina (Pyza and Meinertzhagen, 1997a). This situation is reversed at the end of night. The vertical migration of the screening pigment in *Musca* shows a circadian rhythmicity but in addition the number of pigment granules in DD dramatically decreases compared with their number in LD. This movement, it has been suggested, could indicate a possible function of the screening pigment granules in light adaptation at the terminal (Pyza and Meinertzhagen, 1997a), but direct evidence for this is lacking. The terminals of photoreceptors also contain two type of invaginations; one from the epithelial glial cells, so-called capitate projections, and a second one from neighbouring photoreceptor terminals. The capitate projections which exist in two forms, shallow and deep (Fig. 1B-D), are possibly involved in synaptic transmission at tetrad synapses in the photoreceptors (Meinertzhagen and Hu, 1996) but their number is stable during the day and night. By contrast, the number of inter-receptor invaginations oscillates in LD and also in DD, but the functional significance of these changes is still unknown.

3. A POSSIBLE ORIGIN OF THE CIRCADIAN INPUT DRIVING THE LAMINA RHYTHMS

The origin of circadian input to the lamina is not yet clear. A study of L1 and L2 in flies with lesions that separate the lamina and the second optic neuropile, the medulla, from the rest of the optic lobe and central brain, shows that circadian rhythms in L1 and L2 are abolished (Balys and Pyza, 1999). Only weak day/night oscillations are observed after lesions in flies held under LD conditions. The results of these experiments show that in *Musca* input from the central brain and/or the contralateral optic lobe is needed for rhythmic changes in the sizes of L1 and L2 to occur.

In another study carried out on the null allele of the *per* gene in *Drosophila*, the arrhythmic mutant per^{01} which does not show the size change rhythms in L1 and L2 even in LD, an insertion of a 7.2: 9 kb fragment of *per*, inducing PER expression in the clock neurons and photoreceptors, fails to completely rescue these rhythms (Pyza and Meinertzhagen, 1999c). Thus a third element which controls the sizes of L1 and L2 has to be involved in this process. It may originate from the epithelial glia. These glial cells express PER in *Drosophila* and disruption of their metabolism in *Musca* results in changes in the sizes of L1 and L2. The results obtained in both fly species indicate that the lamina's structural rhythms might be driven by three elements, a direct input from the retina, circadian input from clock neurons in the brain, and involvement of the epithelial glial cells. The epithelial glial cells surrounding cartridges may between them synchronise rhythms in the lamina through gap junctions (Saint Marie and Carlson, 1983). For example, closing gap junctions by injecting octanol results in the blocking of the spread of injected fluorescent dye in the lamina (Shaw S.R., unpublished data) and in disruption of the daily changes in L1 and L2's axon sizes (Pyza et al., 1997).

4. CORRELATION BETWEEN RHYTHMS IN THE VISUAL SYSTEM AND BEHAVIOUR

Although the function of daily and circadian rhythms in the sizes of L1 and L2 is still not known, such changes might correlate with the cells' activity. A ^{3}H-deoxyglucose study in the lamina has shown that during the day the uptake of ^{3}H-deoxyglucose is higher in L1 and L2 than in glial cells while during the night the relative uptake levels are reversed (Bausenwein, 1994). Moreover, studies on three fly species show a correlation between the fly's activity and the sizes of the axonal cross-sectional areas of its L1 and L2 cells. In two diurnal species, *Musca domestica* (Pyza and Meinertzhagen, 1995) and *Calliphora vicina* (Pyza and Cymborowski, 1999), the sizes of L1 and L2 are larger during the day than during the night, exactly mirroring their locomotor activity patterns, with higher activity levels during the day than during the night. A similar correlation was also confirmed in a third fly species, *Drosophila melanogaster*. The fruit fly has two activity peaks in its locomotor behaviour, one in the morning and a second one, which is higher, in the evening. Measurements of L1 and L2 cells in *Drosophila* show that both cells are largest at the beginning of night but a second peak appears at the beginning of day (Pyza and Meinertzhagen, 1999a). Further study of daily rhythms in both locomotor activity and size changes in L1 and L2 in *Calliphora vicina* under LD and DD conditions, has shown that both rhythms are correlated in phase in LD but not in DD, however (Pyza and Cymborowski, 1999). In DD both rhythms are actually reversed in phase. This suggests that daily changes of light and darkness synchronise both rhythms but that in DD the two rhythms run independently. Not all flies in DD show rhythmicity in their locomotor activity, however, and those which are arrhythmic have scattered activity throughout the entire 24 h period. These flies may have either a low or a high level of activity, as well as differences in the sizes of their L1 and L2 cells. Highly active flies have larger cell sizes than arrhythmic ones showing low activity (Pyza and Cymborowski, 1999), again indicating a correlation between anatomy and behaviour.

Recently, L1 and L2 were also measured in *Musca* after stimulating the fly's eyes using a moving periodic grating of vertical black and white stripes, or stimulating both the eyes and its motor system by forcing flies to fly continuously for one hour. After 1 h of stimulation of the visual system at the beginning of day, when the cells are the largest during the 24 h period, L1 and L2 were still larger by 12% and 7%, respectively in stimulated flies than in control flies. After stimulating both visual and motor systems, however, L1 and L2 were both larger by 25% than in control flies (Kula E. and Pyza E., unpublished data). All these results obtained so far support the correlation between a high activity level in flies and the large sizes of their L1 and L2 monopolar cells in the lamina. Our results also confirm in a more general way, old findings from the last century, of differences in cell size in the morning and in the evening in the brains of honeybees and birds, made by Hodge (1892) who tried to explain the changes he observed as resulting from the animal's daily pattern of fatigue. It is possible, however, that there exists some sort of feedback influence from the motor system which affects the sizes of neurons, since after stimulating flies to fly intensively for a period of 1 h their L1 and L2 cells were significantly larger than in resting flies. Stimulation of the visual system results in increased transmission of visual information at the tetrad synapses, at which histamine is a neurotransmitter (Hardie, 1987) between R1-R6 and L1 and L2. Increased transmission may derive from the formation of additional tetrad synapses, for which L1 and L2 provide extra postsynaptic sites, and it is possible that this is causally linked to the enlargement in the girth of their axons. The formation of multiple synapses has been observed in the rat supraoptic nucleus as a result of increases in cell size (Modney and Hatton, 1989). A correlation in the number of synaptic boutons with cell size was also observed in *Xenopus* cardiac ganglion cells (Sargent, 1983). A similar process may exist in the lamina monopolar cells which show circadian and induced changes of the cross-sectional area of their axons because of formation of the new synaptic contacts but it is not confirmed yet by experimental data.

5. MECHANISMS INVOLVED IN THE RHYTHM OF CHANGES IN STRUCTURE OF NEURONS

The lamina of flies is invaded by two sets of tangential   neurons, one immunoreactive to serotonin (Nässel, 1987) and a second one which immunostains with an antibody against pigment-dispersing factor (PDF) (Nässel et al., 1991), a neuropeptide which belongs to pigment-dispersing hormone (PDH) family of peptides first discovered in crustaceans (Rao and Riehm, 1989). In crabs PDHs regulate pigment movement in the integumentary chromatophores as well as circadian rhythms in the eye (Rao, 1985). In insect PDF, a peptide homologue of crustacean PDH (Rao and Riehm, 1993) has been suggested as a neurotransmitter of circadian information from the clock to target neurons (Helfrich-Förster, 2000). In *Drosophila* PDF co-localizes with PER in some lateral neurons (Helfrich-Förster, 1995) and the null mutant of the *pdf* gene is arrhythmic in locomotor activity (Renn et al., 1999). PDF also appears to regulate changes in the sizes of L1 and L2. Injecting PDF during the day or during the night increases the sizes of both cells (Pyza and Meinertzhagen, 1996, 1999b). PDF seems to be cyclically released from terminals of PDF neurons in the lamina and medulla with timing compatible with increasing sizes of L1 and L2. The PDF varicosities in both optic neuropiles, lamina and medulla, change their sizes possibly as a result of cyclical changes in the transport, accumulation and/or release of PDF (Pyza and Meinertzhagen, 1997b). PDF might be not the only peptide affecting the rhythms of size change in L1 and L2, however, since some PDF-positive cells and/or neighbouring cells in *Musca* also exhibit immunoreactivity to FMRF-amide peptides (Pyza E. and Meinertzhagen I.A., unpublished data). Moreover injections of several peptides which belong to the FMRF-amide family into the optic lobe of *Musca* induce changes which are opposite to those observed after PDH injection (Pyza E. and Meinertzhagen I.A., unpublished data). In a circadian system in a mollusc it has been reported that FMRF-amide plays a role in suppressing circadian pacemaker activity (Jacklet et al., 1987).

Changes in L1 and L2 sizes can be also mimicked by injecting other neurotransmitters that have been detected in the lamina (Nässel, 1991). Injections of serotonin (5-HT) during the day increase both cell sizes but a significant increase was only observed for L1. Unlike daytime injections, injections of 5-HT during the night also increase both cell sizes but significantly so only for L2. This suggests that both cells depend on 5-HT in the lamina but at different concentrations during the day and night. Likewise, injections of a specific serotoninergic toxin, 5,7-dihydroxytryptamine (5,7-DHT) show that a decrease of 5-HT during the day decreases cell size, but significantly so only for one cell, L2, again indicating a differential sensitivity of the two cells to altered 5-HT levels. A more drastic shrinkage of both cells was observed after treatment with reserpine, which completely depletes 5-HT from enlargement in the girth of their axons. The formation of multiple synapses has been observed in the rat supraoptic nucleus as a result of increases in cell size (Modney and Hatton, 1989). A correlation in the number of synaptic boutons with cell size was also observed in *Xenopus* cardiac ganglion cells (Sargent, 1983). A similar process may exist in the lamina monopolar cells which show circadian and induced changes of the cross-sectional area of their axons because of formation of the new synaptic contacts but it is not confirmed yet by experimental data.

5. MECHANISMS INVOLVED IN THE RHYTHM OF CHANGES IN STRUCTURE OF NEURONS

The lamina of flies is invaded by two sets of tangential neurons, one immunoreactive to serotonin (Nässel, 1987) and a second one which immunostains with an antibody against pigment-dispersing factor (PDF) (Nässel et al., 1991), a neuropeptide which belongs to pigment-dispersing hormone (PDH) family of peptides first discovered in crustaceans (Rao and Riehm, 1989). In crabs PDHs regulate pigment movement in the integumentary chromatophores as well as circadian rhythms in the eye (Rao, 1985). In insect PDF, a peptide homologue of crustacean PDH (Rao and Riehm, 1993) has been suggested as a neurotransmitter of circadian information from the clock to target neurons (Helfrich-Förster, 2000). In *Drosophila* PDF co-localizes with PER in some lateral neurons (Helfrich-Förster, 1995) and the null mutant of the *pdf* gene is arrhythmic in locomotor activity (Renn et al., 1999). PDF also appears to regulate changes in the sizes of L1 and L2. Injecting PDF during the day or during the night increase the sizes of both cells (Pyza and Meinertzhagen, 1996, 1999b). PDF seems to be cyclically released from terminals of PDF neurons in the lamina and medulla with timing compatible with increasing sizes of L1 and L2. The PDF varicosities in both optic neuropiles, lamina and medulla, change their sizes possibly as a result of cyclical changes in the transport, accumulation and/or release of PDF (Pyza and Meinertzhagen, 1997b). PDF might be not the only peptide affecting the rhythms of size change in L1 and L2, however, since some PDF-positive cells and/or neighbouring cells in *Musca* also exhibit immunoreactivity to FMRF-amide peptides (Pyza E. and Meinertzhagen I.A., unpublished data). Moreover injections of several peptides which belong to the FMRF-amide family into the optic lobe of *Musca* induce changes which are opposite to those observed after PDH injection (Pyza E. and Meinertzhagen I.A., unpublished data). In a circadian system in a mollusc it has been reported that FMRF-amide plays a role in suppressing circadian pacemaker activity (Jacklet et al., 1987).

Changes in L1 and L2 sizes can be also mimicked by injecting other neurotransmitters that have been detected in the lamina (Nässel, 1991). Injections of serotonin (5-HT) during the day increase both cell sizes but a significant increase was only observed for L1. Unlike daytime injections, injections of 5-HT during the night also increase both cell sizes but significantly so only for L2. This suggests that both cells depend on 5-HT in the lamina but at different concentrations during the day and night. Likewise, injections of a specific serotoninergic toxin, 5,7-dihydroxytryptamine (5,7-DHT) show that a decrease of 5-HT during the day decreases cell size, but significantly so only for one cell, L2, again indicating a differential sensitivity of the two cells to altered 5-HT levels. A more drastic shrinkage of both cells was observed after treatment with reserpine, which completely depletes 5-HT from the optic lobe, decreasing L1 and L2 by 50% and 42%, respectively (Pyza and Meinertzhagen, 1999b). The effect of other lamina neurotransmitters: histamine, glutamate, and GABA on L1 and L2 have been also reported (Pyza and Meinertzhagen, 1996). When injected into the optic lobe in *Musca*, histamine affects only L1, increasing its size. This result shows that injected histamine is able to mimic the effects of direct light exposure on monopolar cells because after such a treatment L1 is more affected than L2 (Pyza and Meinertzhagen, 1996). Two other neurotransmitters, glutamate and GABA, have a small but consistent effect on axon size in L1 and L2. They both decrease L1 and L2's sizes when injected into the optic lobe of *Musca* during the day. Not all neurotransmitters affect L1 and L2 cells when injected, however. For example, injections of melatonin, which mimic night changes in the ERG, do not have a significant effect in the lamina (Pyza E. and Shaw S.R., unpublished data). Since the sizes of L1 and L2 seem to be affected by many factors one thing that seems clear is that several neurotransmitters take part in this process, transmitting light, circadian, activity and other inputs to L1 and L2 cells. The concentration of the lamina's neurotransmitters might also be regulated by the epithelial glial cells. For example, in the medulla of *Drosophila,* terminals of PDF cells are in a close contact with PER-expressing glial cells and have been suggested to mediate between generating circadian information PDF cells and expressing circadian changes lamina neurons (Helfrich-Förster, 1995; Meinertzhagen and Pyza, 1996).

Neurotransmitters affect the sizes of neurons but a cellular mechanism for this neuronal plasticity is still unknown. To address this question, and gain access to experimental manipulation of the underlying mechanisms, we have examined whether processes which determine cell morphology and regulate volume changes in biological systems more generally are also involved in the size changes in L1 and L2. We have examined two processes which are usually involved in maintaining cell morphology and volume. Both these parameters depend on a cytoskeleton (Kaech et al., 1996) and ionic and fluid transport (Strange et al., 1996). To examine the role of the cytoskeleton in changing the size of neurons we have used colchicine injections to disrupt the microtubules (Schafer and Reagan, 1981; Dasheiff and Ramirez, 1985). The colchicine injections have shown, however, that this toxin only has an effect on L1 and L2's sizes when it is injected during the night. After colchicine, the cells are not able to shrink during the night and have similar sizes to those during the day (Balys M. and Pyza E., unpublished data). Thus day/night changes in the axon cross-sectional areas of L1 and L2 are abolished. A similar effect was obtained after bafilomycin injections into the housefly's optic lobe (Pyza E., Giebultowicz J.M. and Meinertzhagen I.A., unpublished data). Bafilomycin is a specific blocker of the V-ATPase proton pump (Gagliardi et al., 1999). V-ATPase is present in the brains of *Musca* and *Drosophila* probably in glial cells, and seems to be active in the lamina during the night when the epithelial glial cells increase their sizes and show higher metabolic activity. This suggests that the glial cell proton V-ATPase may participate in the mechanism of L1 and L2's shrinkage during the night.

6. CONCLUDING REMARKS

The lamina's rhythms provide an output system of the circadian clock, which can be studied at the neuronal level. Moreover, the studies on lamina circadian rhythms described above, especially on the rhythm in neuron size changes, have shown that this process may be controlled by oscillators originating from three or more different levels. From the local to central level these are: a local function in the lamina which may arise through *per* expressing glial cells; from the retina, which also regulates the lamina's rhythms; and, finally, lamina rhythms are controlled by a pacemaker located centrally, outside the optic lobe.

ACKNOWLEDGEMENTS

Supported by the International Programme of the Howard Hughes Medical Institute (HHMI 75195-543102).

REFERENCES

Balys. M., and Pyza, E., 1999. Circadian clock in the brain controls circadian rhythms in the fly's optic lobe. Acta Neurobiologiae Experimentalis 59, 226.

Bausenwein, B., 1994. Struktur-Funktionskartierung im Nervensystem von *Drosophila*. Habilitationschrift der Fakultät für Biologie der Albert-Ludwigs-Universität Freiburg.

Barlow, R.B. Jr, Chamberlain, S.C., and Lehman H.K., 1989. Circadian rhythms in the invertebrate retina. In: Stavenga, D.G. and Hardie, R.C. (eds.), Facets of vision. Springer-Verlag, Berlin, pp. 257-280.

Behrens, U.D., Wagner, H.-J., 1996. Adaptation-dependent changes of bipolar cells terminals in fish retina: effects on overall morphology and spinule formation in Ma and Mb cells. Visual Research 36, 3901-3911.

Block, G.D., and Wallace, S.F., 1982. Localization of a circadian pacemaker in the eye of a mollusc, *Bulla*. Science 217, 155-157.

Braitenberg, V., 1967. Patterns of projection in the visual system of the fly. I. Retina-lamina projections. Experimental Brain Research 3, 271-298.

Burkhardt, W., Braitenberg, V., 1976. Some peculiar synaptic complexes in the first visual ganglion of the fly, *Musca domestica*. Cell and Tissue Research 173, 287-308.

Chen, B., Meinertzhagen, I.A., and Shaw, S.R., 1999. Circadian rhythms in light-evoked responses of the fly's compound eye, and the effects of neuromodulators 5-HT and the peptide PDF. Journal of Comparative Physiology A 185, 393-404.

Chen, D.-M., Christianson, J.S., Sapp, R.J., and Stark, W.S., 1992. Visual receptor cycle in normal and *period* mutant *Drosophila*: microspectrophotometry, electrophysiology, and ultrastructural morphometry. Visual Neuroscience 9, 125-135.

Cheng, Y., and Hardin, P.E., 1998. *Drosophila* photoreceptors contain an autonomous circadian oscillator that can function without *period* mRNA cycling. Journal of Neuroscience 18, 741-750.

Colwell, C.S., and Page, T.L., 1989. The electroretinogram of the cockroach *Leucophaea maderae*. Comparative Biochemistry and Physiology 92A, 117-123.

Dasheiff, R.M., and Ramirez, L.F., 1985. The effects of colchicine in mammalian brain from rodents to rhesus monkeys. Brain Research Review 10, 47-67.

Ewer, J., Frisch, B., Hamblen-Coyle, M.J., Rosbash, M., Hall, J.C., 1992. Expression of the *period* clock gene within different cell types in the brain of *Drosophila* adults and mosaic analysis of these cells' influence on circadian behavioral rhythms. Journal of Neuroscience 12, 3321-3349.

Ferrel, B.R., and Reitcheck, B.G., 1993. Circadian changes in cockroach ommatidial structure. Journal of Comparative Physiology A 173, 549-555.

Frisch, B., Hardin, P.E., Hamblen-Coyle, M.J., Rosbash, M., and Hall, J.C., 1994. A promoterless *period* gene mediates behavioral rhythmicity and cyclical *per* expression in a restricted subset of the Drosophila nervous system. Neuron 12, 555-570.

Frixione, E., and Aréchiga, H., 1979. Photomechanical migrations of pigment granules along the retinula cells of the crayfish. Journal of Neurobiology 10, 573-590.

Gagliardi, S., Rees, M., and Farina, C., 1999. Chemistry and structure activity relationships of bafilomycin A_1, a potent and selective inhibitor of the vacuolar H^+-ATPase. Current Medicinal Chemistry 6, 1197-1212.

Gunning, R., and Shipp, E., 1976. Circadian rhythm in endogenous nerve activity in the eye of *Musca domestica* L. Physiological Entomology 1, 241-248.

Hardie, R.C., 1987. Is histamine a neurotransmitter in insect photoreceptors? Journal of Comparative Physiology A 161, 201-213.

Helfrich-Förster, C., 1995. The period clock gene is expressed in central nervous system neurons which also produce a neuropeptide that reveals the projections of circadian pacemaker cells within the brain of *Drosophila melanogaster*. Proceedings of the National Academy of Sciences USA 92, 612-616.

Helfrich-Förster, C., 2000. Ectopic expression of the neuropeptide pigment-dispersing factor alters behavioral rhythms in *Drosophila melanogaster*. Journal of Neuroscience 20, 3339-3353.

Hodge, C.F., 1892. A microscopical study of changes due to functional activity in nerve cells. Journal of Morphology 2, 151-168.

Hunter-Ensor, M., Ousley, A., and Sehgal, A., 1996. Regulation of the *Drosophila* protein *timeless* suggests a mechanism for resetting the circadian clock by light. Cell 84, 677-685.

Jacklet, J.W., and Rolerson, C., 1982. Electrical activity and structure of retinal cells of the *Aplysia* eye. II. Photoreceptors. Journal of Experimental Biology 99, 381-395.

Jacklet, J.W., Klose, M., Goldberg, M., 1987. FMRFamide-like immunoreactive efferent fibers and FMRF-amide suppression of pacemaker neurons in eyes of *Bulla*. Journal of Neurobiology 18, 433-449.

Jacklet, J. W., 1991. Photoresponsiveness of *Aplysia* eye is modulated by the ocular circadian pacemaker and serotonin. Biological Bulletin 180, 284-294.

Kaech, S., Ludin, B., and Matus, A., 1996. Cytoskeletal plasticity in cells expressing neuronal microtubule-associated proteins. Neuron 17, 1189-1199.

Meinertzhagen, I.A., and Hu, X., 1996. Evidence for site selection during synaptogenesis: The surface distribution of synaptic sites in photoreceptor terminals of the flies *Musca* and *Drosophila*. Cellular and Molecular Neurobiology 16, 677-698.

Meinertzhagen, I.A., and Pyza, E., 1996. Daily rhythms in cells of the fly's optic lobe: taking time out from the circadian clock. Trends in Neurosciences 19, 285-291.

Modney, B.K., and Hatton, G.I., 1989. Multiple synapse formation: a possible compensatory mechanism for increased cell size in rat supraoptic nucleus. Journal of Neuroendocrinology 1, 21-27.

Nässel, D.R., 1987. Serotonin and serotonin-immunoreactive neurons in the nervous system of insects. Progress in Neurobiology 30, 1-85.

Nässel, D.R., Shiga, S., Wikstrand, E.M., and Rao, K.R., 1991. Pigment-dispersing hormone-immunoreactive neurons and their relation to serotonergic neurons in the blowfly and cockroach visual system. Cell and Tissue Research 266, 511-523.

Nicol, D., and Meinertzhagen, I.A., 1982. An analysis of the number and composition of the synaptic population formed by photoreceptors of the fly. Journal of Comparative Neurology 207, 29-44.

Pyza, E., and Meinertzhagen, I.A., 1993. Daily and circadian rhythms of synaptic frequency in the first visual neuropile of the housefly's (*Musca domestica* L.) optic lobe. Proceedings of the Royal Society London B 254, 97-105.

Pyza, E., and Meinertzhagen, I.A., 1995. Monopolar cell axons in the first optic neuropil of the housefly, *Musca domestica* L., undergo daily fluctuations in diameter that have a circadian basis. Journal of Neuroscience 15, 407-418.

Pyza, E., and Meinertzhagen, I.A., 1996. Neurotransmitters regulate rhythmic size changes amongst cells in the fly's optic lobe. Journal of Comparative Physiology A, 178, 33-45.

Pyza, E., Górska, J., Czekaj, A., 1997. The role of glial cells in regulating a circadian rhythm of changes in monopolar cell size in the housefly's first visual neuropile. In: Elsner, N., Wässle, H. (eds). Göttingen Neurobiology Report 1997. Georg Thieme Verlag Stuttgart, New York, p. 301.

Pyza, E., and Meinertzhagen, I.A., 1997a. Circadian rhythms in screening pigment and invaginating organelles in photoreceptor terminals of the housefly's first optic neuropile. Journal of Neurobiology 32, 517-529.

Pyza, E., and Meinertzhagen, I. A., 1997b. Neurites of *period*-expressing PDH cells in the fly's optic lobe exhibit circadian oscillations in morphology. European Journal of Neuroscience 9, 1784-1788.

Pyza, E., and Cymborowski, B., 1999. Circadian rhythms in locomotor activity and in the visual system of the blowfly. Acta Neurobiologiae Experimentalis 59, 226.

Pyza, E., and Meinertzhagen, I.A., 1999a. Daily rhythmic changes of cell size and shape in the first optic neuropil in *Drosophila melanogaster*. Journal of Neurobiology 40, 77-88.

Pyza, E., and Meinertzhagen, I.A., 1999b. Serotonin modulates the action of pigment dispersing factor on a structural circadian rhythm in the fly's visual system. IBRO Abstracts p. 162.

Pyza, E., and Meinertzhagen, I.A., 1999c. The role of clock genes and glial cells in expressing circadian rhythms in the fly's lamina. Cold Spring Harbor Laboratory Drosophila Meeting Abstracts, p. 145.

Rao, K.R., 1985. Pigmentory effectors. In: Bliss, D.E., Mantel, L.H. (eds). The biology of Crustacea, vol 9, Integument, pigments and hormonal processes. Academic Press, Orlando, pp. 395-462.

Rao, K.R., and Riehm, J.P., 1989. The pigment-dispersing hormone family: Chemistry, structure activity relations, and distribution. Biological Bulletin 177, 225-229.

Rao, K.R., and Riehm, J.P., 1993. Pigment-dispersing hormones. Annals New York Academy of Sciences 680, 78-88.

Remé, C.E., Wirz-Justice, A., and Terman, M., 1991. The visual input stage of the mammalian circadian pacemaking system: I. Is there a clock in the mammalian eye? Journal of Biological Rhythms 6, 5-29.

68

Renn, S.C.P., Park, J.H., Rosbash, M., Hall. J.C., and Taghert, P.H., 1999. A *pdf* neuropeptide gene mutation and ablation of PDF neurons each cause severe abnormalities of behavioral circadian rhythms in *Drosophila*. Cell 99, 791-802.

Ro, A.I., and Nilsson, D.-E., 1994. Circadian and light-dependent control of the pupil mechanism in tipulid flies. Journal of Insect Physiology 40, 883-891.

Saint Marie, R.L., and Carlson, S.D., 1983. The fine structure of neuroglia in the lamina ganglionaris of the housefly, *Musca domestica* L. Journal of Neurocytology 12, 213-241.

Sapp, R.J., Christianson, J., and Stark, W.S., 1991. Turnover of membrane and opsin in visual receptors of normal and mutant *Drosophila*. Journal of Neurocytology 20, 597-608.

Sargent, P.B., 1983. The number of synaptic boutons terminating on *Xenopus* cardiac ganglion cells is directly correlated with cell size. Journal of Physiology 343, 85-104.

Schafer R., and Reagan, P.D., 1981. Colchicine reversibly inhibits electrical activity in arthropod mechanoreceptors. Journal of Neurobiology 12, 155-166.

Siwicki, K.K., Eastman, C., Petersen, G., Rosbash, M., Hall, J.C., 1988. Antibodies to the *period* gene product of *Drosophila* reveal diverse tissue distribution and rhythmic changes in the visual system. Neuron 1, 141-150.

Strange, K., Emma, F., Jackson, P.S., 1996. Cellular and molecular physiology of volume-sensitive anion channels. American Journal of Physiology 270, C711-C730.

Strausfeld, N.J., and Campos-Ortega, J.A., 1977. Vision in insects: Pathways possibly underlying neural adaptation and lateral inhibition. Science 195, 894-897.

Tosini, G., and Menaker, M., 1996. Circadian rhythms in cultured mammalian retina. Science 272, 419-421.

Trujillo-Cenóz, O., 1965. Some aspects of the structural organization of the intermediate retina of dipterans. Journal of Ultrastructure Research 13, 1-33.

Williams, D.S., 1982. Rhabdom size and photoreceptor membrane turnover in a muscoid fly. Cell and Tissue Research 226, 629-639.

Wills, S.A., Page, T.L., and Colwell, C.S., 1986. Circadian rhythms in the electroretinogram of the cockroach. Journal of Biological Rhythms 1, 25-37.

Zerr, D.M., Hall J.C., Rosbash, M., Siwicki, K.K., 1990. Circadian fluctuations of *period* protein immunoreactivity in the CNS and the visual system of *Drosophila*. Journal of Neuroscience 10, 2749-2762.

Insect Timing: Circadian Rhythmicity to Seasonality
D.L. Denlinger, J. Giebultowicz and D.S. Saunders (Editors)
© 2001 Elsevier Science B.V. All rights reserved.

Anatomy and functions of the brain neurosecretory neurons with regard to reproductive diapause in the blow fly *Protophormia terraenovae*

Sakiko Shiga and Hideharu Numata

Department of Bio- and Geosciences, Graduate School of Science, Osaka City University, Sumiyoshi, Osaka 558-8585, Japan

Hormonal control of diapause has been studied in many insects. However, neural mechanisms controlling endocrine events involved in diapause are poorly understood. In this paper anatomical study and micro-lesions of neurons projecting to the retrocerebral complex were performed to examine their possible roles in the induction of reproductive diapause of the female blow fly, *Protophormia terraenovae* (Diptera: Calliphoridae). Retrograde-filling through the cardiac-recurrent nerve labeled three groups of neurons in the brain/subesophageal ganglion: (1) paramedial clusters of the pars intercerebralis, (2) neurons in each pars lateralis, and (3) neurons in the subesophageal ganglion. The dendritic arborization of the pars intercerebralis and pars lateralis neurons is restricted to the superior protocerebral neuropil and to the anterior neuropil of the subesophageal ganglion. Retrograde-filling from the corpus allatum indicated that the pars lateralis neurons and a few pars intercerebralis neurons project to the corpus allatum, but that the neurons in the subesophageal ganglion do not. Female adults developed their ovaries under diapause-averting conditions (LD 18:6, 25 °C), whereas their ovarian development was suppressed under diapause-inducing conditions (LD 12:12, 20 °C). Under both conditions, the ovaries invariably failed to develop when the pars intercerebralis was removed. When the pars lateralis were removed bilaterally, the ovaries developed in most of the females, irrespective of the rearing conditions. Removal of the pars lateralis prevented females from entering reproductive diapause. These results show that certain neurosecretory neurons in the pars intercerebralis are necessary for vitellogenesis and that the pars lateralis contains inhibitory neurons that suppress vitellogenesis during reproductive diapause. Based on these results, the neuroendocrine mechanisms controlling reproductive diapause in female *P. terraenovae* were discussed compared with those of several other species.

1. INTRODUCTION

Neurons projecting to the retrocerebral complex (RC), which includes the corpora cardiaca (CC) and corpora allata (CA), play important roles in the endocrine control of various physiological and developmental processes (Raabe 1989). Most of them are known as neurosecretory neurons that release hormones from the CC and CA, but the terminals of some of them serve to control the release of hormones from intrinsic endocrine cells in the CC and CA. In the brain, mainly two regions, pars intercerebralis (PI) and pars lateralis (PL), have somata of neurons projecting to the RC.

The neurosecretory neurons producing the prothoracicotropic hormone in the tobacco hornworm, *Manduca sexta,* have somata in the PL and their terminals in the CA that serve as neurohemal release sites rather than as the target of the hormone (Agui et al. 1980). In the PL of the cockroach, *Diploptera punctata,* there are neurosecretory somata of a different type. These somata produce a peptide, allatostatin, and extend fibers to the CC and CA (Stay et al. 1992). Allatostatin released within and from the CA accumulates and acts as a paracrine factor that inhibits release of juvenile hormone (JH) from the intrinsic endocrine cells of the CA. In the migratory locust, *Locusta migratoria,* neurons projecting to the CA have nerve-mediated inhibition of JH production in the CA, because JH is released immediately after transection of the nervi corporis allati (NCA) (Horseman et al. 1994; Okuda and Tanaka 1997). The subesophageal ganglion also contains neurons that terminate in the RC (Nijhout 1994). In the silk moth, *Bombyx mori,* three clusters of neurosecretory cells produce the pheromone biosynthesis-activating neurohormone, which is released from the CC to stimulate the release of sex pheromone in female moths (Ichikawa et al. 1995).

These studies have well characterized hormonal or neuronal outputs of the neurons terminating the RC to certain target organs by combinations of ingenious bioassays, neuroanatomy and electrophysiology. However, upstream mechanisms controlling the neurosecretory neurons have not been studied. Endocrine activities are not always endogenous but internal or external environmental cues can switch the physiological status to ensure the organism's adaptations to the environment.

Facultative diapause is a good model for studying the brain neuroendocrine systems that switch a physiological status. Adult diapause is defined as a reproductive arrest and in diapause females ovarian development is suppressed. This response is controlled by certain environmental cues such as photoperiod and temperature. Photoperiodic or temperature information perceived through a receptor is integrated and stored in a certain part of the brain. This information is translated into endocrine events that control various developmental programs affecting diapause, growth rate and seasonal morphs (Beck 1980; Saunders 1982). In the whole physiological process for adult diapause, the endocrine events have been examined in many insect species.

First, De Wilde and co-workers examined the hormonal control of adult diapause in the Colorado potato beetle, *Leptinotarsa decemlineata* (e. g., De Wilde and De Boer 1961, 1969; Schooneveld et al. 1977), and then the leading role of the CA in adult diapause was confirmed in many other insects. Inactivity of the CA secreting JH has been regarded as a key factor (see Denlinger 1985 for review). The role of ecdysteroids has also been suggested (Briers and De Loof 1981; Lefevere 1989).

In contrast, neural control mechanisms of endocrine events are poorly understood. Projection patterns of brain neurosecretory neurons suggest that they play important roles in conversion of integrated neural information to humoral signals. Hence, understanding the functional roles of neurons in the PI and PL in the physiological mechanism of reproductive diapause is of a great interest. In the linden bug, *Pyrrhocoris apterus,* and the grasshopper, *Tetrix undulata*, neurons in the PI inhibit ovarian development during diapause (Hodková 1976; Poras 1982), whereas in the migratory locust, *Locusta migratoria,* these neurons support the CA activity during vitellogenesis (Poras et al. 1983). The role of neurons in the PL in reproductive diapause, however, has not been well characterized. Most of the findings were obtained from cautery or surgical removal of certain somata without a careful histological examination. Cauterization of the PI can cause interruption of the axonal pathways from the PL neurons to the CA (Poras et al. 1983; Rüegg et al. 1983). As the loci of these somata are close to each other, the surgery must be restricted to a small area of the brain and should be accompanied by a histological examination to distinguish between the PI and PL neurons. Furthermore, anatomical mapping of neurons is also helpful in investing the signal pathway.

Many flies have seasonal adult diapause with suppression of ovarian development (e. g., Stoffolano and Matthysse 1967; Lumme et al. 1974; Stoffolano 1974; Vinogradova 1987; Saunders and Gilbert 1990). The leading role of the CA in the control of adult diapause has been demonstrated in females of several flies (Kambysellis and Heed 1974; Saunders et al. 1990; Agui et al. 1991; Burks et al. 1992; Kim and Krafsur 1995). Female adults in a Japanese population of the blow fly, *Protophormia terraenovae* (Diptera: Calliphoridae) enter diapause with suppression of ovarian development under short-day and low temperature conditions (Numata and Shiga 1995). We have been studying the brain neurosecretory systems using *P. terraenovae* with the aim of investigating the signal cascade switching the reproductive status by environmental stimuli. In this paper we focus on the brain neurosecretory neurons with regard to the reproductive diapause in *P. terraenovae*.

2. EXPERIMENTS AND RESULTS

2.1. Neural control of the corpus allatum by the brain in diapause

Fig. 1 shows a schematic illustration of the brain, subesophageal ganglion and retrocerebral complex of the adult blow fly. In flies, the CC and CA are unpaired and the CC is fused to the hypocerebral ganglion (HG) to form a complex of CCHG (Shiga et al. 2000). A single pair of nervi corporis cardiaca (NCC) extends posteriorly from the brain and merges with the recurrent nerve (RN) to form the cardiac recurrent nerve (CRN). The CRN extends posteriorly, beneath the aorta and above the esophagus, to the CCHG. From the CCHG, a pair of NCA extends dorsally around the aorta to enter the CA. Female adults of *P. terraenovae* develop their ovaries promptly after emergence under LD 18:6 (18-h light and 6-h dark cycles) at 25 °C, although they enter diapause under LD 12:12 at 20 °C (Numata and Shiga 1995).

To examine roles of the CA in adult diapause, first we transplanted the CA (Matsuo et al. 1997). The CA were transplanted from reproductive females 5 days after adult emergence (day 5) under LD 18:6 at 25 °C to females on day 5 under LD 12:12 at 20 °C. The recipients were reared under LD 12:12, 20 °C and dissected to examine ovarian stages on day 10. When one CA was transplanted, yolk deposition was observed in about 70% of the adults and about 30% of the adults had fully matured ovaries (Fig. 2). When two CA were implanted, yolk depositions were observed in all females and fully matured ovaries were observed in about half of them. These results show that humoral substances from the CA in reproductive females are able to induce ovarian development in diapause females. Matsuo et al. (1997) showed that application of JH analogue induced ovarian development in a dose-dependent manner in diapause females of *P. terraenovae*, suggesting that a reduction of JH secretion from the CA is a primary cause of adult diapause. When the NCA were transected on day 5 under LD 12:12 at 20 °C, yolk deposition occurred in all females and about 50% females had fully developed ovaries on day 10 (Fig. 2). Transection of the NCA relieved the inhibition of ovarian development. The results indicate that the brain suppresses CA activity by neurons sending axons in the NCA to inhibit ovarian development under diapause-inducing conditions.

72

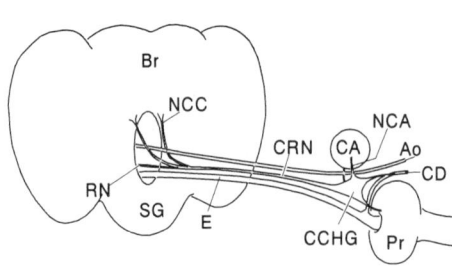

Figure. 1 A schematic illustration of the brain (Br), subesophageal ganglion (SG) and retrocerebral complex in adults of *Protophormia terraenovae*. Br and SG are in posterior view. The retrocerebral complex is in sagittal view. Bilateral nervi corporis cardiaci (NCC) posteriorly exit from the brain and run with the recurrent nerve (RN) as a single nerve, the cardiac-recurrent nerve (CRN). The CRN runs above the esophagus (E) and beneath the aorta (Ao) to a complex of the corpus cardiacum and hypocerebral ganglion (CCHG), which is connected to the corpus allatum (CA) via bilateral nervi corporis allati (NCA) (CD, crop duct, Pr, proventriculus) (from Shiga et al. 2000).

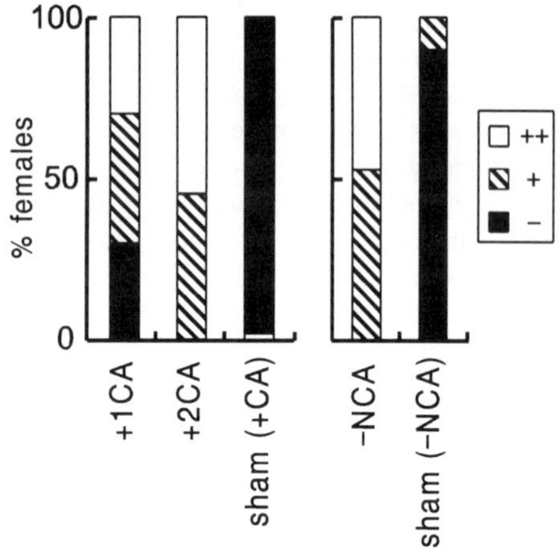

Figure. 2 Effects of implantation of the corpus allatum or transection of the nervi corporis allati on ovarian development in adults of *Protophormia terraenovae* reared under LD 12:12 at 20 °C. Operations were made on day 5 and the ovarian stage was examined on day 10. ++, stage with mature eggs; +, vitellogenic stage; –, previtellogenic stage. +CA, implantation of the corpus allatum from reproductive females reared under LD 18:6 at 25 °C; sham (+CA), sham operation for implantation of the corpus allatum; –NCA, transection of the nervi corporis allati; sham (–NCA), sham operation for transection of the nervi corporis allati. n = 17–41 (altered from Matsuo et al. 1997).

2.2. Brain neurons projecting to the retrocerebral complex

To examine neurons that are involved in suppression of ovarian development during diapause, we performed neuroanatomy of neurons projecting to the retrocerebral complex (Shiga et al. 2000). To label neurons having terminals in the retrocerebral complex including the CCHG and CA, the CRN was first backfilled with $NiCl_2$ (500 mM, supplied with 1% BSA) to the brain and subesophageal ganglion. Labeled neurons were further processed for silver intensification.

Fig. 3 shows the neurons stained through the CRN after the right NCC had been severed. Mainly three groups of somata were stained. The three groups were designated as the PI, PL and subesophageal ganglion (SEG) neurons according to the location of their somata. Through processes extending to the left NCC, seven large and four small somata in the left (ipsilateral) PL and two somata close to the optic lobe in the right (contralateral) PL were stained. On the basis of their projections, we classified the ipsilateral PL neurons as PL-i and the contralateral PL neurons as PL-c. Processes from the two PL-c somata project contralaterally to merge fibers from PL-i neurons and give rise to fine branches along the way. PL neurons exit through the NCC via posterior lateral tract (PLT). The PI somata were stained in the contralateral side to the NCC via a large median bundle (MB) in the anterior portion. The somata of the SEG neurons were also found in the contralateral side (Fig. 3). In these three groups, some neurons extend processes to the CA through the NCA and must be involved in suppression of the CA during diapause.

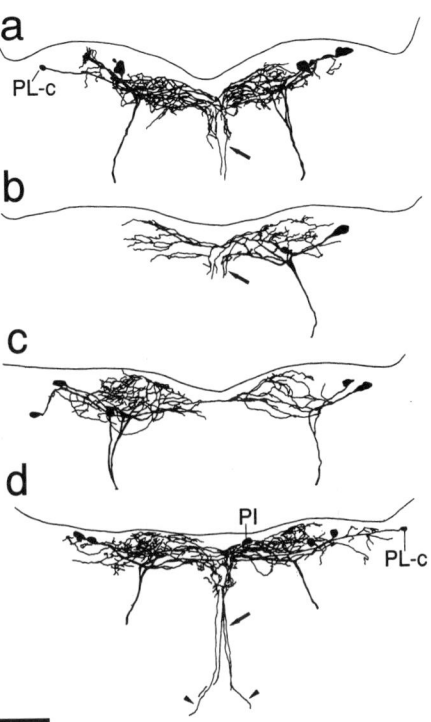

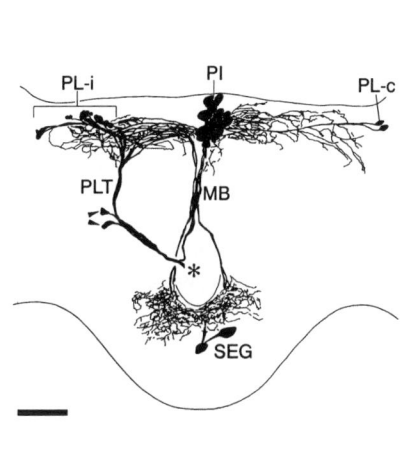

Figure. 3   A drawing of neurons in the brain/subesophageal ganglion backfilled from the cardiac-recurrent nerve after the right nervus corporis cardiaci had been cut in adults of *Protophormia terraenovae* in frontal view. The nervus corporis cardiaci was stained only in the left side (asterisk). Distinct fiber tracts are seen in the posterior lateral tract (PLT) and in the median bundle (MB). Somata of PL-i neurons and on the PLT (arrowheads) were stained ipsilaterally. Somata of PL-c, PI and SEG neurons were stained contralaterally. *Scale* 100 μm (from Shiga et al. 2000).

Figure. 4   Four examples of neurons in the brain backfilled from the corpus allatum in adults of *Protophormia terraenovae* in frontal view. Arrows show processes stained in the median bundle. Small processes further extend into the subesophageal ganglion (arrowheads). *Scale* 100 μm (altered from Shiga et al. 2000).

Then, in the next staining, NiCl$_2$ solution was directly injected into the CA to stain only neurons sending processes to the CA. Fig. 4 shows four examples of this staining. In Fig. 4a, one PL-c soma on the left and PL-i somata on both sides with fibers in the MB were stained. In Fig. 4b, only two PL-i neurons were stained with fibers in both hemispheres branching in the vicinity of the MB. Fig. 4c shows only a few fibers extending between the hemispheres, from PL-i neurons. In Fig. 4d, one PI neuron and processes extending into the subesophageal ganglion through the MB were stained in addition to the PL neurons. Among 14 individuals, the maximum number of somata stained was five of the PL-i, two of the PL-c and one of the PI neurons in a hemisphere. This indicates that most PL neurons and a few PI neurons extend their processes in the CA.

2.3. Removal of neurosecretory cells

Anatomical studies showed that inhibitory neurons that suppress ovarian development during diapause must be included in the PL neurons. Also, a possibility of a few PI neurons as inhibitory neurons is not excluded. In the next experiment, we examined effects of surgical removal of the PL or PI neurons on adult diapause (Shiga and Numata 2000). *P. terraenovae* was reared from eggs under LD 12:12 at 20 °C or LD 18:6 at 25 °C. Operation was made on day 1. Bluish-white neurosecretory cells visible in the PI or PL were removed with a sharpened tungsten hook. After the operations, 10 to 15 females were isolated in a plastic container with sucrose, beef liver and water. On day 11 under LD 12:12 at 20 °C or on day 6 under LD 18:6 at 25 °C, flies were first subject to histological examination in which the neurosecretory neurons were stained by the NiCl$_2$ backfills through the CRN or by paraldehyde-thionin / paraldehyde-fuchsin (PT-PF). After staining, their ovarian stages were examined.

In intact and sham-operated groups, most females had previtellogenic ovaries under LD 12:12 at 20 °C, whereas under LD 18:6 at 25 °C, most females had fully matured ovaries (Fig. 5). When the PI was removed, the ovaries invariably failed to develop, irrespective of the rearing conditions. Under LD 18:6 at 25 °C, the proportion of previtellogenic females was significantly higher in the PI-removal group than in the other three groups (intact, sham-operated and PL-removal). When the PL were removed bilaterally, the ovaries developed in 76% of the females reared under LD 12:12 at 20 °C and in 84% of the females reared under LD 18:6 at 25 °C (Fig. 5). Under LD 12:12 at 20 °C, the proportion of females with previtellogenic ovaries was significantly lower in the PL-removal group than in the other three groups (intact, sham-operated and PI-removal). Removal of the PL prevented females from entering reproductive diapause. Removal of the PL did not affect ovarian development under LD 18:6 at 25 °C.

Fig. 6 shows backfills of intact and experimental brains. In intact brains the backfill labeled the PI and PL neurons bilaterally (Fig. 6A). When the PI was removed, no somata in the PI were labeled (Fig. 6B) but their prominent fibers remained (not shown). Most of the PL neurons were labeled (Fig. 6B). When bilateral regions of the PL were removed, no somata in the PL were labeled, but their axons and fibers were stained even 10 days after the operation (Fig. 6C). Somata in the PI and their fibers were stained as in intact brains (Fig. 6D). Results of ablation experiments clearly indicate that PL neurons inhibit ovarian development during adult diapause and PI neurons are necessary to promote ovarian development.

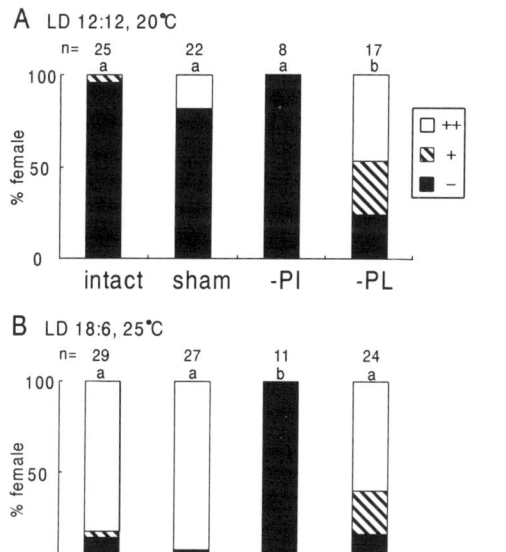

A LD 12:12, 20°C

B LD 18:6, 25°C

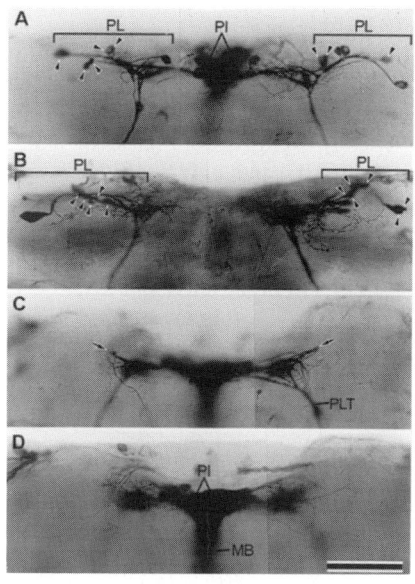

Figure. 5 Effects of surgical removal of the pars intercerebralis (-PI) or pars lateralis (-PL) on ovarian development in adults of *Protophormia terraenovae*. Operations were made on day 1 and the ovarian stage was examined on day 11 under LD 12:12 at 20 °C (**A**), and on day 6 under LD 18:6 at 25 °C (**B**). ++, stage with mature eggs; +, vitellogenic stage; –, previtellogenic stage. Removal of the PL prevented females from entering reproductive diapause. Columns with the same letter show no significant difference in the rate of females with previtellogenic ovaries in respective conditions (Tukey-type multiple comparison test for proportions, P>0.05) (altered from Shiga and Numata 2000).

Figure. 6 Photomicrographs of whole-mount preparations of brain neurons backfilled from the cardiac-recurrent nerve in adults of *Protophormia terraenovae*. Dorsal part of the brain is shown in frontal view. **A** Intact brain. A paramedial cluster of somata in the pars intercerebralis (PI) and 7-8 pairs of somata in the pars lateralis (PL) are stained. Arrowheads indicate out of focus somata. **B** The brain with the PI removed. Note that only PL neurons were labeled. **C** The brain with the PL removed. The PL somata were not labeled but their fibers and axons in the superior protocerebrum (arrows) and in the posterior lateral tract (PLT) still remained even 10 days after the operation. **D** The same preparation as **C** but in a different focal plane. Somata in the PI and axons in the median bundle (MB) were labeled as in the intact brain. *Scale* 100 μm (from Shiga and Numata 2000).

3. DISCUSSION

3.1. Neuroendocrine control of reproductive diapause in female adults of *P. terraenovae*

Fig. 7 summarizes the projection patterns of PL and PI neurons to the retrocerebral complex and the effects of surgical operations on vitellogenesis. Comparison of staining patterns by backfills from the CRN and CA indicates that most of PI neurons have terminals in the CCHG and a few of them extend processes to the CA (not shown in Fig. 7). Forwardfills with Lucifer Yellow from the PI to the RC also confirmed a few processes on a small surface region of the CA (Shiga et al. 2000). In contrast, most of the PL neurons have their terminals on the CA, because almost the same number of somata were stained in the PL by backfills from the CRN and CA.

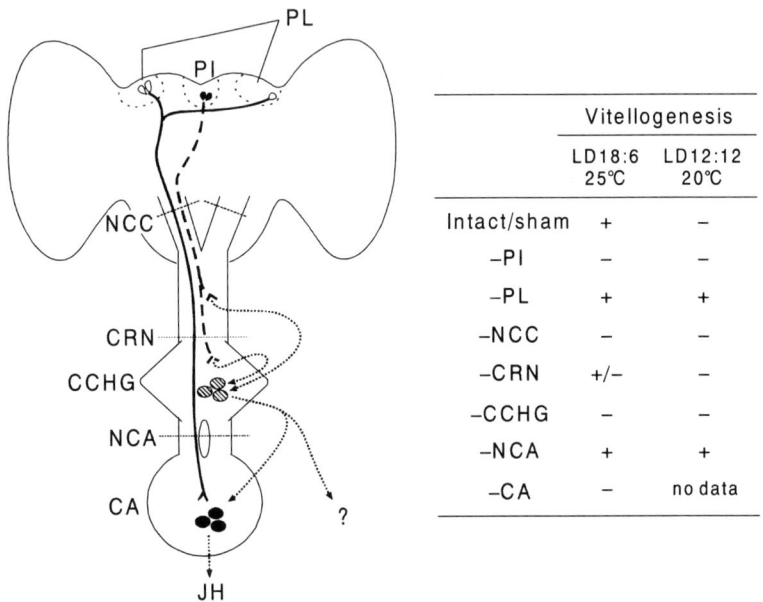

	Vitellogenesis	
	LD18:6 25℃	LD12:12 20℃
Intact/sham	+	−
−PI	−	−
−PL	+	+
−NCC	−	−
−CRN	+/−	−
−CCHG	−	−
−NCA	+	+
−CA	−	no data

Figure. 7 Effects of various surgical operations on ovarian development in adults of *Protophormia terraenovae*. -PI, removal of the pars intercerebralis; -PL, removal of the pars lateralis; -NCC, transection of the nervi corporis cardiaci; -CRN, transection of the cardiac recurrent nerve; -CCHG, removal of the fused corpus cardiacum and hypocerebral ganglion; -NCA, transection of the nervi corporis allati; -CA, removal of the corpus allatum. +, vitellogenesis was observed in most females; -, previtellogenic ovaries were observed in most females.

Together with our previous studies we have made seven kinds of surgical operation to examine neuroendocrine mechanisms controlling the adult diapause (Fig. 7). When the NCC were bilaterally severed, most females failed to develop ovaries irrespective of the rearing conditions (Toyoda et al. 1999). Because the NCC contains both PI and PL axons to the RC, this operation completely disconnected the terminal sites of these neurons. When the NCA were severed, females developed the ovaries irrespective of the rearing conditions (Matsuo et al. 1997). This operation caused the same effects as microlesion of the PL (Shiga and Numata 2000). Severance of the NCA removed terminal sites of most PL neurons and lesion of the PL removed somata of all PL neurons. In both types of operations, physiological outputs of PL neurons must be eliminated and PI neurons were functionally intact to release neurosecretory materials for ovarian development. These operations clearly indicate that even under diapause-inducing conditions PI neurons are able to produce neurosecretory materials for ovarian development.

When the CRN was severed close to the CCHG, less females developed ovaries than sham-operated ones under LD 18:6 at 25 °C and ovarian development was suppressed in most females as in the sham-operated females under LD 12:12 at 20 °C (Toyoda et al. 1999). We interpret the results that the release sites of neurosecretory materials of the PI neurons distribute not only in the CCHG but also in the CRN (Fig. 7). By PT-PF staining, some of the PI somata were revealed with fine processes mainly in the CCHG (Shiga et al. 2000). In addition, on the way to the CCHG, PT-PF positive fibers were also found in the CRN with bearing varicose processes (unpublished). This observation suggests that PI neurons give rise to the secretory region also in the CRN and in some females after transection of the CRN secretory materials from this region was enough to promote ovarian development under diapause-averting conditions but not under diapause-inducing conditions.

Toyoda et al. (1999) also examined the effects of removal of the CCHG. In this operation the CA remained without neural connections to the brain. If neurosecretory materials from the PI neurons directly affect the CA or ovaries to promote vitellogenesis, the results of CCHG removal should be the same as those of the CRN transection. However, this was not the case. After the operation, females completely failed to develop the ovaries irrespective of the rearing conditions (Toyoda et al. 1999). This suggests the presence of cells in the CCHG that mediate actions of PI neurons to promote ovarian development. Probably these cells in the CCHG are intrinsic endocrine cells that release humoral factors increasing JH production in the CA or promote another cascade leading the ovarian development (Fig. 7).

Without the CA no females developed the ovaries under LD 18:6 at 25 °C, clearly indicating that ovarian development is mediated by the CA. Application of a JH analogue caused ovarian development in females in which the CA had been removed. This result strongly suggests that JH from the CA is a key factor for vitellogenesis (Matsuo et al. 1997). Under diapause-inducing conditions, these neuroendocrine cascades are inhibited by PL neurons possibly suppressing the CA to produce JH.

3.2. Comparison with male adults of *P. terraenovae*

Male adults of *P. terraenovae* also have reproductive diapause in which mating behavior is suppressed under short-day conditions. In contrast to females, male adults responded to photoperiod and showed mating behavior even after the CA was removed. The mating behavior is controlled predominantly by a factor or factors other than the CA with respect to diapause (Tanigawa et al. 1999). Thus, the endocrine control of diapause in males is different from that in females in which the CA plays a predominant role. Differences of control mechanisms of adult diapause between the two sexes have also been reported in some other species (Zdárek 1968; Thibout 1982; Hodková and Wagner 1994; see also Pener 1992 for review).

3.3. Comparison with other species

Transection of the NCA induced vitellogenesis in *P. terraenovae* under diapause-inducing conditions, and it was therefore proposed that the brain reduces the endocrine activity of the CA through the NCA in diapause adults (Matsuo et al. 1997). Neural inhibition of the CA by the brain has been reported in several other species with reproductive diapause, e.g., *Pyrrhocoris apterus* (Hodková 1976, 1977), *Eurygaster integriceps* (Panov and Kryuchkova 1977), *Tetrix undulata* (Poras 1982), *Leptinotarsa decemlineata* (Khan et al. 1983), *Locusta migratoria* (Poras et al. 1983; Okuda and Tanaka 1997), *Plautia crossota stali* (Kotaki and Yagi 1989), and *Riptortus clavatus* (Morita and Numata 1997).

The cells responsible for this neural inhibition during reproductive diapause were localized in the PI in *P. apterus* and *T. undulata* (Hodková 1976; Poras 1982), but in the PL in *L. migratoria*, *L. decemlineata* and *P. terraenovae* (Poras et al. 1983; Khan et al. 1986; Shiga and Numata 2000). In *P.* terraenovae and *L. migratoria*, the neurosecretory neurons in the PI are necessary to promote ovarian development. This role of the PI in flies is completely different from that in *P. apterus* and *T. undulata* in which the ovaries can develop without PI neurons (Hodková 1976, 1977; Poras 1982).

In *Anacridium aegyptium*, the hemolymph of diapause adults lacks the stimulatory factor from the brain neurosecretory cells (Girardie et al. 1974). In *L. decemlineata* and *P. apterus*, humorally mediated stimulation and inhibition have been reported, in addition to nervous inhibition by the brain (De Wilde and De Boer 1969; Khan et al. 1983; Hodková 1976, 1977, 1992). In *P. terraenovae*, two humoral stimulating factors have been proposed (see Fig. 7). One is produced by the PI neurons and released from the CCHG and the CRN. The other is probably produced and released by the CCHG intrinsic cells. The pathways of these factors to stimulate ovarian development are unclear, but we suggest the following three possibilities. (1) The factor(s) stimulate the activity of the CA to secrete JH. Such a factor is shown in *L. migratoria* (Tobe et al. 1982). The primary structure of allatotropin, i.e., a peptide stimulating the activity of the CA to secrete JH, was first identified in *Manduca sexta* (Kataoka et al. 1989), and its presence has been shown in several other species (see Tobe and Stay 1985). (2) The factor(s) stimulate the ovaries to secrete ecdysteroids, which are necessary to vitellogenin synthesis in *P. terraenovae* (Huybrechts and De Loof 1982). The primary structure of ovary ecdysteroidogenic hormone, which is a peptide that stimulates the ovaries to secrete ecdysteroids, was recently identified in *Aedes aegypti*. The peptide is localized immunocytochemically in two or three pairs of PI neurons (Brown et al. 1998). (3) The factor(s) are necessary for vitellogenesis without the intervention of JH or ecdysteroids.

3.4. Future perspectives

The radiochemical assay is a good tool to investigate JH biosynthesis and release, and has been applied to many insect species (Pratt and Tobe 1974; see Tobe and Stay 1985 for review). The difference in JH synthetic activity of the CA in diapause and nondiapause adults was shown by radiochemical assays in *L. decemlineata*, *L. migratoria*, and *P. c. stali* (Khan et al. 1982; Okuda et al. 1996; Kotaki 1999). In *P. terraenovae*, it is necessary to measure JH biosynthetic activity of the CA in diapause and nondiapause adults. Furthermore, measurements of JH biosynthesis after surgical operations could give an answer to the questions: (1) Whether the ovary stimulating pathways mentioned above are mediated by the CA? (2) Whether PL neurons directly inhibit JH synthesis or mediate other pathways to inhibit ovarian development under diapause-inducing conditions? With these aims, we have started measurement of JH biosynthesis by the CA in *P. terraenovae*.

The anatomical study has shown that the PI and PL neurons have dendritic arborization in the superior protocerebral neuropil in *P. terraenovae*. In flies, the superior protocerebral neuropil is invaded by the second to fourth order neurons of the visual, chemosensory and mechanosensory systems (Strausfeld 1976). Shiga and Numata (1997) demonstrated that in *P. terraenovae* the compound eyes perceive the short-day photoperiod, which is responsible for the induction of reproductive diapause. Temperature is another stimulus to induce reproductive diapause. In no species, however, temperature receptors for diapause induction have been identified. Although information about where and how photoperiod and temperature signals are processed is completely lacking at this time, it seems likely that these signals are conveyed to PI and PL neurons to control endocrine events for reproductive diapause. As the next step, we survey neurons from which the PI or PL neurons receive information to switch the physiological status with regard to reproductive diapause.

REFERENCES

Agui, N., Bollenbacher, W. E., Granger, N. A., Gilbert, L. I., 1980. Corpus allatum is release site for insect prothoracicotropic hormone. Nature 285, 669-670.

Agui, N., Mihara, M., Kurahashi, H., 1991. Effect of juvenile hormone analogue on ovarian development of the reproductive-diapausing parasitic fly, *Melinda pusilla*. Japanese Journal of Sanitary Zoology 42, 311-317.

Beck, S. D., 1980. Insect Photoperiodism, 2nd ed. Academic Press, New York.

Briers, T., De Loof, A., 1981. Moulting hormone activity in the adult Colorado potato beetle, *Leptinotarsa decemlineata* Say in relation to reproduction and diapause. International Journal of Invertebrate Reproduction 3, 145-155.

Brown, M. R., Graf, R., Swiderek, K. M., Fendley, D., Stracker, T. H., Champagne, D. E., Lea, A. O., 1998. Identification of a steroidogenic neurohormone in female mosquitoes. Journal of Biological Chemistry 273, 3967-3971.

Burks, C. S., Evans, L. D., Kim, Y., Krafsur, E. S., 1992. Effects of precocene and methoprene application in young adult *Musca autumnalis*. Physiological Entomology 17, 115-120.

De Wilde, J., De Boer, J. A., 1961. Physiology of diapause in the adult Colorado beetle—II Diapause as a case of pseudo-allatectomy. Journal of Insect Physiology 6, 152-161.

De Wilde, J., De Boer, J. A., 1969. Humoral and nervous pathways in photoperiodic induction of diapause in *Leptinotarsa decemlineata*. Journal of Insect Physiology 15, 661-675.

Denlinger, D. L., 1985. Hormonal control of diapause. In: Kerkut, G. A., Gilbert, L. I. (Eds.), Comprehensive Insect Physiology, Biochemistry, and Pharmacology, Vol. 8, Pergamon Press, Oxford, pp. 353-412.

Girardie, A., Moulins, M., Girardie, J., 1974. Rupture de la diapause ovarienne d'*Anacridium aegyptium* par stimulation électrique des cellules neurosécrétrices médianes de la pars intercerebralis. Journal of Insect Physiology 20, 2261-2275.

Hodková, M., 1976. Nervous inhibition of corpora allata by photoperiod in *Pyrrhocoris apterus*. Nature 263: 521-523.

Hodková, M., 1977. Function of the neuroendocrine complex in diapausing *Pyrrhocoris apterus* females. Journal of Insect Physiology 23, 23-28.

Hodková, M., 1992. Storage of the photoperiodic information within the implanted neuroendocrine complexes in females of the linden bug *Pyrrhocoris apterus* (L.) (Heteroptera). Journal of Insect Physiology 38, 357-363.

Hodková, M., Wagner, R. M., 1994. Neuroendocrine control of diapause in males of *Pyrrhocoris apterus*. In: Borkovec, A. B., Loeb, M. J. (Eds.), Insect Neurochemistry and Neurophysiology 1993, CRC Press, Boca Raton, pp. 271-274.

Horseman, G., Hartmann, R., Virant-Doberlet, M., Loher, W., Huber, F., 1994. Nervous control of juvenile hormone biosynthesis in *Locusta migratoria*. Proceedings of the National Academy of Sciences of the United States of America 91, 2960-2964.

Huybrechts, R., De Loof, A., 1982. Similarities in vitellogenin and control of vitellogenin synthesis within the genera *Sarcophaga*, *Calliphora*, *Phormia* and *Lucilia* (Diptera). Comparative Biochemistry and Physiology 72B, 339-344.

Ichikawa, T., Hasegawa, K., Shimizu, I., Katsuno, K., Kataoka, H., Suzuki, A., 1995. Structure of neurosecretory cells with immunoreactive diapause hormone and Pheromone biosynthesis activating neuropeptide in the silkworm, *Bombyx mori*. Zoological Science 12, 703-712.

Kambysellis, M. P., Heed, W. B., 1974. Juvenile hormone induces ovarian development in diapausing cave-dwelling *Drosophila* species. Journal of Insect Physiology 20, 1779-1786.

Kataoka, H., Toschi, A., Li, J. P., Carney, R. L., Schooley, D. A., Kramer, S. J., 1989. Identification of an allatotropin from adult *Manduca sexta*. Science 243, 1481-1483.

Khan, M. A., Doderer, A., Koopmanschap, A. B., De Kort, C. A. D., 1982. Improved assay conditions for measurement of corpus allatum activity *in vitro* in the adult Colorado potato beetle, *Leptinotarsa decemlineata*. Journal of Insect Physiology 20, 279-284.

Khan, M. A., Koopmanschap, A. B., De Kort, C. A. D., 1983. The relative importance of nervous and humoral pathways for control of corpus allatum activity in the adult Colorado potato beetle, *Leptinotarsa decemlineata* (Say). General and Comparative Endocrinology 52, 214-221.

Khan, M. A., Romberg-Privee, H. M., Koopmanschap, A. B., 1986. Location of allatostatic centers in the pars lateralis regions of the brain of the Colorado potato beetle. Experientia 42, 836-838.

Kim, Y., Krafsur, E. S., 1995. *In vivo* and *in vitro* effects of 20-hydroxyecdysone and methoprene on diapause maintenance and reproductive development in *Musca autumnalis*. Physiological Entomology 20, 52-58.

Kotaki, T., 1999. Relationship between JH-biosynthetic activity of the corpora allata *in vitro*, their size and adult diapause in a stink bug, *Plautia crossota stali* Scott. Entomological Science 2, 307-313.

Kotaki, T., Yagi, S., 1989. Hormonal control of adult diapause in the brown-winged green bug, *Plautia stali* Scott (Heteroptera: Pentatomidae). Applied Entomology and Zoology 24, 42-51.

Lefevere, K. S., 1989. Endocrine control of diapause termination in the adult female Colorado potato beetle, *Leptinotarsa decemlineata*. Journal of Insect Physiology 35, 197-203.

Lumme, J., Oikarinen, A., Lakovaara, S., Alatalo, R., 1974. The environmental regulation of adult diapause in *Drosophila littoralis*. Journal of Insect Physiology 20, 2023-2033.

Matsuo, J., Nakayama, S., Numata, H., 1997. Role of the corpus allatum in the control of adult diapause in the blowfly, *Protophormia terraenovae*. Journal of Insect Physiology 43, 211-216.

Morita, A., Numata, H., 1997. Role of the neuroendocrine complex in the control of adult diapause in the bean bug, *Riptortus clavatus* Archives of Insect Biochemistry and Physiology 35, 347-355.

Nijhout, H. F., 1994. Insect Hormones. Princeton University Press, Princeton.

Numata, H., Shiga, S., 1995. Induction of adult diapause by photoperiod and temperature in *Protophormia terraenovae* (Diptera: Calliphoridae) in Japan. Environmental Entomology 24, 1633-1636.

Okuda, T., Tanaka, S., 1997. An allatostatic factor and juvenile hormone synthesis by corpora allata in *Locusta migratoria*. Journal of Insect Physiology 43, 635-641.

Okuda, T., Tanaka, S., Kotaki, T., Ferenz, H.-J., 1996. Role of the corpora allata and juvenile hormone in the control of imaginal diapause and reproduction in three species of locusts. Journal of Insect Physiology 42, 943-951.

Panov, A. A., Kryuchkova, E. E., 1977. Control of activity of corpora allata in the bug *Eurygaster integriceps* with obligate imaginal diapause. Doklady Biological Sciences 232, 71-74.

Pener, M. P., 1992. Environmental cues, endocrine factors, and reproductive diapause in male insects. Chronobiology International 9, 102-113.

Poras, M., 1982. Le contrôle endocrinien de la diapause imaginale des femelles de *Tetrix undulata* (Sowerby, 1806) (Orthoptere, Tetrigidae). General and Comparative Endocrinology 46, 200-210.

Poras, M., Baehr, J. C., Cassier, P., 1983. Control of corpus allatum activity during the imaginal diapause in females of *Locusta migratoria* L. International Journal of Invertebrate Reproduction 6: 111-122.

Pratt, G. E., Tobe, S. S., 1974. Juvenile hormones radiobiosynthesized by corpora allata of adult female locusts *in vitro*. Life Sciences 14, 575-586.

Raabe, M., 1989. Recent Developments in Insect Neurohormones. Plenum Press, New York.

Rüegg, R. P., Lococo, D. J., Tobe, S. S., 1983. Control of corpus allatum activity in *Diploptera punctata*: roles of the pars intercerebralis and pars lateralis. Experientia 39, 1329-1334.

Saunders, D. S., 1982. Insect Clocks, 2nd ed. Pergamon Press, Oxford.

Saunders, D. S., Gilbert, L. I., 1990. Regulation of ovarian diapause in *Drosophila melanogaster* by photoperiod and moderately low temperature. Journal of Insect Physiology 36, 195-200.

Saunders, D. S., Richard, D. S., Applebaum, S. W., Ma, M., Gilbert, L. I., 1990. Photoperiodic diapause in *Drosophila melanogaster* involves a block to juvenile hormone regulation of ovarian maturation. General and Comparative Endocrinology 79, 174-184.

Schooneveld, H., Otazo Sanchez, A., De Wilde, J., 1977. Juvenile hormone-induced break and termination of diapause in the Colorado potato beetle. Journal of Insect Physiology 23, 689-696.

Shiga, S., Numata, H., 1997. Induction of reproductive diapause via perception of photoperiod through the compound eyes in the adult blow fly, *Protophormia terraenovae*. Journal of Comparative Physiology A 181, 35-40.

Shiga, S., Numata, H., 2000. The roles of neurosecretory neurons in the pars intercerebralis and pars lateralis in reproductive diapause of the blow fly, *Protophormia terraenovae*. Naturwissenschaften 87, 125-128.

Shiga, S., Toyoda, I., Numata, H., 2000. Neurons projecting to the retrocerebral complex of the adult blow fly *Protophormia terraenovae*. Cell and Tissue Research 299, 427-439.

Stay, B., Chan, K. K., Woodhead, A. P., 1992. Allatostatin-immunoreactive neurons projecting to the corpora allata of adult *Diploptera punctata*. Cell and Tissue Research 270, 15-23.

Stoffolano, J. G. Jr., 1974. Influence of diapause and diet on the development of the gonads and accessory reproductive glands of the black blowfly, *Phormia regina* (Meigen). Canadian Journal of Zoology 52, 981-988.

Stoffolano, J. G. Jr., Matthysse, J. G., 1967. Influence of photoperiod and temperature on diapause in the face fly, *Musca autumnalis* (Diptera: Muscidae). Annals of the Entomological Society of America 60, 1242-1246.

Strausfeld, N. J., 1976. Atlas of an Insect Brain. Springer-Verlag, Berlin.

Tanigawa, N. A., Shiga, S., Numata, H., 1999. Role of the corpus allatum in the control of reproductive diapause in the male blow fly, *Protophormia terraenovae*. Zoological Science 16, 639-644.

Thibout, E, 1982. Le comportement sexual du doryphore, *Leptinotarsa decemlineata* Say. et son possible controle par l'hormone juvenile et les corps allates. Behaviour 80, 199-217.

Tobe, S. S., Stay, B., 1985. Structure and regulation of the corpus allatum. Advances in Insect Physiology 18, 305-432.

Tobe, S. S., Girardie, J., Girardie, A., 1982. Enhancement of juvenile hormone biosynthesis in locusts following electrostimulation of cerebral neurosecretory cells Journal of Insect Physiology 28: 867-871.

Toyoda, I., Numata, H., Shiga, S., 1999. Role of the median neurosecretory cells in the ovarian development of the blow fly *Protophormia terraenovae*. Zoological Science 16, 187-191.

Vinogradova, E. B., 1987. Characteristics and regulation of imaginal diapause in the blackbottle fly *Protophormia terraenovae* R.-D. (Diptera, Calliphoridae). Soviet Journal of Ecology 3, 163–167.

Zdárek, J., 1968. Le comportement d'accouplement à la fin de la diapause imaginale et son contrôle hormonal dans le cas de la punaise *Pyrrhocoris apterus* L (Pyrrhocoridae, Heteroptera). Annales d'Endocrinologie, Paris 29, 703-707.

Insect Timing: Circadian Rhythmicity to Seasonality
D.L. Denlinger, J. Giebultowicz and D.S. Saunders (Editors)
© 2001 Elsevier Science B.V. All rights reserved.

Photoperiodism and seasonality in aphids

Jim Hardie

Aphid Biology Group, Department of Biology, Imperial College at Silwood Park, Ascot, Berks SL5 7PY, United Kingdom

The aphid photoperiodic response controls the mode of reproduction, parthenogenetic in long days and sexual in short days. Parthenogenesis is associated with viviparity and offspring are genetically identical to the mother while sexual reproduction is associated with the laying of diapausing eggs. Different adult forms are associated with these different reproductive strategies and the phenotypic differences between these insects with identical genotypes are extreme. Recently there has been progress in detailing the photoperiodic mechanisms. A putative photoperiodic photoreceptor region of the brain has been located using antibody technology and an opsin-based photoreceptor proposed. A recent model of the photoperiodic response involves separate long-night and short-night mechanisms that are both based on circadian oscillators having different properties. This development arose from contrasting results of studies using light-dark regimes to detect circadian influences and from the observations that long-night accumulation appeared to be temperature independent while short-night accumulation was not. A possible role for melatonin as a neuroendocrine effector molecule has also been examined. In mammals, melatonin levels rise during darkness and fall during light. By feeding or injecting melatonin these high titres can be extended in long days (i.e. short nights) to produce short-day effects. Similar experiments with aphids indicate that melatonin can promote short-day (i.e. long-night) effects in long-day conditions but whole-body titres of melatonin were not significantly different between light and dark.

1. INTRODUCTION

Aphids are not only the major insect pest of agriculture and horticulture in temperate climates but were the first animals shown to respond to day length (Marcovitch 1924). At that time it was known that many aphid species show cyclical parthenogenesis, that is, they reproduced asexually during summer but sexually in autumn. Parthenogenesis is associated with vivipary while sexual females lay diapausing, overwintering eggs and different adult forms (morphs) are associated with the alternative modes of reproduction. Marcovitch (1924) demonstrated that by artificially decreasing day length during summer he could prematurely induce the appearance of sexual forms in the strawberry aphid, *Aphis forbesi*. When day length was extended in the autumn, parthenogenetic reproduction continued and sexual forms did not appear.

Since that pioneering study there has been a great interest in insect photoperiodism with much attention remaining on aphids. This present chapter concentrates on recent findings and ideas on the mechanism of seasonality in two aphid species, the vetch aphid, *Megoura viciae*, and the pea aphid, *Acyrthosiphon pisum.*

The life cycles of aphids are complex and they show a high degree of polyphenism. Not only are different forms associated with the different modes of reproduction but winged and wingless parthenogenetic adults develop in different environmental conditions. In host-alternating species, which move between a woody, primary host plant in autumn and one or more herbaceous, secondary hosts during summer, there are also winged forms that undertake the migration. Some ten percent of the known 4,500 aphid species are host-alternating. The other species, like *M. viciae* and *A. pisum,* do not host alternate and remain on the same host plant species all year round. Their life cycle is, therefore, less complex and is illustrated in Figure 1.

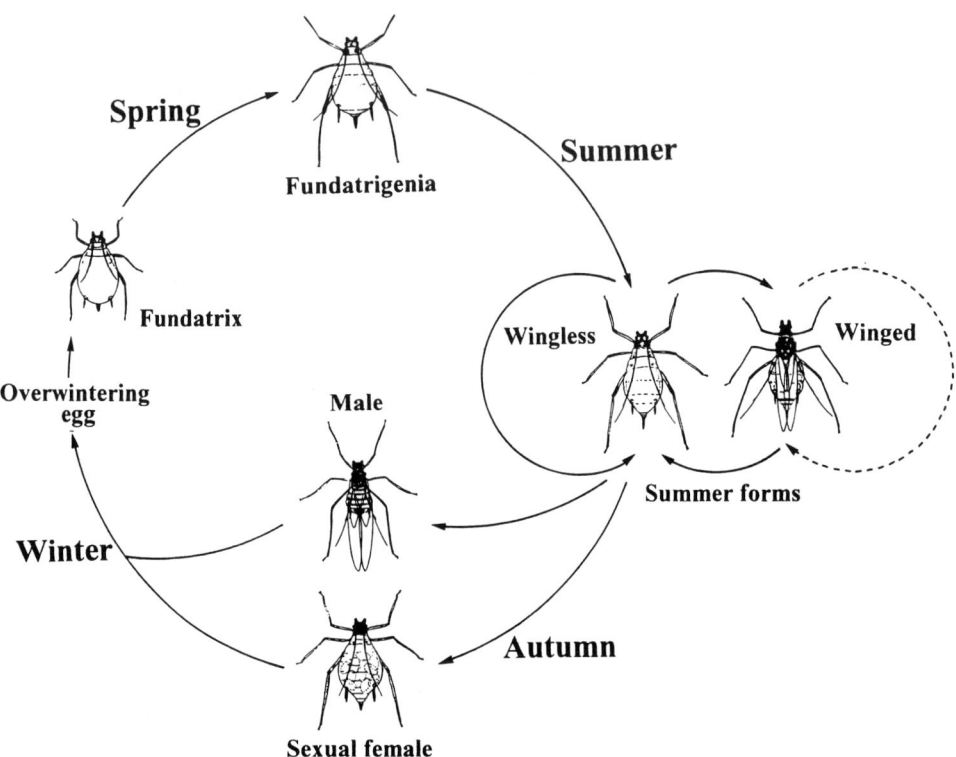

Figure 1. Life cycle of the vetch aphid, *Megoura viciae,* showing the different adult forms occurring throughout the year.

In the laboratory the life cycle complexities can be reduced. By retaining asexual aphids in long-day and uncrowded conditions, wingless asexual females continue to be produced generation after generation. This mode of reproduction also ensures that mothers, daughters and granddaughters are genetically identical and clonal lineages can be set up. The clones described here have been cultured in this way for over 40 years. Nevertheless, when insects are transferred to short-day conditions, sexual forms are induced.

All photoperiodic organisms require
i)   a photoreceptor mechanism to distinguish day from night,
ii)   a clock mechanism to measure day and/or night length,
iii)   a photoperiodic counter or memory to accumulate information from successive light-dark cycles,
iv)   a neuroendocrine effector mechanism to direct the relevant physiological processes.
We have information on each of these mechanisms in aphids but this chapter will concentrate on the more recent findings.

2. THE PHOTOPERIODIC PHOTORECEPTOR

Studies which extended day length over restricted regions of the aphid body, localised illumination experiments, and surgical techniques showed that the photoperiodic photoreceptor of *M. viciae* lies in the brain rather than the compound eyes (Lees, 1964). More recently immunocytochemical techniques have shown that a specific region of the anterio-ventral part of the protocerebral neuropile is labelled with antibodies raised to a variety of visual proteins (Gao et al., 1999). From a selection of 20 antibodies, 7 consistently labelled the same region of protocerebral neuropile. These had been raised to the opsin moiety of photoreceptor pigments from vertebrates and insects (Figure 2), to vertebrate arrestin, to cellular retinoid binding protein and to a protein mixture. The area labelled precisely coincides with that indicated by Lees' local illumination studies and is a prime candidate for the photoperiodic photoreceptor region. No cell bodies were labelled and the position of these remains unknown.
The antibody labelling of this putative photoperiodic photoreceptor region supported an earlier suggestion by Lees based on action spectra studies between wavelengths of 365 and 638 nm (Lees, 1981). He showed that early- and late-night interruptions with a 1-h light pulse indicated a pigment with maximum sensitivity to light with wavelength of 450-470 nm. However, the early-night interruptions produced a more defined peak while the later illumination provided a much broader spectrum with some sensitivity in the yellow/red (570-600 nm). Experiments with monochromatic light forming the main photoperiod produced results between these extremes. The results were compatible with the idea that the photoperiodic photoreceptor pigment was a caroteno-protein and different sensitivities were the result of different forms of the pigment. The shape of the early-night interruption spectrum is also compatible with a retinaldehyde-based photopigment and approximates to the shape of a Dartnell nomogram (Lythgoe, 1979). It is, therefore, similar to visual photoreceptor pigments which are based upon a retinal chromophore and an opsin protein (Gartner and Towner, 1995). This observation, coupled with the labelling of the putative photoreceptor region in the protocerebrum with antibodies raised to opsins, strongly suggest that the photoperiodic photoreceptor pigment is opsin based. Attempts to identify the opsin have involved sequencing opsins from *M. viciae* but, so far, only two have been identified and these are expressed in the eye, and not the brain, indicating that they are associated with vision (Gao et al., 2000). The flattened, broader-based action spectra require further explanation/investigation.

88

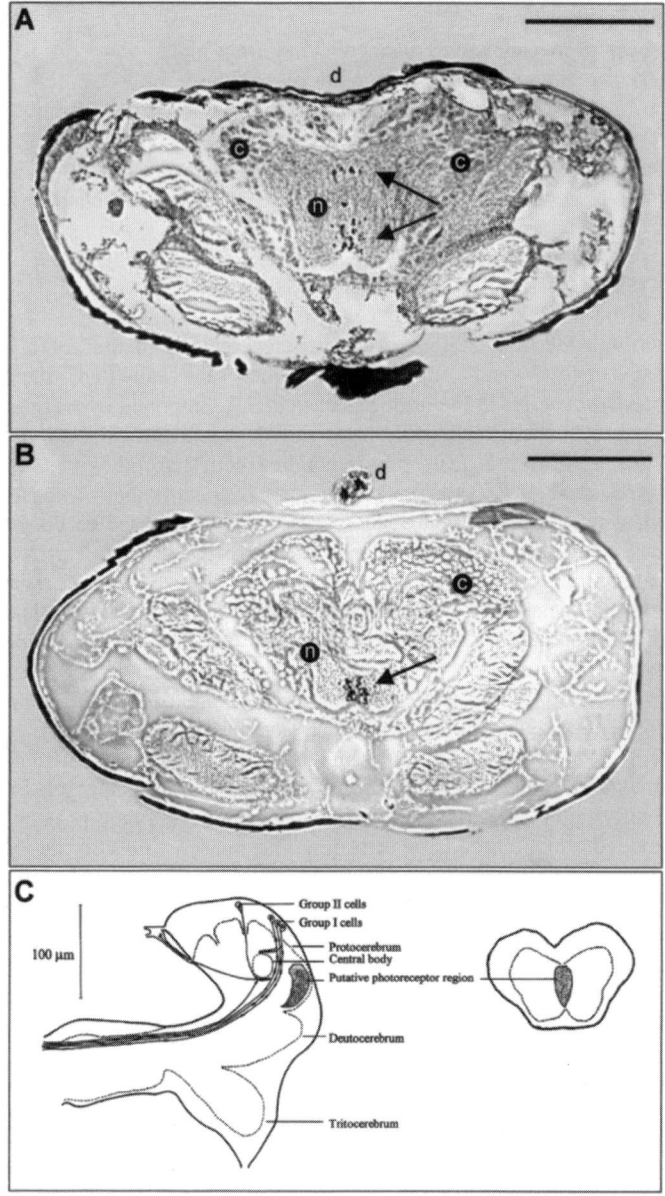

Figure 2. Immunocytochemical labelling (arrows) of the protocerebral neuropile in vertical sections through the head capsule of *M. viciae*. A) antibody raised to the C-terminal of *Drosophila* rhodopsin 1; B) a monoclonal antibody raised to a long-wavelength opsin from chicken cones, c = cell bodies, d = dorsal, n = neuropile, scale bars = 100 μm; C) diagram of the putative photoreceptor region, in longitudinal section, left, and transverse section, right (after Gao et al., 1999).

An alternative to the idea that the photoperiodic photoreceptor pigment is opsin based is that it could be based on a cryptochrome such as those involved in the entrainment of endogenous circadian rhythms to light-dark cycles in *Drosophila* (Stanewsky et al., 1998). Cryptochrome-based pigments were originally identified as photoreceptors in plants and absorb shorter wavelengths of the biological spectrum, blue and ultraviolet light, when bound to flavins/pterin chromophores (Ahmad and Cashmore, 1996). The basic absorption spectra of cryptochrome-based photoreceptors are very different to a retinaldehyde-opsin pigment and they appear an unlikely candidate (Lucas and Foster, 1999). The light-filtering action of tissue overlying a brain photoreceptor can, of course, alter the light spectrum reaching the receptor and thus affects the observed whole-animal response. However, even when the light transmission through the head capsule of *M. viciae* is accounted for, the action spectrum still differs from that predicted by cryptochrome-based photopigments (Hardie et al., 1981). Nevertheless, the possibility of crytochrome-based photoreception deserves further attention in aphids.

3. THE CLOCK AND COUNTER MECHANISMS

The clock mechanism and photoperiodic counter in aphids have recently been reviewed (Hardie and Vaz Nunes, 2000). The latest revision of ideas has taken place with detailed data collected on different clones of *M. viciae* and the photoperiodic induction of winged forms in the host-alternating black bean aphid, *Aphis fabae*. For many years classical experiments designed to reveal whether photoperiodic time measurement was based upon a circadian mechanism failed with *M. viciae* and the timing mechanism was believed to be based on an interval timer/hourglass type of mechanism (Hillman, 1973; Lees, 1973, 1986). With *A. fabae* there was evidence of a circadian involvement at 20 but not at 15 $^{\circ}$C (Hardie, 1987). More recent investigations using a Veerman-Vaz Nunes protocol (Veerman and Vaz Nunes, 1987) have changed ideas on aphid photoperiodic clocks and indicate that circadian rhythms are indeed involved in both species at 15 $^{\circ}$C (Vaz Nunes and Hardie, 1993; Hardie and Vaz Nunes, 2000).

Studies on the photoperiodic counter of these aphid species have also revealed surprises. It appears that the accumulation of long-night information is temperature compensated, whereas short-night accumulation is temperature sensitive (Hardie, 1990; Vaz Nunes and Hardie, 1999, 2000). These findings indicate that there is a fundamental difference between long- and short-night measurement/accumulation and the studies have led to the development of a new computer model of photoperiodic responses (Vaz Nunes, 1998). The model had different long-night and short-night measuring mechanisms and was called the 'double circadian oscillator' model. Here the oscillatory mechanisms for measuring long and short nights are separate, a new concept for photoperiodic time measurement. The oscillators can be ascribed different properties, such as period, damping, temperature sensitivity, and the model can be used to predict photoperiodic responses not only for aphids, but for other insects. It also provides explanations of previously inconsistent results, for example, the *M. viciae* data from Nanda-Hamner protocols indicated an hourglass-like clock while Veerman-Vaz Nunes protocols indicated a circadian component (Vaz Nunes, 1998; Hardie and Vaz Nunes, 2000).

4. THE NEUROENDOCRINE MECHANISM

It has been known for some time that the neuroendocrine control of the photoperiodic response in aphids involves brain neurosecretory cells and the corpus allatum (Hardie, 1984). Destruction of the medial neurosecretory cells (also known as the Group I cells in aphids) results in short-day reared aphids not responding to a transfer to long days by switching from the production of sexual to asexual progeny. After such surgery, long-day reared insects will spontaneously switch to the production of sexual females, as if they had been transferred to short-days (Steel and Lees, 1977). The original hypothesis was that these cells produced a neurohormone that acted directly upon developing embryos and promoted the asexual, long-day phenotype. Later work, however, indicated a role for juvenile hormone in effecting the long-day response. Topical application of juvenile hormone agonists promoted the production of asexual forms (Hardie, 1981; Corbitt and Hardie, 1985; Hardie and Lees, 1985). It now seems possible that the Group I cells play a role in the control of juvenile hormone synthesis by the corpora allata (Hardie, 1987). Whatever the mechanisms, it is evident that a link is required from light/dark detection in the brain, via measurement of night length and accumulation of the photoperiodic information, which is assumed to be a central mechanism, to the maternal ovaries. It is here that the final influence on the phenotypic development of embryos takes place.

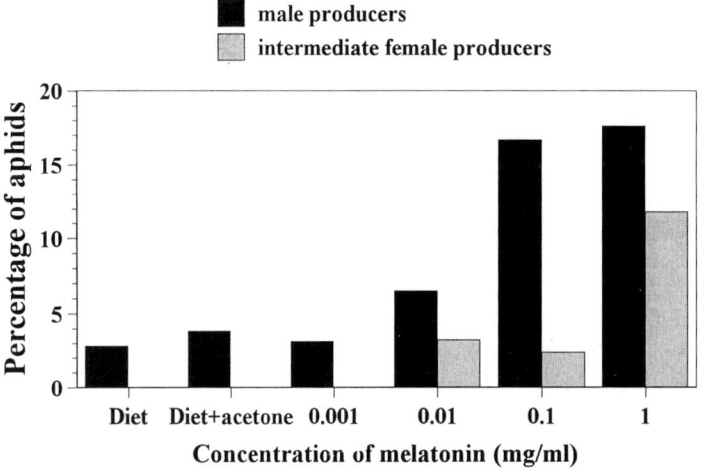

Figure 3. Percentage of aphids reared on artificial diet containing different concentrations of melatonin giving birth to sexual/asexual intermediate females or males (from Gao and Hardie, 1997).

Melatonin was first isolated from vertebrate pineal gland and identified as an indoleamine by Lerner et al. (1958). Melatonin levels tend to rise during darkness and fall during the day. The amplitude, phase and duration of the daily melatonin cycle contains information that could provide calendar cues and identify the seasonal cycle. In addition, injection and pinealectomy indicate that melatonin mediates the effect of day length on reproduction in a range of mammals (Tamarkin et al., 1985; Arendt, 1995).

Melatonin has also been reported in insects; in locusts, *Locusta migratoria*, (Vivien-Roels et al., 1984), in the face fly, *Musca autumnalis*, (Wetterberg et al., 1987), in the fruit fly, *Drosophila* spp., (Finocchiaro et al., 1988) in two species of damselfly (Tilden et al., 1994) and the silk worm *Bombyx mori* (Itoh et al., 1995). In *M. autumnalis*, one damselfly, *Ischnura verticalis*, and *B. mori* melatonin levels were higher during the night than during the day and it is tempting to speculate that melatonin could be involved in insect photoperiodism. Certainly, topical application of melatonin to the linden bug, *Pyrrhocoris apterus*, which enters reproductive diapause in short days, did delay the onset of oviposition but did not induce diapause in diapause-averting day lengths (Hodkova, 1989). Neither did melatonin application prevent the termination of diapause when diapausing *P. apterus* females were placed in diapause-averting photoperiods.

Recently we looked at the possibility of a role for melatonin in aphid photoperiodism (Gao 1996; Gao and Hardie, 1997). A green biotype of the pea aphid, *A. pisum*, was used and daily topical applications were attempted 4 h before lights off in long-day (16L:8D) conditions. This strategy should ensure a prolonged period of high melatonin titres if the titres are naturally high during the dark period. However, there was a high mortality when treatment started in first- or second-stadium aphids, so treatment was delayed until the start of the third stadium and continued throughout life. At 8 ng doses in acetone, melatonin reduced fecundity and almost all offspring died. Doses of 2 ng also reduced fecundity but two 2 male progeny (of 1047) were induced while controls produced only asexual females (Gao, 1996). As short days promote the appearance of males it appeared that a short-day (long-night) effect had been induced even though sexual females were not observed.

Although topical application allows the precise timing of melatonin treatment, the delay in starting treatment until the third stadium will have affected the results. Aphids are photoperiodically responsive from late embryo right through to the adult stage and an improved treatment would begin at birth to examine for effects on the next generation. So, insects were also fed on an artificial diet containing various concentrations of melatonin and 1% acetone in long-day conditions. Newly-born nymphs were transferred from bean seedlings to the diet and left to develop into adults prior to returning them to plants. Control insects were reared on stock diet or diet with 1% acetone. The progenies from diet-reared insects were collected every 2-3 days and, in turn, left to develop until they could be identified as long-day asexual females, as short-day sexual females or males, or as intermediate females whose ovaries contain both asexually developing diploid embryos and haploid yolky eggs. Again, melatonin treatment reduced fecundity and extended development time (significantly at 1 mg/ml) (Gao and Hardie, 1997). None of the progeny born to treated mothers became sexual females but a number were intermediate females and some were males (Figure 3). The numbers of males and intermediate females increased with concentration of melatonin and indeed it appears that melatonin can evoke short-day (long-night) effects. Stock diet and acetone diet controls produced no intermediate females but a few males were found in the progeny.

It is not feasible to measure hormone titres in the haemolymph of aphids but whole animal extracts are possible. Adult pea aphids reared in long days and short days were collected at either mid-photophase or mid-scotophase and extracted with chloroform. After drying, the extracts were assayed for melatonin using radioimunoassay procedures (Fraser et al., 1983).

The presence of melatonin was confirmed by estimating a series of diluted aphid samples against the standard melatonin curve and showed that they ran parallel. In addition, the Rf value of immunoreactivity of aphid samples on TLC plates was precisely the same as that for tritiated melatonin (Gao and Hardie, 1997). In both long days and short days the titres during the scotophase were higher than during the photophase (Figure 4) but despite using clonal insects, the variation was high and the results were not significantly different (Gao and Hardie, 1997).

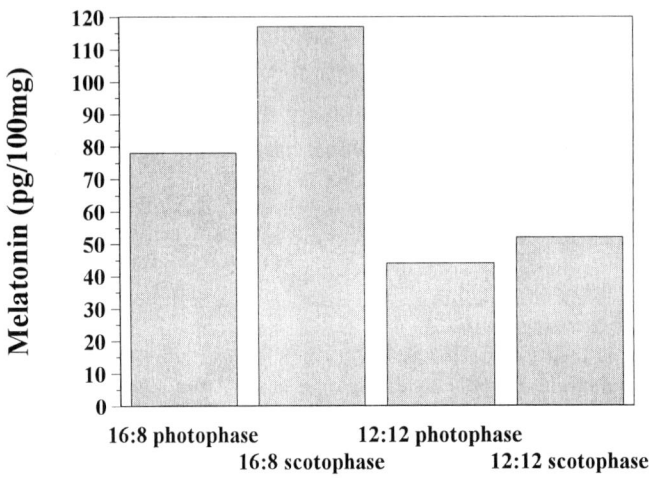

Light-dark cycle and time of sampling

Figure 4. Melatonin titres in pea aphid, *Acyrthosiphon pisum*, expressed as pg/100 mg wet weight, in long (LD 16:8) and short (LD 12:12) days. Insect samples were collected at mid-photophase and mid-scotophase. There were no significant differences between photophase and scotophase titres (from Gao and Hardie, 1997).

Despite the fact that melatonin offers a signal for dark period measurement in mammals the precise mechanism for translation of this into seasonal reproduction is not known (Arendt, 1995). It acts via reproductive hormones, most probably through the central nervous system, but peripheral actions are also present. If melatonin is involved in the aphid photoperiodic response and has a long-night type action, i.e. extended high titres of melatonin are read as long nights, then its action is opposite to that of juvenile hormones which promote short-night responses (see above). Melatonin has been reported in a number of invertebrate species and it has been considered as an evolutionarily-conserved molecule that transduces day-length messages (Vivien-Roels and Pevet, 1986). The involvement of melatonin in insect photoperiodism remains a distinct possibility but requires further elucidation. It is possible, for example, that control could be effected by changes in melatonin levels in localised regions of the brain. If this were the case, then the results of our melatonin titre assessments are not surprising.

REFERENCES

Ahmad M., Cashmore A.R., 1996. Seeing blue: The discovery of cryptochrome. Plant Molecular Biology 30, 851.

Arendt, J., 1995. Melatonin and the Mammalian Pineal Gland. Chapman & Hall, London.

Corbitt, T.S., Hardie, J., 1985. Juvenile hormone effects on polymorphism in the pea aphid, *Acyrthosiphon pisum*. Entomologia experimentalis et applicata 38, 131-135.

Finocchario, L. Callebert, J. Launay J.M., Jallon, J.M., 1988. Melatonin biosynthesis in *Drosophila*: Its nature and and its effects. Journal of Neurochemistry 50, 382-387.

Fraser, S. Cowen, P. Franklin, M. Franey C., Arendt, 1983. Direct radioimmunoassay for melatonin in plasma. Clinical Chemistry 29, 396-397.

Gao, N., 1996. Physiological aspects of form determination in aphids. PhD Thesis, University of London.

Gao N., Foster R.G., Hardie J., 2000. Two opsin genes from the vetch aphid, *Megoura viciae*. Insect Molecular Biology 9, 197-202.

Gao, N., Hardie, J., 1997. Melatonin in the pea aphid, *Acyrthosiphon pisum*. Journal of Insect Physiology 43, 615-620.

Gao, N., von Schantz M., Foster R.G., Hardie J., 1999. The putative brain photoperiodic photoreceptors in the vetch aphid, *Megoura viciae*. Journal of Insect Physiology 45, 1011-1019.

Gartner, W., Towner, P., 1995. Invertebrate visual systems. Photochemistry and Photobiology 62, 1-16.

Hardie, J., 1981. Juvenile hormone and photoperiodically controlled polymorphism in *Aphis fabae*: postnatal effects on presumptive gynoparae. Journal of Insect Physiology 27, 347-355.

Hardie, J., 1984. A hormonal basis for the photoperiodic control of polymorphism in aphids. In Photoperiodic Regulation of Insect and Molluscan Hormones, (Pitman, London), 240-252. (Ciba Symposium 104, 1983).

Hardie, J., 1987. The corpus allatum, neurosecretion and photoperiodically controlled polymorphism in an aphid. Journal of Insect Physiology 33, 201-205.

Hardie, J., 1987. The photoperiodic control of wing development in the black bean aphid, *Aphis fabae*. Journal of Insect Physiology 33, 543-549.

Hardie, J., 1990. The photoperiodic counter, quantitative day-length effects and scotophase timing in the vetch aphid *Megoura viciae*. Journal of Insect Physiology 36, 939-949.

Hardie, J., Lees, A.D., 1985. The induction of normal and teratoid viviparae by juvenile hormone and kinoprene in two aphid species. Physiological Entomology 10, 65-74.

Hardie, J., Lees, A.D. Young, S., 1981. Light transmission through the head capsule of an aphid, *Megoura viciae*. Journal of Insect Physiology 27, 773-777.

Hardie, J., Vaz Nunes, M., 2000. Aphid photoperiodic clocks. Journal of Insect Physiology, in press.

Hillman, W.S., 1973. Non-circadian photoperiodic timing in the aphid *Megoura viciae*. Nature 242, 128-129.

Hodkova, M., 1989. Indication of the role of melatonin in the regulation of reproduction in *Pyrrhocoris apterus* (Heteroptera). Acta Entomologia Bohemoslovakia 86, 81-85.

Itoh, M.T., Hattori, A., Nomura, T., Sumi Y., Suzuki, T., 1995. Melatonin and arylalkylamine *N*-acetyltransferase activity in the silkworm. Molecular and Cellular Endocrinology 115, 59-64

94

Lees, A.D., 1964. The location of the photoperiodic receptors in the aphid *Megoura viciae* Buckton. Journal of Experimental Biology 41, 119-133.

Lees, A.D., 1973. Photoperiodic time measurement in the aphid *Megoura viciae*. Journal of Insect Physiology 19, 2279-2316.

Lees, A.D., 1981. Action spectra for the photoperiodic control of polymorphism in the aphid *Megoura viciae*. Journal of Insect Physiology 27, 761-771.

Lees, A.D., 1986. Some effects of temperature on the hourglass photoperiod timer in the aphid *Megoura viciae*. Journal of Insect Physiology 32, 79-89.

Lerner, A.B.. Case, J.C., Takahashi, Y., 1958. Isolation of melatonin, pineal factor that lightens melanocytes. Journal of the American Chemical Society, 80, 2587.

Lucas, R.J., Foster, R.G., 1999. Photoentrainment in mammals: A role for cryptochrome. Journal of Biological Rhythms, 14, 4-10.

Lythgoe, J., 1979. The Ecology of Vision, Oxford University Press, Oxford.

Marcovitch, S., 1924. The migration of the Aphididae and the appearance of the sexual forms as affected by the relative length of daily light exposure. Journal of Agricultural Research 27, 513-522.

Stanewsky, R., Kaneko, M., Emery, P., Berretta, B., Wagener-Smith, K., Kay, S., Rosbash M., Hall, J., 1998. The cry^b mutation identifies cryptochrome as a circadian photoreceptor in *Drosophila*. Cell 95, 681-692.

Steel, C.G.H., Lees, A.D., 1977. The role of neurosecretion in the photoperiodic control of polymorphism in the aphid *Megoura viciae*. Journal of Experimental Biology 67, 117-135.

Tamarkin, L. Baird C.J., Almeida, O.F.X., 1985. Melatonin: A coordinating signal for mammalian reproduction. Science, 227l, 714-720.

Tilden, A.R., Anderson W.J., Hutchison, V.H., 1994. Melatonin in two species of damselfly *Ischnura verticalis* and *Enallagma civile*. Journal of Insect Physiology 40, 775-780.

Vaz Nunes, M., 1998. A double circadian oscillator model for quantitative photoperiodic time measurement in insects and mites. Journal of Theoretical Biology 194, 299-311.

Vaz Nunes, M., Hardie, J., 1993. Circadian rhythmicity is involved in photoperiodic time measurement in the aphid *Megoura viciae*. Experientia 49, 711-713.

Vaz Nunes, M., Hardie, J., 1999. The effect of temperature on the photoperiodic 'counters' for female morph and sex determination in two clones of the black bean aphid, *Aphis fabae*. Physiological Entomology 24, 339-345.

Vaz Nunes, M., Hardie, J., 2000. The effect of temperature on the photoperiodic 'clock' and 'counter' of a Scottish clone of the vetch aphid, *Megoura viciae*. Journal of Insect Physiology 46, 727-733.

Veerman, A., Vaz Nunes, M. 1987. Analysis of the operation of the photoperiodic counter provides evidence for hourglass time measurement in the spider mite *Tetranychus urticae*. Journal of Comparative Physiology A 160, 421-430.

Vivien-Roels, B., Pevet, P., 1986. Is melatonin an evolutionarily conservative molecule involved in transduction of photoperiodic information in all living organisms. Advances in Pineal Research 1, 61-68.

Vivien-Roels, B., Pevet, P., Beck, O., Fevre-Montange, M., 1984. Identification of melatonin in the compound eyes of an insect, the locust (*Locusta migratoria*), by radioimmunoassay and gas chromatography-mass spectrometry. Neuroscience Letters 49, 153-158.

Wetterberg, L., Hayes, D.K., Halberg, F., 1987. Circadian rhythms of melatonin in the brain of the face fly, *Musca autumnalis* De Geer. Chronobiologia, 14, 377-381.

Insect Timing: Circadian Rhythmicity to Seasonality
D.L. Denlinger, J. Giebultowicz and D.S. Saunders (Editors)
© 2001 Elsevier Science B.V.   All rights reserved.

Photoperiodic time measurement and shift of the critical photoperiod for diapause induction in a moth

Sinzo Masaki[a] & Yuji Kimura[b]

[a]12-13 Matsubara Higashi-1, Hirosaki 036-8141, Japan
[b]Aomori Agric.Exp. Sta., 1-1 Sakaimatsu, Kuroishi 036-0389, Japan

Mamestra brassicae (Lepidoptera: Noctuidae) may take one of the three different channels of pupal development in response to photoperiod in the larval stage: long winter diapause in short photoperiods (≤13L11D, 13-h light: 11-h dark), short summer diapause in long photoperiods around 16L8D and nondiapause in intermediate photoperiods close to 14L10D. This response is based on dark-time measurement, and shows either an hourglass or a circadian-oscillator feature depending on experimental conditions. The photoperiod in the egg and early larval stages modifies the critical photoperiod for diapause induction. Under stationary conditions, the critical photoperiod is close to 14L10D, but shifted to 13.5L10.5D and 15L9D by a preliminary exposure to 10L14D and 18L6D, respectively. This flexibility suggests that the critical photoperiod is not a fixed phase point of the photoperiodic clock but that it might be a reflection of the physiological threshold for determining the developmental patterns in the effector component of the photoperiodic response system.

1. INTRODUCTION

1.1. Critical photoperiod

Photoperiodism as a means of seasonal adaptation depends on the appropriate timing of diapause and other seasonal responses. This timing is mainly accomplished by the critical photoperiod, which is the key parameter for life-cycle regulation [for comprehensive reviews see Danilevskii (1965), Beck (1980), Saunders (1982), Tauber et al. (1986), Danks (1987), Zaslavski (1988)]. Owing to its adaptive importance, the critical photoperiod should have been under stringent selection as exemplified by many parallel clines along the latitudinal or altitudinal gradients of climatic conditions (see the reviews cited above; Masaki, 1999). In many species, the critical photoperiod tends to be longer in the northern districts with cooler and shorter growing seasons than in the warmer south.

The genetic nature of such variations have been elucidated by crossing between different geographic populations (Danilevskii, 1965; Riihimaa & Kimura, 1989; Cambell & Bradshaw, 1992; Hard et al., 1993; Kimura & Yoshida, 1995). Consequently, the critical photoperiod might be regarded as a genetically fixed time lapse from a given signal (dawn or dusk) for the photoperiodic clock to assess the natural daylength (or nightlength in many cases) to be below or above the inductive level.

Clock models of various types, including circadian oscillators, hourglasses and complexes of these two types, have been proposed to explain or simulate photoperiodic responses under various conditions (for review, see Vaz Nunes & Saunders, 1999). In the external coincidence version of oscillator models, the critical photoperiod is postulated to be a special light-sensitive phase ϕi (Pittendrigh & Minis, 1964; Saunders, 1979). The internal coincidence version postulates that the induction of diapause depends on whether the active phases of the dawn and dusk oscillators overlap or not (Danilevsky et al. 1970); therefore the critical photoperiod is again determined by the special phase points starting and ending the active phases of the two oscillators. In hourglass models, the critical nightlength is related to the particular stage in the course of dark-time measurement (Lees, 1973), so that this stage is comparable to ϕi of the oscillator model. In these cases, a daylength (nightlength) is assumed to be perceived in a qualitative way, i.e. shorter or longer than the critical photoperiod.

On the other hand, insects might measure photoperiodic time quantitatively based on the product of the time-measuring process, and Zaslavski (1988) ascribed the critical photoperiod to interaction of two antagonistic responses and the thresholds for their activities. A similar but more sophisticated feedback system model consisting of two independent circadian oscillators has been constructed, which successfully simulates responses under various photoperiodic regimes (Vaz Nunes, 1998).

1.2. The study insect

We used the noctuid moth *Mamestra brassicae* which is a widely distributed pest of vegetable crops in Eurasia. Its photoperiodic response has been studied for more than 40 years (Masaki, 1956; Danilevskii, 1965; Goryshin & Tyshchenko, 1969, 1973, 1976; Bonnemaison, 1976; Poitout & Bues, 1977; Goryshin et al., 1979; Furunishi et al., 1982; Grüner & Sauer, 1988; Kimura & Masaki, 1992, 1993). Despite these long-continued efforts, an important feature of the photoperiodic response had been overlooked.

We have recently found that the critical photoperiod can be shifted during the course of its development (Kimura & Masaki, 1998). This discovery poses a problem for our understanding of photoperiodism in insects, particularly the critical photoperiod. Here, we review first an account of the photoperiodic response of this moth based on data obtained in our laboratory, and then give evidence for the shift of the critical photoperiod and discuss its implication for understanding insect photoperiodism.

1.3. Terminology

Unfortunately, none of the many monographs hitherto published (cited in **1.1** above) try to standardize the format for photoperiodic regimes. Some authors carefully avoid to abbreviate photoperiodic regimes. For simplicity and clarity, however, we use L and D following numbers to indicate hours of light and dark; e.g. 12L12D for a cycle of 12 h light and 12 h dark; 1L10D1L12D indicates a skeleton photoperiod formed by a pair of two 1-h light pulses separated by 10 or 12 h dark. By this means, we can show the duration of both light and dark components of any regime, even in a complicated one consisted of four or more components.

'Photoperiod' is used to represent a cycle with light and dark components, and not to indicate only the light period in a daily cycle in spite of its original meaning. For convenience, however, 'long' or 'short' is used to characterize a photoperiod by referring only to its photophase.

2. PHOTOPERIODIC TIME MEASUREMENT IN *MAMESTRA*

2.1. Response to stationary photoperiod

This noctuid moth undergoes basically a bivoltine life cycle regulated by three options for the pupal development in Japan   (Masaki, 1956) as well as in western Europe (Poitout & Bues, 1977; Sauer and Grüner, 1988; Grüner & Masaki, 1994). The distribution of pupae reaching the eyespot stage of adult morphogenesis clearly indicates three different types of pupae (Fig. 1). At 25 °C, a sharp peak of developing pupae may occur within 10 days after pupation, representing nondiapause development. This peak may be followed by a lower and wider peak, extending from day 20 to day 60 after pupation. This short developmental delay is termed summer diapause, for it is induced in early summer in the field as shown below and terminated in autumn (Masaki, 1956; Masaki & Sakai, 1965). After the completion of adult eclosion from the summer pupae, many pupae may still remain in diapause for more than three months at 25°C. They are in winter diapause (Kimura & Masaki, 1992).

The percentages of winter diapause, nondiapause and summer diapause pupae vary as

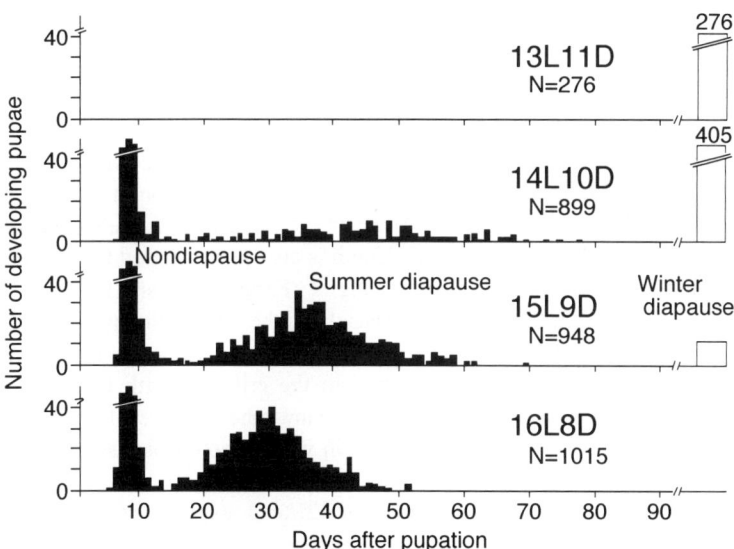

Fig. 1. Three types of pupal development at 25°C in *Mamestra brassicae* reared as larvae at 13L11D (13-h light, 11-h dark) to 16L8D (Kimura & Masaki, 1992, modified). Open histograms indicate the number of persisting pupae after three months.

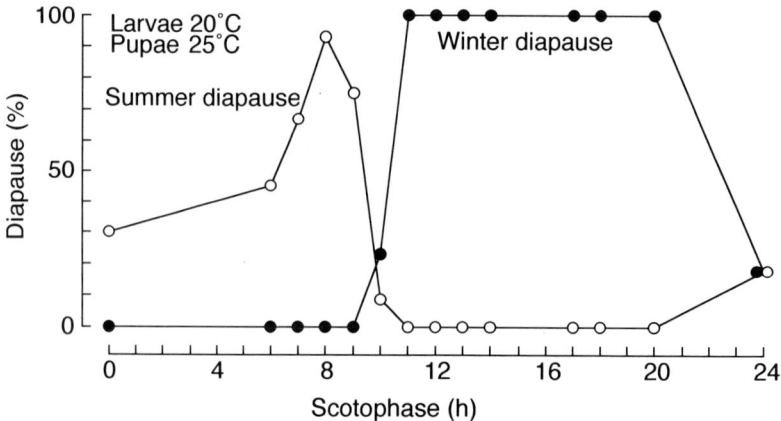

Fig. 2. Effect of larval photoperiod on the developmental pattern in pupae of *M. brassicae* at 25°C (Furunishi et al., 1982, modified). Closed circles: winter diapause. Open circles: summer diapause.

functions of the larval photoperiod (Fig. 2; Furunishi et al., 1982). If only the incidence of winter diapause is taken into account, the critical photoperiod is about 14L10D, at which the induction curves for summer diapause and winter diapause cross each other and the frequency of nondiapause pupae reaches a maximum. Although the occurrence of the two different types of diapause complicates the response of this moth, the induction curve for winter diapause is a common type among insects.

2.2. Hourglass seems to measure nightlength

In a French population of *M. brassicae*, Bonnemaison (1976) obtained high incidence of diapause by 10 h or longer dark periods irrespective of the length of combined light periods ranging from 7 to 20 h. This result indicates the importance of dark-time measurement in the photoperiodic response. The importance of nightlength is also demonstrated by the effect of a brief light pulse (1 h) interrupting a long night (Fig. 3; Furunishi et al., 1982). When the interval between dusk (the end of the main photophase) and the interrupting pulse was 9 h, i.e. 1 h shorter than the critical nightlength, the long night effect was more or less suppressed. Therefore, a dark period should continue longer than the critical length to induce winter diapause. In this respect, *M. brassicae* is similar to many other species of arthropods (see reviews cited in **1.1**) except for *Pyrrhocoris apterus* in which photophase is more important than scotophase (Saunders, 1987).

We extended night-interruption using 12L36D and 12L60D, experimental designs known as Bünsow protocol. Only one depression in the percentage winter diapause was observed by a light pulse again 9 h after dusk. Such an effect was not repeated by any later interruption, suggesting that the dark time measurement depends on an hourglass mechanism under these experimental conditions (Fig. 4; Kimura & Masaki, 1993).

The importance of dark-time measurement is also indicated in the following experiments.

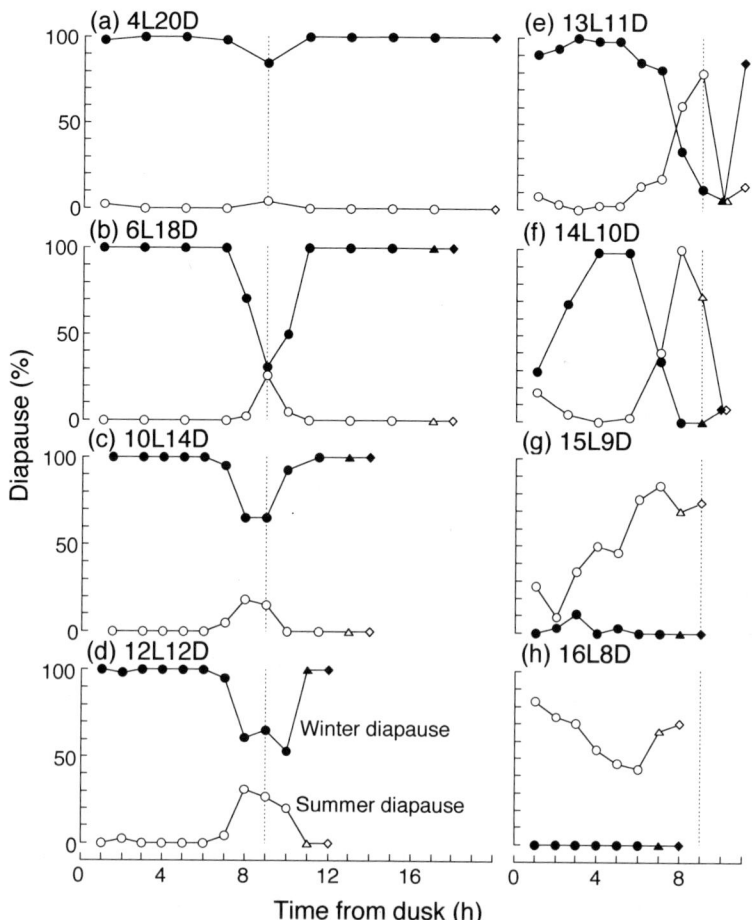

Fig. 3. Effects of night interruption by a 1 h light pulse on the incidence of summer (open circles) and winter diapause in various photoperiods (Furunishi et al., 1982). Horizontal axis indicates the lights-on time of the interrupting pulse, taking dusk as hour 0. Diamonds and triangles represent controls without night interruption. The main photophase was 1 h shorter for triangles than indicated at the top of each panel.

When scotophases ranging from 8 to 12 h are combined with photophases ranging from 1 to 36 h, the frequency and the type of diapause mainly depend on the scotophase duration (Fig. 5). Long scotophases tend to induce winter diapause and short ones, summer diapause. The frequencies of the two types of diapause are mirror images of each other. This tendency is observed over a wide range of combined photophases from 8 to 36 h, strongly supporting the notion that scotophase is the determinant. The total cycle length, whether it is close to multiples of 24 h or not, does not affect the response. This fact suggests that an hourglass, or a

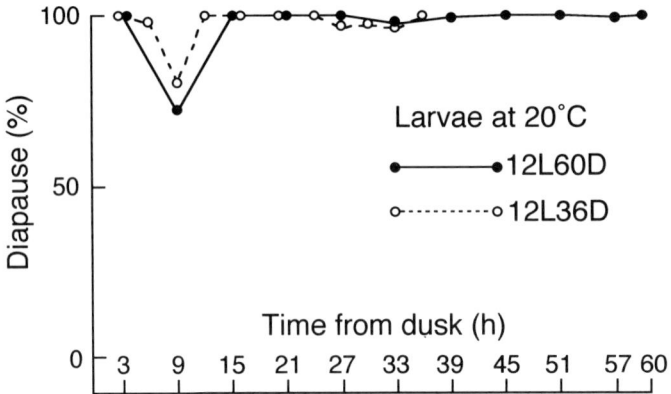

Fig. 4. Diapause response to Bünsow protocols at 20°C (Kimura & Masaki, 1993). The scotophase was scanned systematically by a 2 h light pulse in 12L60D (solid circles) or by 1 h light pulse in 12L36D (open circles). Abscissa indicates hours from the lights-off of the 12-h photophase to the beginning of the light pulse.

circadian oscillator instantly damping in light measures the dark time, as observed in the photoperiodic induction of diapause in *Sarcophaga argyrostoma* (Saunders, 1976) and *Tetranychus urticae* (Vaz Nunes & Veerman, 1986).

These results are in accordance with those of interruption of 12 h scotophase preceded by 12, 24, and 36 h photophase, giving cycle lengths of 24, 36 and 48 h, respectively. Irrespective of the photophase duration, winter diapause was suppressed by a light pulses about 9 h after dusk (Kimura & Masaki, 1992).

However, photophase exerts some effect on the dark-time measurement. The critical nightlength for the induction of winter diapause varied from 9.5 to 11.5 h depending on the combined photophase. It was shortest when the photophase was about 12 h, becoming longer as the combined photophase was either lengthened or shortened (Kimura & Masaki, 1992). There seems to be an optimum length of the preceding photophase for the rate of the dark-time measuring process.

2.3. Involvement of circadian system

We analyzed the photoperiodic response of *M. brassicae* also by resonance protocols, using systematically varied periods of light-dark cycles by combining, for example, a 12 h photophase with scotophases ranging from 12 to 60 h.

In contrast to the result of the tests with a Bünsow protocol, *Mamestra*'s diapause response resonates with light/dark cycles of about 24 h and its multiples when the light phase was 12 h, and winter-diapause peaks appear with cycle lengths of 24, 48 and 72 h and troughs with cycle lengths of 36 and 60 h at 23 and 25°C (Fig. 6A). Therefore, a circadian component is somehow involved in the *Mamestra*'s photoperiodic response.

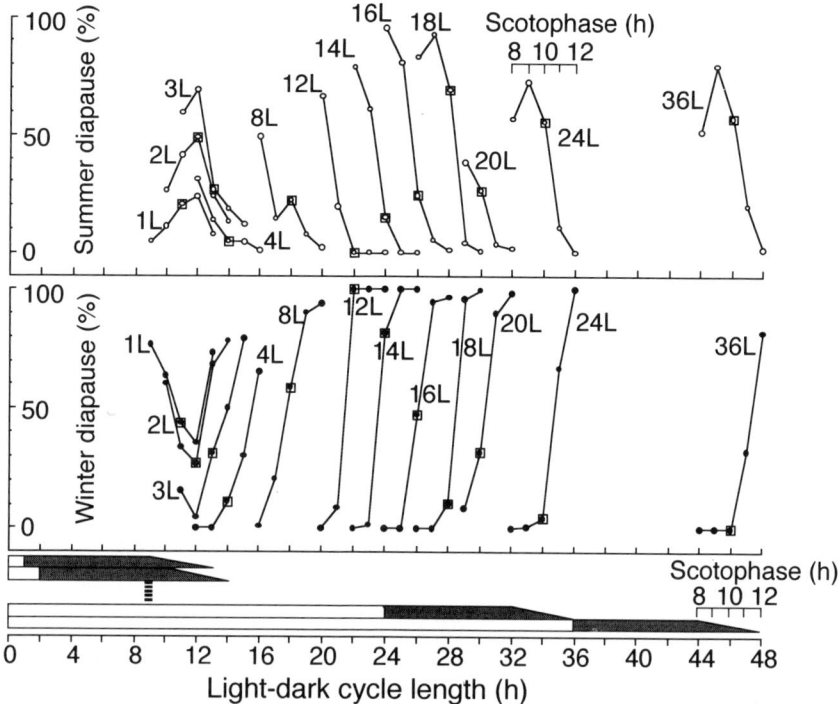

Fig. 5. Effects of photophase (varied from 1 to 36 h, indicated for each response curve), scotophase (from 8 to 12 h for each photophase except 20 h photophase) and cycle length on the diapause responses at 20°C (Kimura & Masaki, 1992). The squares show responses to 10 h scotophase. Bottom bars show examples of light-dark cycles used; open portion, photophase; shaded portion, scotophase with slanted edge indicating the range of scotophase.

However, the expression of this circadian rhythmicity depends on temperature. The diapause incidence remained very low at 28°C and very high at 20°C and does not show circadian fluctuations at these temperatures. Similar effects of temperature on the circadian expression of photoperiodic response have been found in *Sarcophaga argyrostoma* (Saunders, 1973, 1982), *Ostrinia nubilalis* (Takeda & Skopik, 1985), *Pieris brassicae* (Veerman et al., 1988) and *Drosophila auraria* (Pittendrigh et al., 1991).

2.4. Bistability phenomenon

One of the basic properties of most circadian systems is its entrainment to a light dark cycle of about 24 h, and a circadian oscillation maintains a fixed phase relationship with such a driving cycle (e.g. Pittendrigh & Minis, 1964). Under a symmetrical skeleton photoperiod of 1L12D1L10D, such a circadian system may take this regime as either 14L10D or 12L12D, depending on the phase (subjective day or night) of its oscillation when it sees the first 1-h

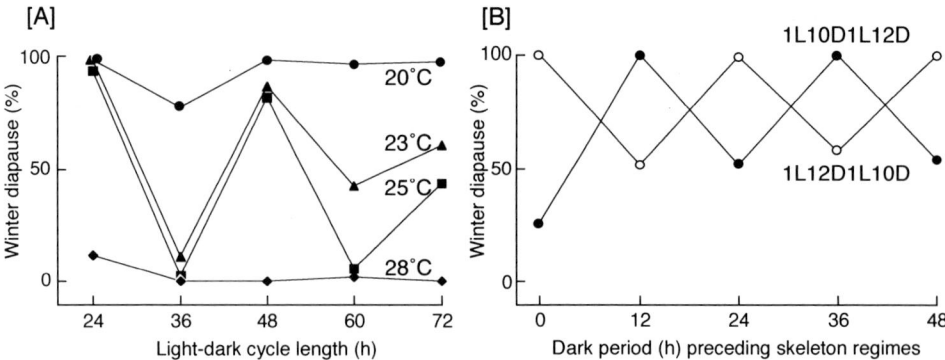

Fig. 6. [A] Winter-diapause response to Nanda-Hamner protocols at different temperatures (Kimura & Masaki, 1993). A photophase of 12 h was combined with different scotophases ranging from 12 to 60 h giving cycle lengths from 24 to 72 h.   [B] Winter-diapause response to the skeleton regimes of 1L10D1L12D (solid circles) and 1L12D1L10D (open circles) at 20°C (Kimura & Masaki, 1993). The larvae were kept in constant light for five days after hatching before transfer to darkness.

light pulse. The response of this kind is known as a bistability phenomenon. Thus, the skeleton photoperiod would provide another means to test whether or not a circadian oscillator underlies the photoperiodic time measurement. A clear bistability phenomenon in photoperiodic induction has been observed for diapause induction for the first time in *Sarcophaga argyrostoma* (Saunders, 1975) and later in *Calliphora vicina* (Vaz Nunes et al., 1990) and also in the wing-form determination of *Dianemobius fascipes* (Masaki & Watari, 1989).

In *M. brassicae*, 10-h night and 12-h night give conspicuously different effects (Fig. 2). A 10-h night is close to the critical length giving about 50% winter diapause, whereas 12-h night is long enough to induce 100% winter diapause. We kept larvae in constant light for 5 days after hatching and then each group was divided into five subgroups, which were exposed to 0, 12, 24, 36 and 48 h dark before transferring to either one of the two skeleton photoperiods that were mirror images of each other, 1L12D1L10D and 1L10D1L12D. Which one, 10D or 12D, is measured as night depends on the clock phase free-running in darkness, if the clock is a self-sustaining circadian oscillator.

The results are given in Figure 6B, which suggests the involvement of a circadian oscillator in the *Mamestra*'s photoperiodic response.   First, the percentage of winter diapause changed as a circadian function of the duration of darkness preceding the skeleton regimes in both series of experiments, i.e. the photoperiodic clock in this moth free-runs in continuous dark, so that the phase of the clock when it sees the first light pulse varies with the preceding dark time. Second, the responses in the two series were exactly mirror images of each other, reflecting their mirror-image skeleton regimes. This result can be accounted for only by assuming a circadian phase change in the photoperiodic system.

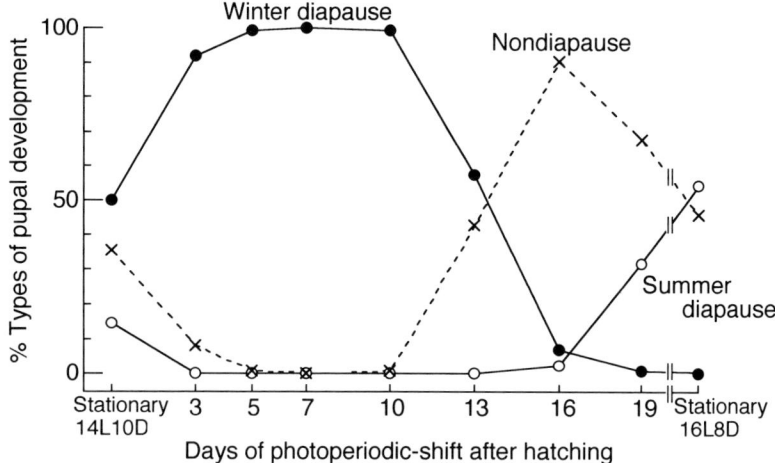

Fig. 7. Effects of shift from 16L8D to 14L10D at 20°C on pupal development (Kimura & Masaki, 1998). The egg stage was kept in darkness.

3. CHANGE IN THE CRITICAL PHOTOPERIOD

3.1. Response to shift in photoperiod

Shift in photoperiod at various developmental stages has been conventionally used to determine the stage sensitive to photoperiod (e.g. Tanaka, 1950) or the number of light-dark cycles required for diapause induction (e.g., Goryshin & Tyshchenko, 1973). Furthermore, the occurrence of stepwise responses (see Zaslavski, 1988) and responses to the direction of photoperiodic change (e.g. Tauber & Tauber, 1970, 1976) require more careful studies on photoperiodic changes. Our approach along this line lead to an unexpected finding.

In *M. brassicae*, the photoperiodic conditions in the early larval stage dramatically modify the subsequent response to daylength. When larvae were transferred from a typical short night of 16L8D to an intermediate night of 14L10D (close to the critical one) at various times after hatching, the diapause incidence remarkably varied with the time of transfer (Fig. 7). Only three days' exposure to 16L8D after hatching modified the intermediate effect (50% winter diapause) of 14L10D and gave 100% winter diapause.

We repeated and expanded transfer experiments, using various combinations of the first and second photoperiods (Fig. 8). The results indicate complicated interactions between the first and second photoperiods. We cannot explain them by any previously proposed model for responses to photoperiodic change. First, the involvement of a two-step response is rejected, simply because *Mamestra*'s pupal development can be fully controlled by stationary photoperiods (Fig. 2). Second, the direction of change in photoperiod itself is not the determinant. Increasing shifts of photoperiod from 10L14D to 12L12D or 13L11D and also from 12L12D to 13L11D, all within the range below the critical photoperiod, failed to prevent

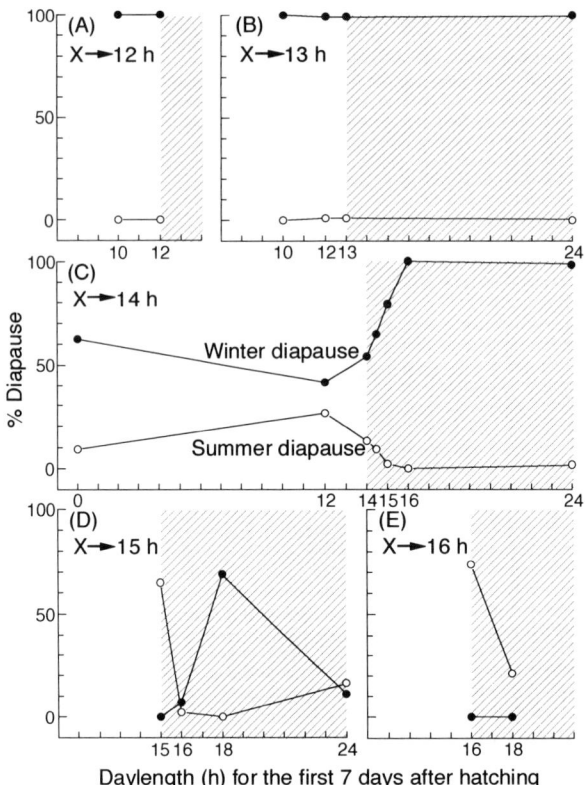

Fig. 8. Effects of photoperiodic shift at day 7 of the larval life at 20°C (Kimura & Masaki, 1998). Larvae were exposed to daylengths indicated on the X-axis for the first 7 days after hatching and transferred to the daylength as shown left in each panel. The egg stage was kept in darkness. The plots in the open portion represent increasing shifts and those in the shaded portion decreasing shifts.

winter diapause. At the same time, some of the decreasing shifts of photoperiod from 16L8D to 15L9D and also from 18L6D to 16L8D, both within the range above the critical photoperiod, prevented winter diapause (Fig. 8). Therefore, we have to look for another model.

3.2. The response curve moves with the early photoperiod

We hypothesize that the three different types of development, winter diapause, summer diapause and nondiapause, are expressions of different levels of a putative physiological quantity, diapause propensity [=liability for overwintering in Kimura & Masaki (1998)], adopting a quantitative genetic model for threshold traits (Falconer, 1981). The reason to use this model has been given previously (Kimura & Masaki, 1998). A hypothetical distribution

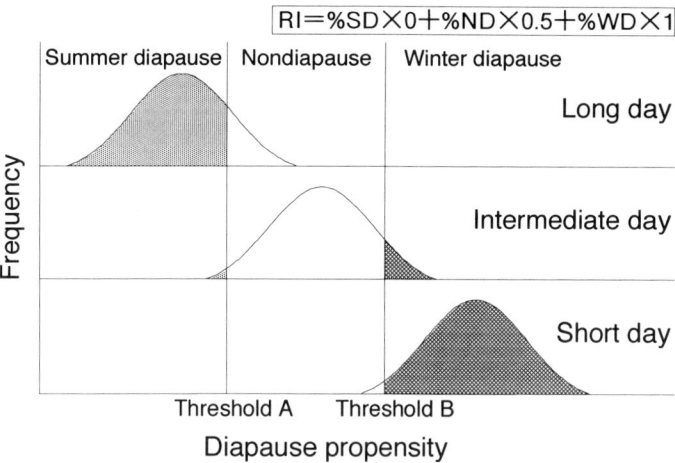

Fig. 9. A threshold model for the determination of three options of pupal development (Kimura & Masaki, 1998). Diapause propensity distributes continuously, but two thresholds divide three phenotypes. The distribution moves along the horizontal axis with photoperiod. The response index defined on the top right roughly represents the relative position of the distribution. For further explanation see Kimura & Masaki (1998).

of diapause propensity with two thresholds, one between winter diapause and nondiapause, and the other between nondiapause and summer diapause is approximately derived from the photoperiodic response curve (Fig. 9). Thus the frequencies of the three types of pupae can be roughly represented by a single variable, the response index (RI), as defined at the top of Figure 9. This index would roughly correspond to the distribution of diapause propensity in a given regime, so that we can assess the photoperiodic response by this single variable (Kimura & Masaki, 1998).

When RI is plotted against the second daylength for each of the first daylength, a group of curves are obtained (Fig. 10A). Although the datum points are not sufficient, these curves look like portions of photoperiodic response curves shifted along the horizontal (photoperiod) axis. They move regularly from left (short) to right (long) as the first photoperiod changes from 12L12D to 18L6D. From this tendency, we might propose that, in *M. brassicae*, the critical photoperiod might be modified by the photoperiod in the early stage of development. This proposition claims that the critical photoperiod is not a genetically fixed phase point of the photoperiodic clock.

3.3. Shift of the critical photoperiod

In view of the importance of this discovery, we tried to determine more precisely the change in the critical photoperiod caused by the early photoperiodic conditions. We kept two groups of larvae under 10L14D and 18L6D, respectively, for 10 days after oviposition, and then each group was separated into five subgroups, which were exposed to five different

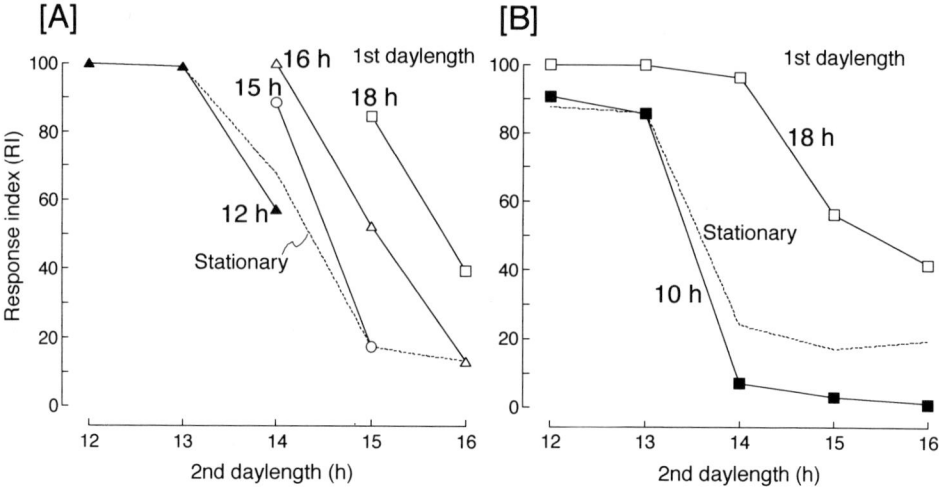

Fig. 10. [A] Response index (see text) calculated from the data given in Figure 8 are plotted as functions of the second daylength for each first daylength (Kimura & Masaki, 1998). The larvae were transferred from the first to the second daylength at day 7 after hatching.

[B] Shift of the critical photoperiod for the determination of pupal development as expressed by response index in *M. brassicae*. The test animals were exposed to 10L14D (closed square) or 18L6D (open square) in the egg and early larval stages for 10 days and transferred to various daylengths indicated on the X-axis. The dotted line indicates the results with stationary photoperiods. The larvae were reared at 20°C and pupae at 25°C.

photoperiods ranging from 12L12D to 16L8D, respectively. A control group was kept under the same five photoperiods from the egg to prepupal stages.

As a result, three clearly separated response curves were obtained, indicating three different critical photoperiods, 13.5L10.5D, 14L10D and 15L9D (Fig. 10B), phenotypically resembling variation in critical photoperiod with latitude (Danilevskii, 1965). Thus, our hypothesis — the critical photoperiod for the induction of pupal diapause is modified under the influence of photoperiod during the early larval life is confirmed.

Such a modifying effect on the critical photoperiod is exerted also in the egg stage (Kimura and Masaki, 1998). When the egg stage (lasting for 6 days at 20°C) was kept under 16L8D, rearing larvae at 14L10D gave about 100% winter diapause. In contrast, the percentage winter diapause decreased to 20-30% under the same photoperiod, when the eggs were kept at 12L12D.

Further evidence for the shift of the critical photoperiod should be observed in the response to night interruption by a brief light pulse. As shown in Figure 3, 1-h interruption of a long night suppresses the induction of winter diapause when it occurs 9 h after dusk. If this suppression of winter diapause is due to the effect of the dark period between dusk and the interrupting pulse shorter than the critical length, then the shortened critical nightlength would reveal a corresponding shift of the effective time of an interrupting light pulse.

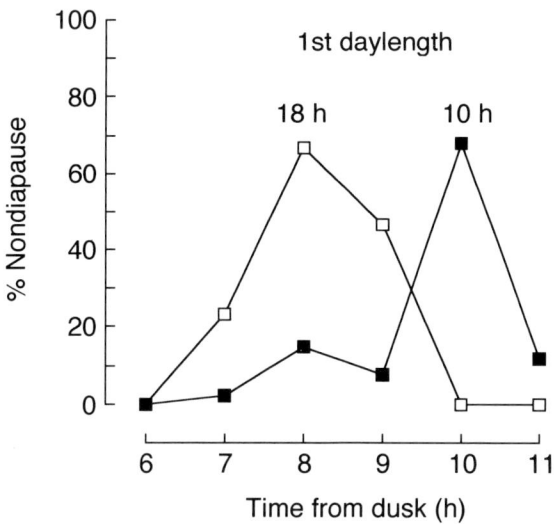

Fig. 11. The responses to 1-h night interruption in 12L12D of two groups exposed to 18L6D and 10L14D, respectively, during the first 10 days of life (the egg and early larval stages) and then transferred to the night interruption regimes in 12L12D at 20°C.

We are now testing this inference by exposing two groups of insects to 18L6D and 10L14D, respectively, for the first 10 days of life (the egg and early larval stages) and then transferred to 12L12D and the 12D component is systematically scanned by a 1-h pulse. At present, the final results have not yet been obtained, because more than three months are necessary to distinguish between the winter and summer diapause pupae. However, the nondiapause pupae have already eclosed as adults, so that the responses of the two groups can be tentatively compared in terms of the percentage nondiapause (Fig. 11). The short night (18L6D) pre-treatment made the peak time of the interrupting effect 2 h earlier than in the long night (10L14D) pre-treatment, and this difference is exactly comparable to the difference in the critical photoperiod caused by such pre-treatments (Fig. 10B).

4. CLOSING REMARKS

M. brassicae shows a photoperiodic response typical for insects in terms of the incidence of winter diapause (Fig. 2), although two phenologically different kinds of diapause occur in its life cycle (Fig. 1). This response is based on the dark-time measurement, because the long-night effect inducing winter diapause is reversed by an interrupting light pulse at around the critical nightlength (Fig. 3). The clock measuring dark time seems to work as an hourglass (Figs. 4, 5), but the involvement of a circadian oscillation is indicated in the final expression of the photoperiodic response (Figs. 6A, B). All these features are common in arthropod photoperiodism (see reviews cited in **1.1**). Probably, this moth shares the basic structure of photoperiodism with many other species of insects. Therefore, the observed shift of the critical photoperiod (Figs. 10A, 10B, 11) might be related to the general functional make-up of the system. If so, such a shift may also be found in other species of insects. In fact, the univoltine *Chrysopa downsi* shows a response to photoperiodic shift very similar to what we have observed in *M. brassicae*: the longer the photoperiod before the obligatory diapause stage, the longer is the critical photoperiod for terminating diapause (Tauber & Tauber, 1976).

One of the important issues in insect photoperiodism is whether the photoperiodic time-measurement is qualitative or quantitative. For the qualitative measurement, the photoperiodic clock itself (either oscillator or hourglass types) should have a genetically fixed special phase (ϕi) designating the critical photoperiod. In such a case, it is difficult to conceive that ϕi changes under the influence of the preceding photoperiod. In order to do so, the clock itself must change ϕi while it measures the nightlength by referring to ϕi, which is logically impossible. The shifting critical photoperiod thus excludes the ϕi model. Then we might assume that the critical photoperiod is not calibrated on the clock itself [cf. Zaslavski (1988), Spieth & Sauer (1991), Vaz Nunes (1998)].

The photoperiodic response system includes at least the following components: the photo-receptor, clock, accumulator (counter) and effector (cf. Takeda & Skopik, 1997; Takeda et al. 1999). The substance produced during the time measurement by the clock might be fed via the accumulator to the effector (probably neuroendocrine; Hodkova, 1999), where the level of the accumulated substance is sorted out by a threshold for switching on one or the other developmental pathway, if the concerned trait can take one of the dichotomous character states such as diapause or nondiapause, short-wing or long-wing forms, etc.; or the clock product may be directly transmitted to the effector that incessantly controls a quantitative trait such as growth rate, body size, etc. When a threshold response intervenes in the effecting process, it reflects in the critical photoperiod.

If this hypothesis is accepted, the shift of the critical photoperiod by the preceding photoperiod should not be ascribed to modification of the clock function itself, but to a change in the effector threshold. A change in the threshold can be effected by the early-stage photoperiod independently of the clock, because the effector itself is under the control of the circadian organization that is highly sensitive to photoperiod.

As the photoperiodic response may be manifested in various traits such as diapause, wing form, developmental rate, body colour, body size and shape, behaviour, etc. (see reviews cited in **1.1**) apparently regulated by different physiological mechanisms, there should be more than one effector component deriving the seasonal information from the same clock. There is then the possibility that the different effectors for different traits give their own response curves with different critical photoperiods. The distinction between the qualitative and quantitative responses may not be ascribed to the clock function itself but to the different modes of input to and output from the effector.   This assumption may explain the multiple photoperiodic responses of *Plautia stali* (Numata & Kobayashi, 1994), *Aquarius paldum* (Harada & Numata, 1993) and *Sasakia charonda* (Kato & Hasegawa, 1984). In short, the photoperiodic response can take either qualitative or quantitative expression and also different critical photoperiods for different traits depending on the effector function.

The above remarks may also concern with the hourglass and oscillator aspects of the photoperiodic response (Figs 4, 5, 6A, 6B). As discussed above, the final expression of the photoperiodic response might be determined by the effector component, which is probably under the control of the circadian organization. In *Tetranychus urticae*, the hourglass nature of the clock and the circadian nature of the photoperiodic counter are inferred to be the cause of the hourglass and oscillatory expressions of the photoperiodic response (Veerman & Vaz Nunes, 1987). The interference in the apparent clock function by the circadian system might be exerted directly, for example, by setting the hourglass ready to start in response to light/dark transition, or indirectly by controlling the effector functions. We have as yet no evidence to make further comments on this issue in *M. brassicae*.

ACKNOWLEDGMENTS

This study has been started at Entomological Laboratory, Hirosaki University and continued at Environmental Section, Aomori Agricultural Experiment Station. We are gratefully indebted our colleagues in the two laboratories for helping us in various ways and the Ministry of Education, Science and Culture, Japan, for financial supports. Our sincere thanks are due to D.S. Saunders and H. Numata for inviting us to the symposium "Photoperiodic Induction of Diapause and Seasonal Morphs" at Iguaçu. This symposium has been the impetus for writing the present paper. The following organizations kindly permitted to reproduce figures from their publications: The Entomological Society of Japan (Figs 7, 8, 9, 10A), Japanese Society of Applied Entomology and Zoology (Figs 2, 3), Pergamon Press (Figs 1, 5) and Blackwell Science (Figs 4, 6A, 6B).

We dedicate this paper to Professor D.S. Saunders as a token of our appreciation of his outstanding contributions to our understanding of insect photoperiodism.

REFERENCES

Beck, S.D., 1980. Insect Photoperiodism, 2nd ed. Academic Press New York.

Bonnemaison, L., 1976. Action de la photopériode sur l'induction de la diapause chez *Mamestra brassicae* (Lep., Noctuidae). Annales de la Societé entomologique Française 11, 767-781.

Campbell, M.D. & Bradshaw, W.E., 1992. Genetic coordination in the pitcherplant mosquito, *Wyeomyia smithii* (Diptera: Culicidae). Annals of the Entomological Society of America 85, 445-451.

Danilevskii, A.S., 1965. Photoperiodism and Seasonal Development of Insects. Oliver & Boyd, Edinburgh.

Danilevsky, A.S., Goryshin, N.I. & Tyshchenko, V.P., 1970. Biological rhythmus in terrestrial arthropods. Annual Review of Entomology 15, 201-244.

Danks, H.V., 1987. Insect Dormancy: an Ecological Perspective. Biological Survey of Canada (Terrestrial Arthropods), National Museum of Natural Sciences, Ottawa.

110

Falconer, D.S., 1981. Introduction to Quantitative Genetics, 2nd ed. Longman, New York.

Furunishi, S., Masaki, S., Hashimoto, Y. & Suzuki, M., 1982. Diapause response to photoperiod and night interruption in *Mamestra brassicae* (Lepidoptera: Noctuidae). Applied Entomology and Zoology 17, 398-409.

Goryshin, N.J. & Tyshchenko, G.F., 1969. A comparative analysis of effectiveness of light breaks of darkness in photoperiodic induction of diapause in insects. Zhurnal Obshchei Biologii 29, 481-498. (In Russian with English summary.)

Goryshin, N.J. & Tyshchenko, G.F., 1973. The accumulation of the photoperiodic information during the diapause induction in *Barathra brassicae* L. (Lepidoptera, Noctuidae). Entomologicheskoe Obozrenie 52, 249-255. (In Russian with English summary.)

Goryshin, N.J. & Tyshchenko, G.F. 1976. Some parallel phenomena in insect photoperiodic reaction and in circadian rhythms under light-dark cycles of various duration. Entomologicheskoe Obozrenie 55, 750-762. (In Russian with English summary.)

Goryshin, N.I., Volkovich, T.A. & Saulich, A.H., 1979. The comparative efficiency of constant and changing photoperiods for three species of Lepidoptera. Zoologicheskii Zhurnal 58, 44-53. (In Russian with English summary.)

Grüner, C. & Masaki, S., 1994. Summer diapause in the noctuid moth *Mamestra brassicae*. In Danks, H.V. (ed.), Insect Life-Cycle Polymorphism, 191-204.

Grüner, C. & Sauer, K.P.J., 1988. Aestival dormancy in the cabbage moth *Mamestra brassicae* L. (Lepidoptera: Noctuidae). I. Adaptive significance of variability of two traits: day length threshold triggering aestival dormancy and duration of aestival dormancy. Oecologia 74, 515-523.

Harada, T. & Numata, H., 1993. Two critical day lengths for the determination of wing forms and the induction of adult diapause in the water strider, *Aquarius paludum*. Naturwissenschaften 80, 430-432.

Hard, J.J., Bradshaw, W.E. & Holzapfel, C.M., 1993. The genetic basis of photoperiodism and its evolutionary divergence among populations of the pitcherplant mosquito, *Wyomyia smithii*. American Naturalist 142, 457-473.

Hodkova, M., 1999. Regulation of diapause and reproduction in *Pyrrhocoris apterus* (L.) (Heteroptera)—Neuroendocrine outputs (mini-review). Entomological Science 2, 563-566.

Kato, Y. & Hasegawa, Y., 1984. Photoperiodic regulation of larval diapause and development in the nymphalid butterfly, *Sasakia charonda*. Kontyû (Japanese Journal of Entomology) 52, 363-369.

Kimura, M.T. & Yoshida, T., 1995. A genetic analysis of photoperiodic reproductive diapause in *Drosophila biauraria*. Physiological Entomology 20, 253-256.

Kimura, Y. & Masaki, S., 1992. Effect of light period on dark-time measurement for diapause induction in *Mamestra brassicae*. Journal of Insect Physiology 38, 681-686.

Kimura, Y, & Masaki, S., 1993. Hourglass and oscillator expressions of photoperiodic diapause response in the cabbage moth *Mamestra brassicae*. Physiological Entomology 18, 240-246.

Kimura, Y. & Masaki, S., 1998. Diapause programming with variable critical daylength under changing photoperiodic conditions in *Mamestra brassicae* (Lepidoptera: Noctuidae). Entomological Science 1, 467-475.

Lees, A.D., 1973. Photoperiodic time measurement in the aphid *Megoura viciae*. Journal of Insect Physiology 19, 2279-2316.

Masaki, S., 1956., The local variation in the diapause pattern of the cabbage moth, *Barathra brassicae* Linné, with particular reference to the aestival diapause (Lepidoptera: Noctuidae). Bulletin of Faculty of Agriculture, Mie University 13, 29-46.

Masaki, S., 1999. Seasonal adaptations of insects as revealed by latitudinal diapause clines. Entomological Science 2, 539-549.

Masaki, S. & Sakai, T., 1965. Summer diapause in the seasonal life cycle of *Mamestra brassicae* (Linné) (Lepidoptera: Noctuidae). Japanese Journal of Applied Entomology and Zoology 8, 191-205.

Masaki, S. & Watari, Y., 1989 Response to night interruption in photoperiodic determination of wing form of the ground cricket *Dianemobius fascipes*. Physiological Entomology 14, 179-186.

Numata, H. & Kobayashi, S., 1994. Threshold and quantitative photoperiodic responses exist in an insect. Experientia 50, 969-971.

Pittendrigh, C.S., Kyner, W.T. & Takamura, T., 1991. The amplitude of circadian oscillation: temperature dependence, latitudinal clines, and the photoperiodic time measurement. Journal of Biological Rhythms 6, 299-313.

Pittendrigh, C.S. & Minis, D.H., 1964. The entrainment of circadian oscillation by light and their role as photoperiodic clocks. American Naturalist 98, 261-264.

Pittendrigh, C.S. & Minis, D.H., 1971. The photoperiodic time measurement in *Pectinophora gosypiella* and its relation to the circadian system in that species. In Menaker, M. (ed.), Biochronometry, 212-250. National Academy of Sciences, Washington, D.C.

Poitout, S. & Bues, R., 1977. Quelques aspects génétique de l'hétérogéneité de manifestation de la diapause éstivale dans les populations européenne de deux Lépidoptères Noctuidae Hadeninae (*Mamestra oleracea* L. et *Mamestra brassicae* L.). Annales de Zoologie et Ecologie des Animaux 9, 235-259.

Riihimaa, A.J. & Kimura, M.T., 1989. Genetics of the photoperiodic larval diapause in *Chymomyza costata* (Diptera: Drosophilidae). Hereditas 110, 193-200.

Sauer, K.P. & Grüner, C., 1988. Aestival dormancy in the cabbage moth *Mamestra brassicae* L. (Lepidoptera: Noctuidae). 2. Geographical variation in two traits: daylength thresholds triggering aestival dormancy and duration of aestival dormancy. Oecologia 76, 89-96.

Saunders, D.S., 1973. The photoperiodic clock in the flesh-fly *Sarcophaga argyrostoma*. Journal of Insect Physiology 19, 1941-1954.

Saunders, D.S., 1975. 'Skeleton' photoperiod and the control of diapause and development in the flesh fly, *Sarcophaga argyrostoma*. Journal of comparative Physiology A 97, 97-112.

Saunders, D.S., 1976. Circadian eclosion rhythm in *Sarcophaga argyrostoma*: some comparisons with the photoperiodic "clock". Journal of comparative Physiology A 110, 111-133.

Saunders, D.S., 1979 External coincidence and the photoinducible phase in the *Sarcophaga* photoperiodic clock. Journal of comparative Physiology A 132, 179-189.

Saunders, D.S., 1982. Insect Clocks, 2nd ed. Pergamon Press, Oxford.

Saunders, D.S., 1987. Insect photoperiodism: the linden bug, *Pyrrhocoris apterus*, a species that measures daylength rather than nightlength. Experientia 43, 935-937.

Spieth, H.R. & Sauer, K.P., 1991. Quantitative measurement of photoperiod and its significance for the induction of diapause in *Pieris brassicae* (Lepidoptera, Pieridae). Journal of Insect Physiology 37, 231-238.

Takeda, M., Endo, Y., Ohnishi, H. & Ishikawa, N., 1999. Photoperiodic system in physiological reality. Entomological Science 2, 567-574.

Takeda, M. & Skopik, S.D., 1985. Geographic variation in the circadian system controlling photoperiodism in *Ostrinia nubilalis*. Journal of comparative Physiology A 156, 653-658.

Takeda, M. & Skopik, S.D., 1997. Photoperiodic time measurement and related physiological mechanism in insects and mites. Annual Review of Entomology 42, 323-349.

Tanaka, Y., 1950 Studies on hibernation with special reference to photoperiodicity and breeding of Chinese tussar-silkworm, I. Japanese Journal of Sericultural Science 19, 358-371. (In Japanese.)

Tauber, M.J. & Tauber, C.A., 1970. Photoperiodic induction of diapause in an insect: response to changing daylengths. Science 167, 170.

Tauber, M.J. & Tauber, C.A., 1976. Developmental requirements of the univoltine species *Chrysopa downesi*: photoperiodic stimuli and sensitive stages. Journal of Insect Physiology 22, 331-335.

Tauber, M.J., Tauber, C.A. & Masaki, S., 1986. Seasonal Adaptations of Insects. Oxford Univ. Press, New York.

Vaz Nunes, M., 1998. A double circadian oscillator model for quantitative photoperiodic time measurement in insects and mites. Journal of theoretical Biology 194, 299-311.

Vaz Nunes, M., Kenny, N.A.P. & Saunders, D.S., 1990. The photoperiodic clock in *Calliphora vicina*. Journal of Insect Physiology 36, 61-67.

Vaz Nunes, M. & Saunders, D., 1999. Photoperiodic time measurement in insects: a review of clock models. Journal of Biological Rhythms 4, 84-104.

Vaz Nunes, M. & Veerman, A., 1986. A "dusk" oscillator affects photoperiodic induction of diapause in the spider mite, *Tetranychus urticae*. Journal of Insect Physiology 32, 605-614.

Veerman, V. & Vaz Nunes, M., 1987. Analysis of the operation of the photoperiodic counter provides evidence for hourglass time measurement in the spider mite *Tetranychus urticae*. Journal of comparative Physiology A 160, 421-430.

Veerman, A., Beekman, M. & Veenendal, R.L., 1988. Photoperiodic induction of diapause in the large white butterfly, *Pieris brassicae*: evidence for hourglass time measurement. Journal of Insect Physiology 34, 1063-1065.

Zaslavski, V.A., 1988. Insect Development: Photoperiodic and Temperature Control. Springer-Verlag, Berlin.

Insect Timing: Circadian Rhythmicity to Seasonality
D.L. Denlinger, J. Giebultowicz and D.S. Saunders (Editors)
113

Geographical strains and selection for the diapause trait in *Calliphora vicina*

D. S. Saunders

Institute of Cell, Animal and Population Biology, University of Edinburgh, West Mains Road, Edinburgh EH9 3JT, Scotland, United Kingdom

Like many insects with a large geographical range, the blow fly *Calliphora vicina* shows latitudinal clines in seasonally related traits. Five strains spanning 36° to 65°N were shown to display such clines in critical day length, diapause duration and, perhaps, diapause related cold tolerance. Divergent selection for diapause incidence using a diapause-depleted stock restored both a high incidence of diapause and its complete elimination within less than five generations. Reciprocal crossings between northern (65°N) and southern (51°N) strains showed that males contributed nothing to the diapause status of their immediate offspring and even in later generations and back crosses, the male contribution was less than that of the female. Diapause *duration,* on the other hand, showed contributions from both sexes, suggesting that incidence and duration may be inherited in a different manner. The central problem of photoperiodic time measurement is also reviewed, especially the evidence that night length measurement is a function of the insect's circadian system (Bünning's general hypothesis).

1. INTRODUCTION

The blow fly *Calliphora vicina* is widespread in the Palaearctic and Nearctic regions where it over-winters in a photoperiodically induced diapause at the end of the third larval instar. Like many insects with a large geographical range, it presents latitudinal clines in diapause-related traits, such as critical day length (CDL) (McWatters and Saunders, 1996), diapause duration (Saunders, 2000), and cold tolerance (Saunders and Haywood, 1998). Flies from more northerly latitudes thus show a longer CDL, a more intense (longer lasting) diapause, and an increased resistance to the cold than those from the south.

Although some strains of *C. vicina* show larval sensitivity to day length (Vinogradova, 1986; Saunders et al. 1986), larval diapause is commonly regulated by a maternally operating photoperiodic clock over most of this geographical range (Vinogradova and Zinovjeva, 1972; Saunders, 1987). Thus, in a strain originating from southeastern Scotland, adult flies exposed to the long days of summer produce successive generations of continuously developing, nondiapausing progeny. Those exposed to autumnal day lengths less than the critical value (LD 14½: 9½), however, lay eggs giving rise to larvae that enter diapause after wandering from their food, and just prior to their metamorphosis to the puparium. Diapause development and the metabolic rate of diapause and post-diapause larvae is then a function of temperature (in the soil) (Saunders, 2000); over-wintering insects then emerge as adults in the following spring.

114

This chapter examines what we know about these geographical clines in photoperiodic and diapause traits of *C. vicina*, the genetical basis of their regulation, and the physiological nature of time measurement in the photoperiodic clock.

2. LATITUDINAL STRAINS OF *C. vicina*

2.1. The critical day length

Figure 1 shows photoperiodic response curves (PPRCs) of strains of blow flies originating from four localities in Europe and one in North America. The four European strains were isolated from wild populations in the north of Finland (Nallikari, near Oulu, 65°N), south eastern Scotland (Musselburgh, near Edinburgh, 55°N), southern England (Silwood Park, Ascot, 51°N) and central Italy (Barga, 44°N). The North American strain, established from flies caught at Chapel Hill, North Carolina (36°N), is not of the same latitudinal cline, but is included as part of a north-south series. The Barga strain was established from a single female fly caught at altitude (about 1,000 m); all others were collected at sites at or near sea level.

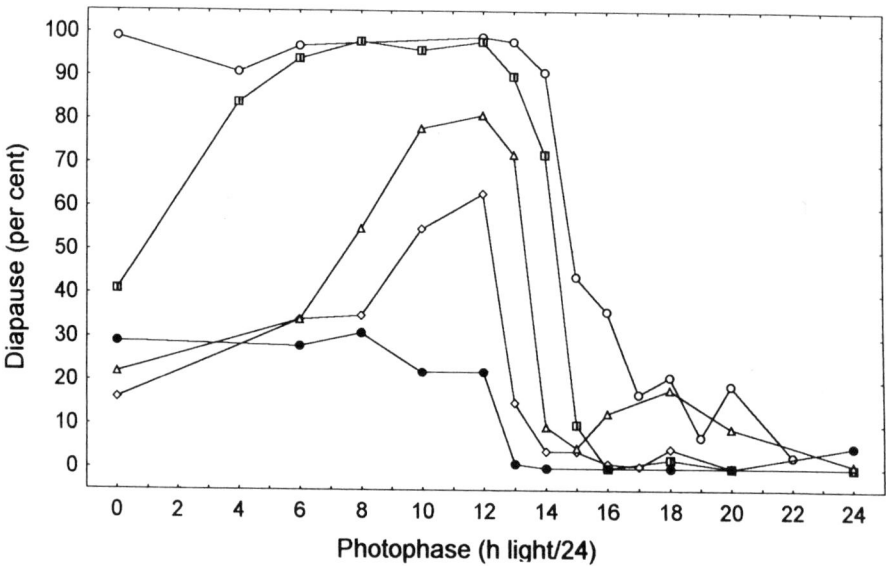

Figure 1. Photoperiodic response curves for blow flies (*Calliphora vicina*) originating from 5 localities (right to left: Nallikari, Finland, 65°N, Edinburgh, Scotland, 55°N, Barga, Italy, 44°N, Silwood park, England, 51°N, Chapel Hill, North Carolina, 36°N). All PPRCs were conducted at 20°C.

The PPRCs shown in Fig. 1 were determined soon after capture, in most cases within one or two generations of laboratory culture. They were established by exposing groups of adult flies to a range of photoperiods (LD 4:20 to LD 22:2), and in some cases to continuous darkness (DD) and continuous light (LL), at a constant temperature of 20°C. Adult flies were supplied with water and sugar *ad libitum* and with fresh meat (beef muscle) on day 4 post-eclosion and at intervals thereafter. Eggs laid on days 12 to 14 were allowed to hatch at 20°C, and the hatchlings established as larval cultures, in darkness at 11 to 12°C, on meat augmented by a yeast, dried milk and agar medium (Saunders, 1987). Fully fed larvae were allowed to wander into dry sawdust for pupariation.

The progeny of 'long day' exposed adults formed puparia promptly (after about 17 to 24 days) whereas those from 'short day' flies showed much delayed pupariation. As with other cases of larval diapause, there are no overt morphological or physiological signs of diapause in *C. vicina*. For this reason, larvae failing to pupariate by day 30 were considered to be in diapause (Saunders, 1987). This tallies well with endocrinological data suggesting that non-release of prothoracicotropic hormone (PTTH) and a developing refractoriness of the larval ring glands to PTTH stimulation appeared by about day 26 (Richard and Saunders, 1987). These events lead directly to the low haemolymph titre of ecdysteroids, and the suspended development and metamorphosis characteristic of the diapause state. In Fig. 1, the proportion of the larvae failing to form puparia by day 30 is presented as the incidence of diapause under that particular photoperiod.

Figure 1 shows that several aspects of the photoperiodic response were associated with the geographical origin of the strain. Firstly, the critical day length (CDL) varied with latitude. It was longest (about 15 h/24) at 65°N, shorter (14½ h/24) at 55°N (Edinburgh), and about 12½ h/24 at 51°N (Silwood Park). At the lowest latitude (Chapel Hill, 36°N) the photoperiodic response was much weaker, although an increased incidence of diapause occurred under photoperiods of less than 12 h/24. The CDL for the higher altitude strain (Barga, 44°N) was about 13½ h/24, greater than that for Silwood Park, and therefore more than that predicted from a simple latitudinal series. The *shape* of the PPRC also depended on geographical origin. At 65°N, diapause incidence under continuous darkness (DD) was almost 100 per cent, whereas that for more southerly strains was 40 per cent or less. In addition, the northernmost strain showed a higher incidence of diapause under photoperiods longer than the CDL, with a proportion of the larvae entering diapause under LD 18:6 to LD 22:2; this may indicate a tendency towards univoltinism at higher latitudes.

The data presented in Fig. 1 were obtained at 20°C. However, it is known that the incidence of larval diapause in *C. vicina* is greater when the adult flies are maintained at a lower temperature (Saunders et al., 1986). McWatters and Saunders (1998) showed, for Silwood Park (51°N) and Nallikari (65°N), that lowering the maternal temperature from 20° to 15°C shifted the CDL to a longer value and raised the incidence of diapause, particularly under ultra-short and ultra-long photoperiods. Lowering the maternal temperature therefore had an effect resembling that of increased latitude. The known interaction between maternal photoperiod and temperature thus suggests that northerly strains enter diapause more readily at higher temperatures, whereas those from more southerly areas tend to avoid diapause unless the temperature falls.

2.2. Diapause duration

The 'intensity' or duration of the larval diapause appearing among the progeny of 'short-day' flies is known to depend on both temperature and the number of inductive cycles experienced by the adult fly before she lays her eggs (Saunders, 1987). In the Edinburgh strain, for example, larval diapause was more intense when parental flies were exposed to short days (LD 12:12) at 15°, rather than 20°C (Saunders et al., 1986), and to more than 10 such cycles before oviposition (Saunders, 1987). Under 'standard' conditions of maternal photoperiod and temperature, larval diapause was also more intense (longer lasting) in strains from more northerly localities (McWatters and Saunders, 1996, 1998). With a maternal regime of 12 pre-oviposition cycles of LD 12:12 at 19°C, larval diapause (in darkness at 11 to 12°C) continued for more than 70 days in some individuals from the Nallikari strain (65°N). Diapause duration under the same conditions was less for flies from Edinburgh (55°N) and only up to about 30 days for flies from Barga (44°N) (Saunders, 2000).

2.3. Cold tolerance

There is also evidence that diapausing larvae of *C. vicina* from more northerly latitudes show a greater cold tolerance than larvae from localities further south (Saunders and Hayward, 1998). Survival to eclosion, after 1 to 18 days exposure to low temperature (down to -8°C), was thus greater for the larval progeny of short-day than long-day females from two northern strains (Nallikari, 65°N, and Edinburgh, 55°N) but not for a southern strain (Barga, 44°N). Diapause-destined larvae from Edinburgh, however, were more cold tolerant than larvae from Nallikari, at both -4° and -8°C, a difference possibly associated with longer lasting snow cover in the more northern locality, which might insulate the over wintering soil microclimate.

The data described above demonstrate that the induction and incidence of larval diapause, the duration (intensity) of the diapause so induced, and probably the associated cold resistance, are all regulated by both environmental (photoperiod, temperature) and genetic factors. Some of this genetic variation has been explored by the selection of high and low diapause lines, and by crossing northern and southern strains.

3. SELECTION FOR HIGH AND LOW DIAPAUSE INCIDENCE

Divergent selection for high and low diapause has been achieved many times. Tauber et al. (1986), for example, list over 40 cases of artificial selection of diapause and photoperiodic traits from a 'parental' stock, often - especially in the case of a diapause free line - after rather few generations. These studies suggest that natural, and even laboratory, insect populations possess a considerable store of genetic variability, and that a multigenic mode of inheritance is involved. In *C. vicina*, selection for high and low diapause strains was prompted by a relentless weakening of its response to short days during years of laboratory culture. After 10 years of culturing the Edinburgh stock under continuous light at 25°C, for example, the incidence of diapause had fallen from over 90 per cent to about 20 per cent, when tested under LD 12:12.

A selection regime was established based upon breeding from the first larvae to pupariate and from the last. Adult flies of the diapause-depleted stock population were exposed to a short day regime (LD 12:12) at 16°C, and their larval progeny raised under continuous darkness at 11 to 12°C. Puparia were collected daily as they formed, each day's batch being maintained separately under the same conditions. The first puparia to form, and the last, then provided the material for the next generation; the selection pressure was about 10 per cent in each case. After just two generations of breeding from the earliest pupariaters, all of the larval progeny showed prompt pupariation characteristic of *C. vicina* under nondiapause (e.g. long day) conditions. In contrast, 4 or 5 generations of breeding from the last to form puparia led to strains in which practically all larvae delayed pupariation beyond the 30 days accepted as the manifestation of larval diapause (Fig. 2).

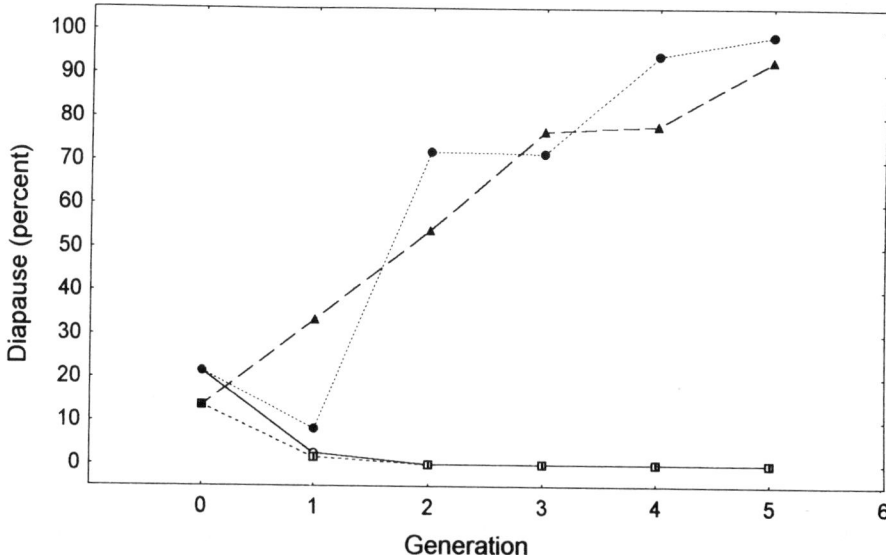

Figure 2. Selection for high incidence of larval diapause (closed points) and low incidence of diapause (open points) (two parallel cultures of each) from a diapause depleted stock of *C. vicina*. All experiments were conducted under a short day (LD 12:12) at 16°C.

After the fifth generation of selection, the selected 'high' and 'low' diapause lines were then compared with the depleted parental stock over a range of photoperiods from LD 4:20 to LD22:2 to produce photoperiodic response curves (figure 3). Diapause was absent in the low diapause strain right across this range, but the PPRC for the high diapause strain had been restored to its previous high level (as in Fig. 1). After very few generations, therefore, diapause incidence had been reduced to zero, or restored to over 90 per cent, thereby approaching that of the stock population shortly after its isolation from the wild.

4. CROSSING NORTHERN AND SOUTHERN STRAINS

Reciprocal crosses between parental flies from a northern (N) strain (Nallikari, 65°N) and a southern (S) strain (Silwood Park, 51°N) were performed in an attempt to unravel the contributions made by maternal, paternal and latitude-related genes (McWatters and Saunders, 1996, 1997). The first series of crosses between N and S was conducted at 15°C where the CDL for N was about 16 h/24, and that for S was about 14½ h/24. Under a photoperiod of 15½:8½, therefore, N flies reacted as though they were in a short day and produced a high incidence of diapause in their progeny, whereas S flies reacted as though they were in a long day regime.

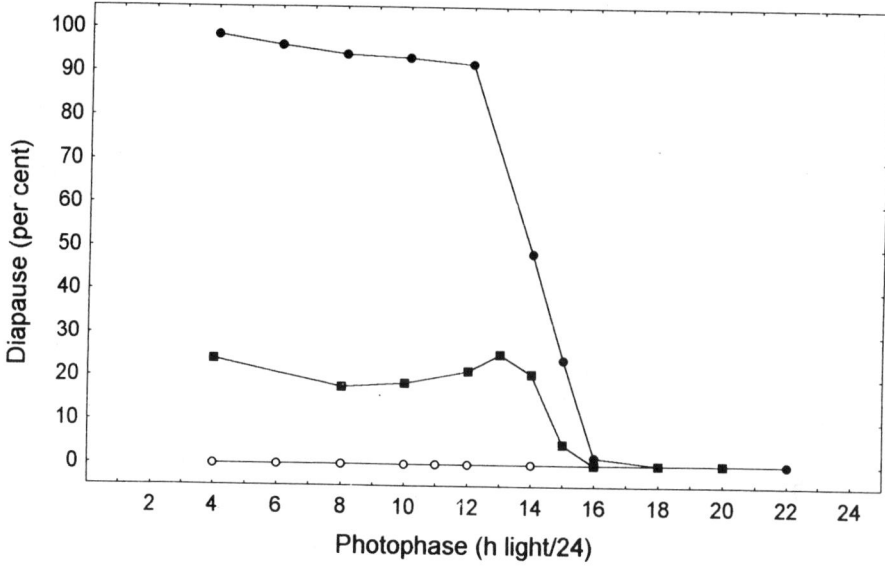

Figure 3. Photoperiodic response curves (PPRCs) for blow flies (*C. vicina*) from the 5th generation of selection for high (closed circles) and low (open circles) incidence of larval diapause, compared with the PPRC for the 'parental' diapause depleted stock (closed squares). All experiments were conducted under short days (LD 12:12) at 16°C.

The results of these crosses showed that the incidence of larval diapause was entirely dependent upon the mother, with the father contributing nothing to the diapause status of his immediate offspring. Thus, crosses between S female x S male, or S female x N male gave a low incidence of diapause (1.6 and 2.9 per cent), whereas those between N female x N male or N female x S male gave 92.7 per cent and 86.9 per cent. The result for diapause *duration*, however, was entirely different. Diapausing larvae from the S x S and S x N crosses showed rapid pupariation or a short, low intensity diapause. Crosses involving northern mothers (N x N and N x S), on the other hand, produced a high initial incidence of larval diapause but the duration of diapause was much lower in the hybrid group. These data suggest that whilst diapause incidence is controlled by the female alone, its duration or intensity is affected by both parents. Subsequent crosses between the F_1 flies, and backcrosses between F_1 and the parental strains (McWatters and Saunders, 1997) showed that indirect effect of inheritance down the male line was weaker than that down the female line. A more intense diapause was observed in larvae with a greater admixture of northern genes, regardless of whether they were maternal or paternal.

5. PHOTOPERIODIC TIME MEASUREMENT

The central problem in photoperiodism has always been, and continues to be, the mechanism of the time measurement involved. In this review we may also ask: How does knowledge of the latitudinal and genetical variation in photoperiodic and diapause-related traits help our understanding of this process?

There is now ample evidence, in both *C. vicina* and many other insects, to suggest that photoperiodic time measurement (PPTM) is a function of the circadian system. Evidence for this - perhaps causal - association has been reviewed many times, most recently by Saunders (1998) and Vaz Nunes and Saunders (1999). In particular, experiments involving the manipulation of the light-dark cycle have highlighted a number of formal similarities between the two systems: both present characteristics typical of circadian rhythms, both use light for entrainment or photoinduction and, in *C. vicina,* both operate simultaneously in the adult female fly. The most widely used of these experiments is the so-called Nanda-Hamner or 'resonance' protocol. In this type of experiment, groups of insects are exposed to a range of cycles in which the light component is held constant, but the dark component is varied widely, so that the period of the cycle (T) ranges over several multiples of the circadian period (τ). Vaz Nunes and Saunders (1999) have recently reviewed responses to such experiments. In cases of 'positive' resonance, peaks of high diapause incidence are interspersed with troughs of low diapause, with the intervals between peaks (or troughs) reflecting τ of the underlying photoperiodic oscillator. For the blow fly, resonance and other experiments indicating a circadian basis for PPTM are described by Vaz Nunes et al. (1990).

Overt behavioural rhythmicity and photoperiodic timing, however, appear to involve separate circadian pacemakers with different characteristics. For example, locomotor rhythmicity of *C. vicina* free-runs in DD with an essentially undamped rhythm and a period shorter than 24 hours ($\tau \sim 22.5$ h) (Hong and Saunders, 1998). Nanda-Hamner experiments, on the other hand, indicate that τ for the photoperiodic oscillator is much closer to 24 h, and that the oscillator is rather heavily damping (Saunders, 1987). In *Drosophila melanogaster,* there is more direct evidence for the separate regulation of overt rhythmicity and PPTM. Photoperiodic regulation of ovarian diapause continues in apparently arrhythmic flies (*per*^{O}), and in flies deprived of the entire *period* locus (*per*‾), albeit with a critical day length that differs from wild type (Saunders, 1990). The current view, therefore, is that overt behavioural rhythms and PPTM, although both circadian based, reflect separate components of the multioscillator circadian system.

Most studies of latitudinal clines in photoperiodic responses show that the critical day length becomes longer as one travels north (e.g. Bradshaw, 1976; this review), whereas the period (τ) of overt rhythms becomes steadily shorter (e.g. Lankinen, 1986). In the spider mite, *Tetranychus urticae,* Vaz Nunes et al. (1990) also demonstrated an inverse correlation between latitude and Nanda-Hamner inter-peak intervals, thereby indicating that the period of the circadian oscillation underlying PPTM also shortens as one moves north. An inverse relationship between endogenous period and critical day length is exactly what one would expect from the application of circadian entrainment theory within a framework of photoperiodic clock models (see Saunders, 1982). However, both Lankinen (1986) and Vaz Nunes et al. (1990) found only weak correlations between CDL and circadian period. One interpretation might be that, although both parameters show the expected south-north trends, they reflect

120

separate components of the multioscillator circadian system and are therefore not *causally* associated.

In summary, *C. vicina,* like many other insects with a large south-north range, exhibits latitudinal clines in critical day length, diapause incidence, diapause duration, and possibly also in cold tolerance. Selection experiments indicate a substantial reservoir of genetic variation for some of these diapause-related parameters. Larval diapause, however, is purely maternal with the males contributing nothing to the incidence of diapause among their immediate offspring. Furthermore, F_1 crosses, and backcrosses between F_1 and parental lines from different geographical stocks, show that the male contribution is always less than the maternal. In contrast, *both* parents contribute to the diapause duration (or diapause intensity) of their immediate offspring.

There is no doubt that the regulation of larval diapause in *C. vicina* is extremely complex. At its simplest it must involve a cascade of events from photoreception through a 'clock' (to measure night length) and a 'counter' (to accumulate successive night lengths to a threshold value) to the final endocrine effectors. In *C. vicina* this cascade is additionally complicated by photoreception and the clock-counter mechanism being maternal, whereas the endocrine effectors regulating diapause occur in the next generation, after many intervening hormonal events employing the same endocrine mechanisms that regulate moulting and metamorphosis. Some of the simple genetical tools used in unravelling, for example, circadian rhythmicity, may be very difficult to use for the dissection of photoperiodic responses. Point mutations affecting diapause incidence or duration, for example, could be attributed to genetic lesions affecting several (or any) of the layers of this concatenation. There is no doubt that much more work will be required before we begin to unravel this complexity.

REFERENCES

Bradshaw, W.E., 1976. Geography of photoperiodic response in a diapausing mosquito Nature, 262, 384-386.

Hong, S-F., Saunders, D.S., 1998. Internal desynchronisation of the circadian locomotor rhythm of the blow fly, *Calliphora vicina,* as evidence for the involvement of a complex pacemaker. Biological Rhythm Research 29, 387-396.

Lankinen, P., 1986. Geographical variation in circadian eclosion rhythm and photoperiodic adult diapause in *Drosophila littoralis.* Journal of Comparative Physiology A 159, 123-142.

McWatters, H.G., Saunders, D.S., 1996. The influence of each parent and geographic origin on larval diapause in the blow fly, *Calliphora vicina.* Journal of Insect Physiology 42, 721-726.

McWatters, H.G., Saunders, D.S., 1997. Inheritance of the photoperiodic response controlling larval diapause in the blow fly, *Calliphora vicina.* Journal of Insect Physiology 43, 709-717.

McWatters, H.G., Saunders, D.S., 1998. Maternal temperature has different effects on the photoperiodic response and duration of larval diapause in the blow fly (*Calliphora vicina)* strains collected at two latitudes. Physiological Entomology 23, 369-375.

Richard, D.S., Saunders, D.S., 1987. Prothoracic gland function in diapause and non-diapause *Sarcophaga argyrostoma* and *Calliphora vicina.* Journal of Insect Physiology 33, 385-392.

Saunders, D.S., 1982. Insect Clocks, second edition. Pergamon Press, Oxford.

Saunders, D.S., 1987. Maternal influence on the incidence and duration of larval diapause in *Calliphora vicina*. Physiological Entomology 12, 331-338.

Saunders, D.S., 1990. The circadian basis of ovarian diapause regulation in *Drosophila melanogaster:* is the *period* gene causally involved in photoperiodic time measurement? Journal of Biological Rhythms 5, 315-331.

Saunders, D.S., 1998. Insect circadian rhythms and photoperiodism. Invertebrate Neuroscience 3, 155-164.

Saunders, D.S., 2000. Larval diapause duration and fat metabolism in three geographical strains of the blow fly, *Calliphora vicina*. Journal of Insect Physiology 46, 509-517.

Saunders, D.S., Haywood, S.A.L., 1998. Geographical and diapause-related cold tolerance in the blow fly, *Calliphora vicina*. Journal of Insect Physiology 44, 541-551.

Saunders, D.S., Macpherson, J.N., Cairncross, K., 1986. Maternal and larval effects of photoperiod on larval diapause in the flies *Calliphora vicina* and *Lucilia sericata*. Experimental Biology 46, 51-58.

Tauber, M.J., Tauber, C.A., Masaki, S., 1986. Seasonal Adaptations of Insects. Oxford University Press, 411 pp.

Vaz Nunes, M., Kenny, N.A.P., Saunders, D.S., 1990. The photoperiodic clock in the blowfly, *Calliphora vicina*. Journal of Insect Physiology 36, 61-67.

Vaz Nunes, M., Koveos, D.S., Veerman, A., 1990. Geographical variation in photoperiodic induction in the spider mite (*Tetranychus urticae*): A causal relation between critical nightlength and circadian period? Journal of Biological Rhythms 5, 47-57.

Vaz Nunes, M., Saunders, D.S., 1999. Photoperiodic time measurement in insects: A review of clock models. Journal of Biological Rhythms 14, 84-104.

Vinogradova, E.B., 1986. Geographical variation and ecological control of diapause in flies. In: F. Taylor and R. Karban (Eds.), The Evolution of Insect Life Cycles, p. 35-47. Springer-Verlag, New York,

Vinogradova, E.B., Zinovjeva, K.B., 1972. Maternal induction of larval diapause in the blow fly *Calliphora vicina*. Journal of Insect Physiology 18, 2401-2409.

Insect Timing: Circadian Rhythmicity to Seasonality
D.L. Denlinger, J. Giebultowicz and D.S. Saunders (Editors)
© 2001 Elsevier Science B.V. All rights reserved.

Evolutionary aspects of photoperiodism in *Drosophila*

M. T. Kimura

Graduate School of Environmental Earth Science, Hokkaido University, Sapporo, Hokkaido 060-0810, Japan

To understand the evolution of photoperiodism in the *Drosophila montium* species subgroup, I examined photoperiodic responses of several species from temperate and subtropical areas. No sign of photoperiodic response was observed in the subtropical species from Taiwan, whereas a distinct photoperiodic response was observed in *D. rufa* from warm-temperate areas. In *D. rufa* and *D. asahinai* from islands located at the boundary of subtropical and warm-temperate regions, the effect of photoperiod on cold tolerance was observed when reared at 11 °C, but was not at 15 °C. However, it is questionable whether the photoperiodic response of *D. asahinai* is functional or not, since mean January (the coldest month) temperature of the islands it occurs is about 15 °C. In this subgroup, photoperiodism might have evolved under circumstances cooler than these islands of the present, and *D. asahinai* might have weakened the photoperiodic response as a result of its adaptation to the present climatic conditions of the islands.

The present paper also reports the evolution of univoltinism in *Drosophila moriwakii*, *D. alpina* and northernmost populations of *Chymomyza costata*. These species and populations showed photoperiodic responses when reared under unusual conditions; i.e., northernmost populations of *C. costata* at high temperatures, *D. alpina* when maintained for very long period, and *D. moriwakii* when flight was retarded. These results indicate that they originated from multivoltine species or populations with photoperiodic diapause. Northernmost populations of *C. costata* evolved a univoltine life cycle by changing the critical temperature for the induction of diapause, and *D. alpina* and *D. moriwakii* did though the acquisition of 'long-day' or 'non-photoperiodic' diapause.

1. THE EVOLUTION OF PHOTOPERIODISM

A large number of studies have been made on physiological and ecological aspects of insect photoperiodism, but little is known on the origin and evolution of photoperiodism (Tauber et al., 1986; Danks, 1987). This is partly because most extant species or populations have developed photoperiodism to fully functional levels or do not have photoperiodism at all and there are few opportunities to study the early stage of its evolution. In this study, I focus on the *Drosophila montium* species subgroup of the *melanogaster* species group. This subgroup is diversified in tropical regions, but only six species are known from temperate areas (Lemeunier et al., 1986; Goto et al., 2000). This indicates that the colonization of this subgroup in temperate areas was rather recent in the evolutionary time scale. Therefore, this subgroup may provide us with an opportunity to study the early stage of evolution of photoperiodism.

The temperate species of this subgroup are subdivided into the cool-temperate (*D. auraria* and its siblings) and warm-temperate species (*D. rufa* and *D. tani*) on the basis of their distributions and cold tolerance (Kimura, 1988; Goto et al., 2000). On the other hand, the species occurring in subtropical areas of Asia (Taiwan and southern China) are subdivided into the highland (*D. trapezifrons, D. constricta* and some others) and lowland species (*D. watanabei* and some others) (Goto et al., 2000). In addition to these species, two sibling species of *D. rufa, D. asahinai* and *D. lacteicornis*, are known from islands located at the boundary of subtropical and temperate areas (Kim et al., 1994) (Fig. 1). We have shown that the cool- and warm-temperate species have photoperiodic reproductive diapause (Kimura, 1988). However, little information has been reported on the photoperiodic responses of the other species. Here, I examined the effects of photoperiod on ovarian development and cold tolerance in several species and populations from subtropical and warm-temperate areas to understand the evolution of photoperiodism in this subgroup. Figure 1 shows the original localities of experimental strains.

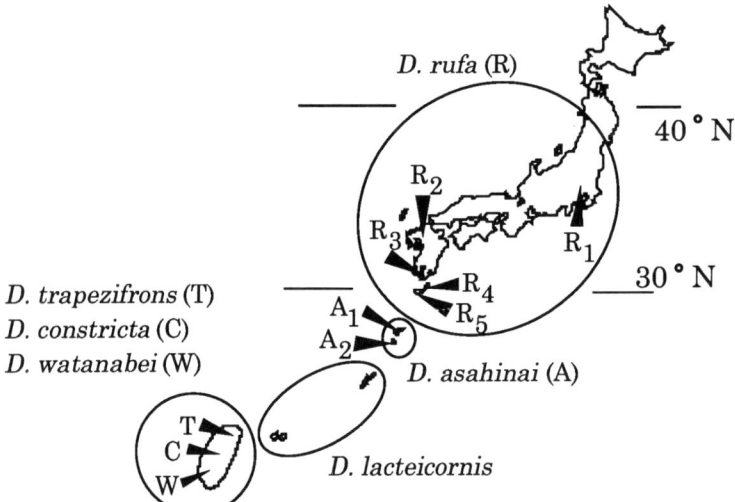

Figure 1. Distributions of experimental species in Japan and Taiwan and original localities of experimental strains.

1.1. Photoperiodic control of ovarian development

Figure 2 shows ovarian development in geographic strains of *D. rufa* and *D. asahinai* under various photoperiods at 15 °C. The effect of photoperiod was apparent in *D. rufa* from northern areas (R_1-R_3), and the critical daylength increased with the increase of latitude. However, the photoperiodic response was not observed in *D. rufa* or *D. asahinai* from southern islands (R_4, R_5, A_1 and A_2).

1.2. Photoperiodic control of cold tolerance

Figure 3 shows the effect of photoperiod on cold tolerance in females of *D. rufa*, *D. asahinai*, *D. trapezifrons*, *D. constricta* and *D. watanabei*. Males had similar cold tolerance with females (data not shown). Flies were reared under a short (10-h-light:14-h-dark) or long (15-h-light:9-h-dark) daylength at 15 °C before eclosion and at 11 or 15 °C after eclosion. They were exposed to 1.5 °C (constant darkness) 16- (for those reared at 15 °C) or 32-days (for those reared at 11 °C) after eclosion and monitored for survival. Time (days) to kill 25, 50 and 75 % of population was obtained by directly reading the survival curves and plotted against the original latitudes. No photoperiodic effect was observed in the

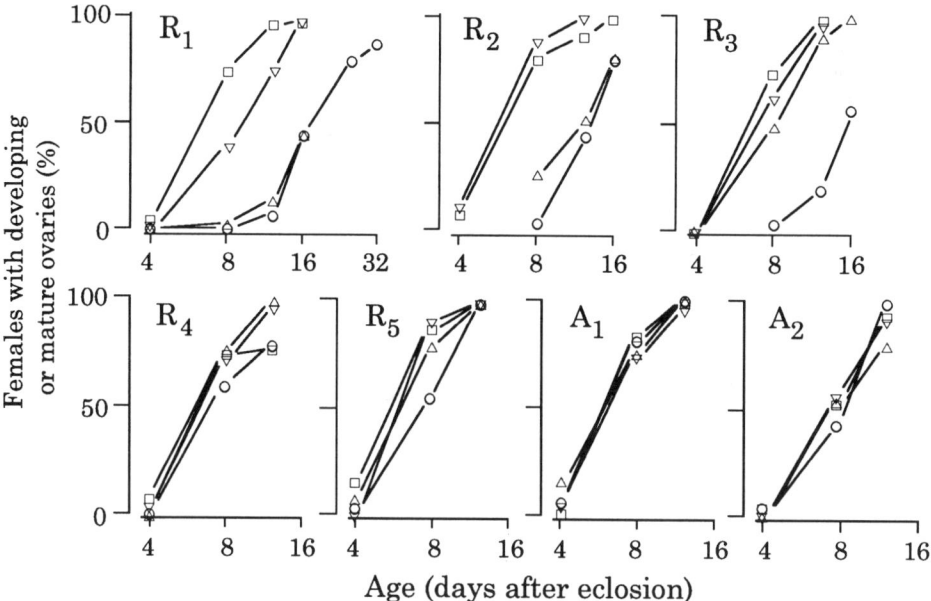

Figure 2. Ovarian development in geographic strains of *D. rufa* (R_1-R_5) and *D. asahinai* (A_1, A_2) under various photoperiods (○: LD 10:14, △: LD 11:13, ▽: LD 12:12, □: LD 13:11) at 15 ℃. Latitudes of the original localities of the strains, see Figure 3.

subtropical species, *D. trapezifrons*, *D. constricta* and *D. watanabei*, irrespective of rearing temperature, whereas a short day condition enhanced cold tolerance not only in northern populations of *D. rufa* but also in southern populations of *D. rufa* and *D. asahinai* at 11 °C, although no or only little effect of photoperiod was observed at 15 °C.

Thus, *D. rufa* and *D. asahinai* from the islands located at the boundary of subtropical and warm-temperate regions retained a weak photoperiodic response. There are two possible explanations for the weak photoperiodic response of these species; 1) it represents the early stage of the evolution of photoperiodism, and 2) it is a relic of firm photoperiodic diapause of the past. If the first explanation is the case, their photoperiodic response would be functional. However, it is questionable whether the weak photoperiodic response of *D. asahinai* is functional or not, since mean January (the coldest month) temperature of the islands where it occurs (A_1 and A_2 in Fig. 1) is about 15 °C, a temperature at which its response to photoperiod became unclear. On the other hand, climates of these islands are assumed to be much cooler during the glacial periods, e.g., about 15-20 thousand years ago. Therefore, if it occurred in these islands during the glacial periods, it might have a firm photoperiodic diapause as the present-day warm-temperate

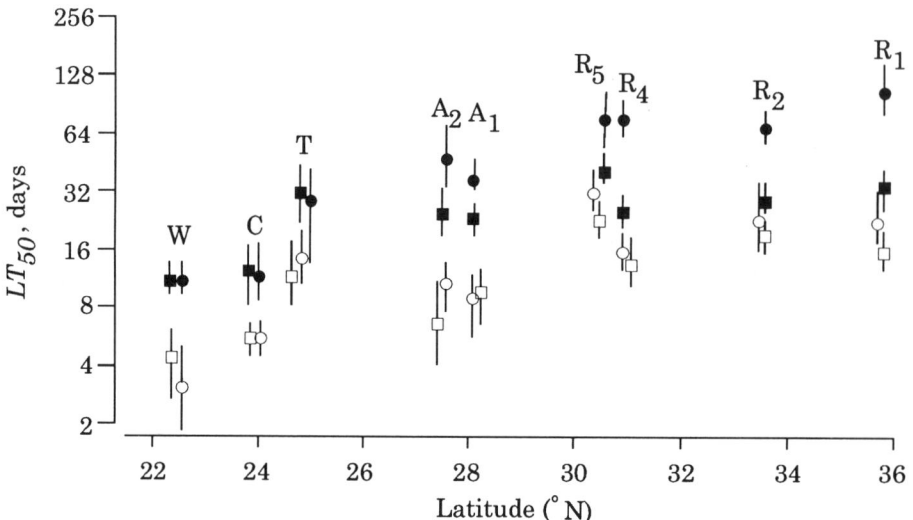

Figure 3. Half lethal time *(LT50)* of adult flies of experimental strains of *D. rufa* (R1-R5), *D. asahinai* (A1, A2), *D. trapezifrons* (T), *D. constricta* (C) and *D. watanabei* (W) at 1.5 ℃. Experimental individuals were reared under LD 10:14 (○, ●) and LD 15:9 (□, ■). Rearing temperature was 15 ℃ before eclosion and 11 (●, ■) and 15 ℃ (○, □) after eclosion. Vertical lines indicate *LT25* and *LT75*.

populations of *D. rufa*. In this subgroup, photoperiodism might have evolved under circumstances cooler than these southern islands of the present, and *D. asahinai* might have weakened the photoperiodic response as a result of its adaptation to the present climatic conditions of the islands. However, our preliminary survey suggests that at least *D. trapezifrons* undergoes seasonal migration between highlands and lowlands (Goto et al., 2000). In a number of insects, it has been shown that seasonal migration is controlled by photoperiodism (Dingle, 1978; Tauber et al., 1986; Danks, 1987). Therefore, there is also a possibility that photoperiodism has already evolved in subtropical highland species as a means to control seasonal migration.

2. THE EVOLUTION OF UNIVOLTINE LIFE CYCLES

A number of boreal and alpine insects pass only one generation in a year. Limited heat sum in boreal or alpine areas would not allow them to produce two generations in a year (Tauber et al., 1986; Danks, 1987). Univoltine life cycles are also known in some insects from temperate areas where heat sum is sufficient for the production of two or more generations. Their univoltine life cycles would be adaptations to seasonal changes in resource conditions, predator abundance or humidity in their habitats. Since univoltine species always enter diapause at certain stages, they need not respond to environmental cues in the induction of diapause. However, many univoltine species respond to environmental cues in the process of diapause. A well studied example is a lacewing *Chrysoperla downesi* which enters reproductive diapause in response to stationary long daylengths or decreasing daylength but reproduces in response to increasing daylength (Tauber and Tauber, 1976). Here, I review our studies on diapause characteristics of univoltine drosophilids, *C. costata*, *D. alpina* and *D. moriwakii*, with addition of new data, and discuss how they have evolved univoltine life cycles.

2.1. *Chymomyza costata*

This species enters larval diapause to pass the winter (Lumme and Lakovaara, 1983; Riihimaa and Kimura, 1988). Its populations from northern Finland (north of 66 °N) have a univoltine life cycle, but those from southern Finland or Japan have multivoltine life cycles (Riihimaa et al., 1996). This species is basically a long day species. The critical daylength is ca. 14 h in a population from Sapporo (northern Japan) and increases with the increase of latitude (Riihimaa and Kimura, 1988; Riihimaa et al., 1996). However, it does not increase over 20 h even in northern Finnish univoltine populations. Therefore, if the induction of diapause is solely determined by photoperiod, larvae growing in June or July may fail to enter diapause in northern Finland, since night becomes very short (Riihimaa et al., 1996). The northern Finnish populations solve this problem by changing their response to temperature. In this species, temperature has been

128

shown to play an important role in the induction of diapause even in southern multivoltine populations. For example, a temperature of 11 °C induces diapause in about 70 % of individuals of a population from Sapporo even under continuous light (Riihimaa and Kimura, 1988). Northern Finnish populations have raised the critical temperature to ca. 22 °C, a temperature seldom met in northern Finland, and therefore they are able to enter diapause even in mid summer (Riihimaa et al., 1996).

2.2. *Drosophila alpina*

D. alpina and its relative, *D. subsilvestris*, enter pupal diapause for overwintering (Beppu et al., 1996; Goto et al., 1999). Here, I compare the photoperiodic responses of *D. alpina* (originated from a highland of central Japan) and *D. subsilvestris* (from Tübingen, Germany). Figure 4 shows eclosion of these species under combinations of short (10 h light-14 h dark) and long (15 h light-9 h dark) daylengths at 15 °C. In *D. alpina*, almost all pupae did not eclosed at least for 5 months if they experienced a short daylength for a certain period during the pupal stage. When individuals of this species were maintained under a long daylength throughout their entire lives, about a half of them eclosed 3-5 months after pupation; i.e., they stayed as pupae at least for two and a half months. On the other hand, *D. subsilvestris* eclosed about a month after being oviposited under a long daylength, much earlier than *D. alpina*. It is therefore considered

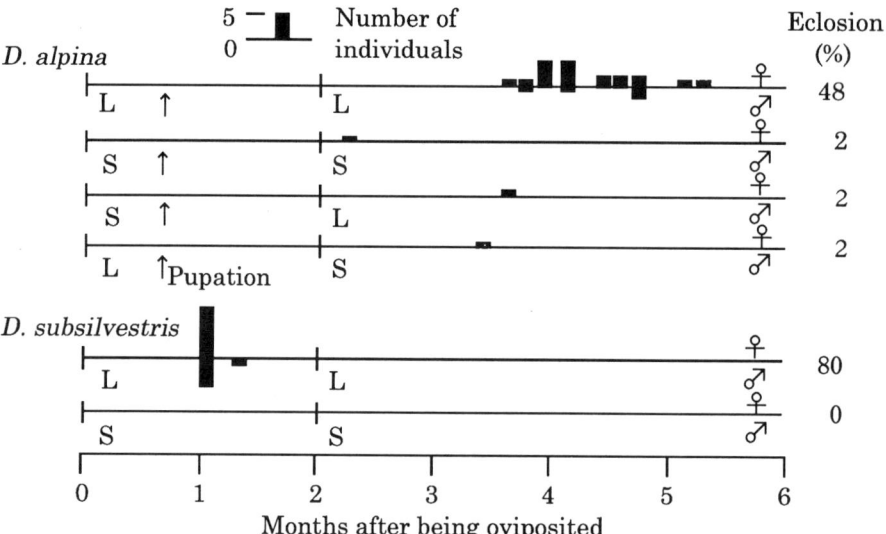

Figure 4. Eclosion of *D. alpina* and *D. subsilvestris* under combinations of long (L: LD 15:9) and short (S: LD 10:14) daylengths.

that the physiological state that arises under a long daylength in *D. alpina* is diapause. However, it can not be determined whether this diapause is 'long-day' diapause (induced only at long daylengths) or 'non-photoperiodic' diapause (induced independent of photoperiod), because it is not known whether this diapause is induced at short daylengths or not. At short daylengths, the expression of this diapause would be masked by the 'short-day' diapause even if it is induced.

2.3. *Drosophila moriwakii*

In northern Japan, this species overwinters in reproductive diapause and reproduces in May and June (Ichijô and Beppu, 1990). Imagoes of the next generation eclose from late June and enter summer-autumnal-winter diapause. In laboratory, imagoes of this species entered diapause irrespective of photoperiod when they were maintained in a cube-shaped cage (27 liter in volume) in which they were able to fly freely, but they showed a long-day type photoperiodic response when confined in a vial (50 ml) in which they are unable to fly freely (Ichijô et al., 1992). In addition, imagoes of which wings were excised showed a long-day type response even if they were maintained in a cage (Ichijô et al., 1992). On the basis of these results, it is assumed that this species has two types of diapause as well as *D. alpina*; i.e., 'short-day' diapause which is resistant to

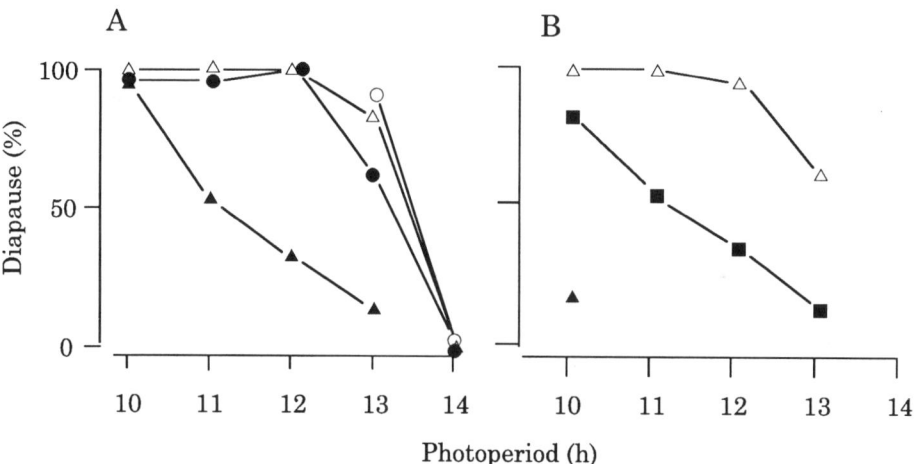

Figure 5. Females of *D. triauraria* in reproductive diapause 16 (A) or 32 days (B) after eclosion under various photoperiods. ○: intact flies reared in cages, △: intact flies reared in vials, ●: flies of which wings were excised on the day of eclosion, ▲: flies of which wings were fixed on the day of eclosion, ■: flies of which wings were fixed 16 days after eclosion. Wing-treated flies were reared in vials. Experimental temperature was 15 ℃.

diapause-averting action of flight suppression and 'long-day' or 'non-photoperiodic' diapause which is averted by suppression of flight. Here, I studied the effect of the retardation of flight on reproductive diapause in a multivoltine species *Drosophila triauraria* for further understanding of the relation between flight and reproductive diapause.

The experimental strain was collected from Onuma (42 °N), northern Japan, and maintained for years under diapause-preventing conditions, 23 °C and continuous illumination. Experimental animals were reared at various photoperiods at 15 °C in milk bottles (200 ml in volume) before eclosion and in 50-ml vials or 27-liter cages after eclosion. In addition, wings were excised or fixed with a small piece of paper (1x3 mm) using a quick-drying glue on the day of eclosion or 16-day after eclosion. Fixation of wings prevents flies from not only flying but also moving flight muscles. The incidence of diapause was checked in females 16 or 32 days after eclosion; those with undeveloped ovaries were determined to be in diapause (Kimura, 1983).

No or small difference was observed in the photoperiodic response between vial and cage populations or between intact and wing-excised individuals (Fig. 6). However, diapause was averted when wings were fixed. At a photoperiod of 10 h, it was averted when wings were fixed for 32 days from eclosion but was not when wings were fixed only for either of earlier or later 16 days (Fig. 6), revealing that longer fixation is more effective to avert diapause. In addition, the effect of wing fixation was more apparent at longer photoperiods. In this species, it has been shown that diapause induced at a longer photoperiod (but shorter than the critical daylength) is shallower (i.e., maintained for a shorter period) than that induced at a shorter photoperiod (Kimura, 1990). These results indicate that shallower diapause is less resistant to diapause-averting action of wing fixation.

Thus, flight (or movement of flight muscles) affected not only the diapause of *D. moriwakii* at long daylengths but also the 'short-day' diapause of a multivoltine species. These results raise a question whether the diapause of *D. moriwakii* at long daylengths is essentially different from the 'short-day' diapause or not. However, it is at least certain that the diapause of *D. moriwakii* at long daylengths is very shallow, since it is averted by excision of wings or confinement into a small vial, the treatments which are not sufficient to avert the 'short-day' diapause of *D. moriwakii* and *D. triauraria*.

The univoltine drosophilid species and populations studied here thus retained photoperiodic responses which are expressed when reared under unusual conditions. This suggests that these species and populations have derived from multivoltine species or populations with photoperiodic diapause. They evolved univoltine life cycles in different ways; northernmost populations of *C. costata* raised the critical temperature for the induction of diapause, and *D. alpina* and *D. moriwakii* acquired 'long-day' or 'non-photoperiodic' diapause.

3. ACKNOWLEDGMENTS

I thank K. Beppu, H. Watabe, N. Ichijô, A. Riihimaa, T. Ohtsu, T. Yoshida, T. Awasaki and S. G. Goto for their help in the course of the present study. This study was supported by a Grant-in-Aid from the Ministry of Education, Science, Sports and Culture of Japan (No. 10836003).

REFERENCES

Beppu, K., Yoshida, T., Kimura, M.T., 1996. Seasonal life cycles and adaptations of four species of *Drosophila* at high altitudes in central Japan. Japanese Journal of Entomology 64, 627-635.

Danks, H.V., 1987. Insect Dormancy: An Ecological Perspective. Biological Survey of Canada, Ottawa.

Dingle, H. (ed.), 1978. Evolution of Insect Migration and Diapause. Springer-Verlag, New York.

Goto, S.G., Yoshida, T., Beppu, K., Kimura, M.T., 1999. Evolution of overwintering strategies in Eurasian species of the *Drosophila obscura* species group. Biological Journal of Linnean Society 68, 429-441

Goto, S.G., Kitamura, H.W., Kimura, M.T., 2000. Phylogenetic relationships and climatic adaptations in the *Drosophila takahashii* and *montium* species subgroups. Molecular Phylogenetics and Evolution 15, 147-156.

Ichijô, N., Beppu, K., 1990. Bionomics of Drosophilidae (Diptera) in Hokkaido. XI. Five species of the *Drosophila robusta* species group. Japanese Journal of Entomology 58, 625-636.

Ichijô, N., Beppu, K., Kimura, M.T., 1992. Aestivo-hibernal reproductive diapause in *Drosophila moriwakii*: flight as a controlling factor. Entomologia Experimentalis et Applicata 62, 23-28.

Kim, B.K., Aotsuka, T., Kitagawa, O., 1993. Evolutionary genetics of *Drosophila montium* subgroup. II. Mitochondrial DNA variation. Zoological Science 10, 991-996.

Kimura, M.T., 1983. Geographic variation and genetic aspects of reproductive diapause in *Drosophila triauraria* and *D. quadraria*. Physiological Entomology 8, 181-186.

Kimura, M.T., 1988. Adaptations to temperate climates and evolution of overwintering strategies in the *Drosophila melanogaster* species group. Evolution, 42, 1288-1297.

Kimura, M.T., 1990. Quantitative response to photoperiod during reproductive diapause in the *Drosophila auraria* species-complex. Journal of Insect Physiology 36, 147-152.

Lemeunier, F., David, J.R., Tsacas L., Ashburner, M., 1986. The *melanogaster* species group. In: Ashburner, M., Carson H.L., Thompson, J.N. Jr., (eds.), The

Genetics and Biology of *Drosophila*, vol. 3e, Academic Press, New York, pp. 147-256.

Lumme, J., Lakovaara, S., 1983. Seasonality and diapause in drosophilids. In: Ashburner, M., Carson H.L., Thompson, J.N. Jr., (eds.), The Genetics and Biology of *Drosophila*, vol. 3d, Academic Press, New York, pp. 171-220.

Riihimaa, A., Kimura, M.T., 1988. A mutant strain of *Chymomyza costata* (Diptera: Drosophilidae) insensitive to diapause-inducing action of photoperiod. Physiological Entomology 13, 441-445.

Riihimaa, A., Kimura, M.T., Lumme, J., Lakovaara, S., 1996. Geographical variation in the larval diapause of *Chymomyza costata* (Diptera, Drosophilidae). Hereditas 124, 151-163.

Tauber, M.J., Tauber, C.A., 1976. Developmental requirements of the univoltine species *Chrysopa downesi*: photoperiodic stimuli and sensitive stages. Journal of Insect Physiology 22, 331-335.

Tauber, M.J., Tauber, C.A., Masaki, S., 1986. Seasonal Adaptations of Insects, Oxford Univ. Press, New York

Insect Timing: Circadian Rhythmicity to Seasonality
D.L. Denlinger, J. Giebultowicz and D.S. Saunders (Editors)
© 2001 Elsevier Science B.V.   All rights reserved.

Molecular analysis of overwintering diapause

S. R. Palli[1], R. Kothapalli[2], Q. Feng[2], T. Ladd[2], S.C. Perera[2], S.-C. Zheng[2],
K. Gojtan[2], A.S.D Pang[2], M. Primavera[2], W. Tomkins[2] and A. Retnakaran[2]

[1]Biotechnology Division, Rohm and Haas Company, 727 Norristown Road, Spring
House, PA 19477, USA

[2]Great Lakes Forestry Centre, Canadian Forest Service
1219 Queen Street East, Sault Ste. Marie, Ontario, Canada P6A 5M7

The spruce budworm, *Choristonuera fumiferana*, overwinters as a diapausing
second-instar larva. Molecular analyses of larvae in prediapause, diapause and
postdiapause stages led to the identification of sugars and proteins that are
related to diapause or overwintering. *C. fumiferana* produces the disaccharide
trehalose and proteins such as diapause associated proteins 1&2 and glutathione
S-transferase that are diapause related. The glycerol and the defensin-like
protein that the larvae produce appear to be related to the overwintering process.
Absence of ecdysteroids in the diapausing second instars prevents molting and
metamorphosis, allowing them to overwinter in this stage.

1. INTRODUCTION

Diapause is a genetically determined, hormonally mediated state of
suppressed development and is an important adaptive mechanism for insect
survival during unfavourable environmental conditions such as low winter
temperatures. The continuous developmental program is switched off and an
alternative diapause-genetic program is initiated. While simple dormancy is a
direct response to adverse physical conditions, diapause is the initiation of a
complex genetic program, which often begins long before the onset of
unfavourable conditions and may not be terminated when such a state ends.
Token stimuli that anticipate future seasonal changes are perceived during
specific stages and then translated into a biological response. Diapause is
considered obligatory, a relatively rare phenomenon, when it is faithfully
repeated as a part of the life history and is considered facultative, a relatively
common occurrence, when it can be changed depending on the environmental
conditions. Diapause can occur at the egg or embryonic (silkmoth, *Bombyx mori*),
larval (rice stem borer, *Chilo suppressalis*), pupal (giant silkworm, *Hyalophora
cecropia*) or adult (Colorado potato beetle, *Leptinotarsa decemlineata*) stages, but
the stage is fixed in each species (Beck, 1968). The spruce budworm, *C.
fumiferana*, for example, undergoes an obligatory diapause as a second-instar
larva (Harvey, 1957).

Although most insects undergo overwintering and diapause simultaneously, these could be two separate events influenced concomitantly by both internal (e.g., hormone) and external factors (e.g., temperature and photoperiod). In this article we will summarise our studies of diapause in the spruce budworm, *C. fumiferana*, showing, *inter alia*, that overwintering and diapause are two distinct physiological events that occur simultaneously in this species.

2. THE SPRUCE BUDWORM LIFE CYCLE

The spruce budworm, *Choristoneura fumiferana* Clemens (Lepidoptera: Tortricidae) is univoltine, has six larval instars and overwinters as a diapausing second-instar larva. In Northern Ontario, the adult moths eclose in July and lay their eggs in masses on the under surface of needles of primarily balsam fir and white spruce. The eggs hatch in approximately eight days at 25°C, and the pale green, newly emerged first-instar larvae crawl toward the base of branches. These first instars, which were once thought to be non-feeding, actually feed by grazing on the surface of the needles (Retnakaran et al. 1999). Each larva spins a hibernaculum, molts into a light brown second-instar and enters into an obligatory diapause until the following spring. In the field, first-instars take approximately three days at 25°C to molt into second-instars, but in the laboratory we were able to prolong it to five days by keeping the larvae at 22°C, 70% relative humidity (RH) and 16L:8D photoperiod. Characteristically, within each hibernaculum, we observed the presence of an ecdysed first-instar head capsule and two frass pellets eliminated by the first-instar larva prior to moulting. The second instars come out of diapause in spring, mine into needles and later enter the buds. They remain inside the buds until the caps fall off and the buds flush in June, at which time most have reached the fourth instar. These larvae feed actively, molt into fifth-instar larvae, then into sixth-instars and pupate. (Prebble, 1975).

Grisdale (1970) has described in detail the laboratory rearing procedure for the spruce budworm. The eggs are collected within one hour after oviposition and are kept at 22°C and 70% RH. Under these conditions the eggs develop, and first-instar larvae hatch in eight days. The first-instar larvae crawl into the mesh of the cheesecloth, spin hibernacula, molt into second-instar larvae in about five days, and enter diapause. The diapausing larvae in cheesecloth sheets are maintained at 16°C for two weeks and at 2°C for 20 to 25 weeks to satisfy the diapause requirement. At the end of this period, the larvae are moved to 20°C. In about four days at this temperature the larvae start crawling out of the cheesecloth, indicating the termination of diapause. At this stage the larvae are transferred to cups containing artificial diet and reared at 22°C, 70% RH under a circadian photoperiodic regimen of 12h light and 12h darkness.

3. PRODUCTION OF POLYOLS AND SUGARS DURING OVERWINTERING DIAPAUSE

Polyhydric alcohols play an important role in both diapause and the overwintering of insects (Storey and Storey, 1991). Glycerol is the most common cryoprotectant used by insects. Other polyols such as sorbitol, mannitol, ribitol, erythritol, threitol and ethylene glycol as well as sugars such as trehalose, sucrose, glucose and fructose are known to act as cryoprotectants in insects (Storey and Storey, 1991). Han and Bauce (1993) reported the presence of high levels of glycerol in diapausing second-instar *C. fumiferana* larvae. We have analysed whole body alcohol extracts of first- and second-instar *C. fumiferana* larvae by HPLC. Two main peaks corresponding to trehalose and glycerol were detected. As shown in Figure 1, both trehalose and glycerol levels were very low on the day of hatching into first-instar larvae. The glycerol levels remained low throughout the first-instar. On the other hand, the trehalose levels started increasing in two-day-old first-instar larvae and continued to increase until the second-instar larvae entered diapause. The high trehalose levels were maintained throughout diapause and started decreasing when the larvae were transferred to 20°C for terminating diapause.

The glycerol levels remained low until the diapausing second-instar larvae were transferred to 2°C. The glycerol levels started increasing at five weeks after the second-instar larvae were transferred to 2°C and reached maximum levels by 10 weeks; these levels were maintained as long as the larvae were kept at 2°C. Glycerol levels decreased 5-fold within four days after transferring the second-instar larvae from 2°C to 20°C.

Thus, glycerol and trehalose in *C. fumiferana* larvae seem to increase in response to different cues. It appears that trehalose levels increase as soon as the first-instar larvae hatch in preparation for the obligatory overwintering diapause. On the other hand, glycerol levels increase only after the second-instar larvae are exposed to cold, indicating that the synthesis of this sugar may be under the control of an environmental cue such as exposure to low temperature. The levels of both sugars decrease as soon as the second-instar larvae are transferred from 2°C to 20°C for diapause termination. In *C. fumiferana*, exposure to 2°C for 20 weeks is sufficient to satisfy the obligatory diapause requirement. When the larvae were kept at 2°C for extended periods up to 30 weeks, both trehalose and glycerol levels remained high (Fig 1). On the other hand, when the second-instar larvae were transferred to 20°C prematurely, at eight weeks after exposure to cold, the levels of both sugars decreased within four days exposure to 20°C (Kothapalli, R. and Palli, S.R. unpublished results).

To verify our hypothesis that glycerol levels increase only after the larvae are exposed to cold treatment, we maintained diapausing second instar larvae at 20°C for up to eight weeks and measured the levels of the sugars in the hemolymph. In these larvae the trehalose levels were high but glycerol levels remained very low (Kothapalli, R. and Palli, S.R. unpublished results). Thus, it appears that the increase in trehalose concentration is related to the obligatory diapause, whereas, the increase in glycerol concentration is related to overwintering diapause in *C. fumiferana.*

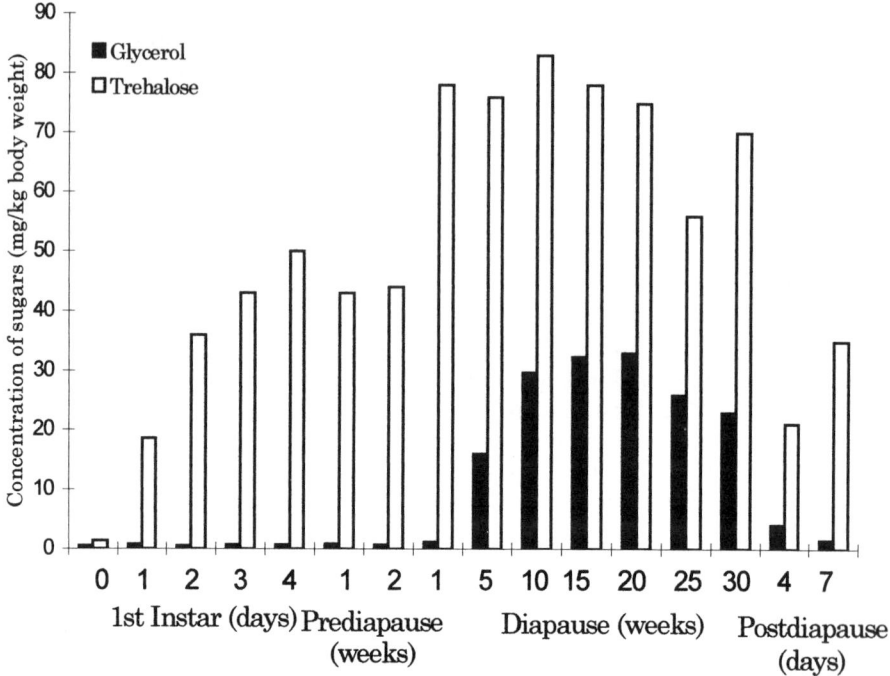

Figure 1. Glycerol and trehalose levels in first-instar larvae and second instars in prediapause, diapause and postdiapause stages. Larval samples were collected at various times after hatching. First-instar larvae were collected at 0, 1, 2, 3, and 4 days after hatching. The second-instar larvae were kept at 16°C for 2 weeks (prediapause), prior to their transfer to 2°C for 20-30 weeks (diapause). The second-instar larvae were sampled at 1 and 2 weeks in prediapause, 1, 5, 10, 15, 20, 25 and 30 weeks in diapause. At the end of the diapause, the larvae were transferred to 20°C, which results in the larvae crawling out of the hibernacula and starting to feed about four days after the transfer to the higher temperature (postdiapause). The second-instar larvae were sampled at 4 and 7 days after the transfer to 20°C. Sugars from the larvae were extracted by homogenisation in 80% ethanol followed by centrifugation. The alcohol was evaporated and the sugars were resuspended in water and injected into a HPLC carbopack PCX-30 anion exchange column. Sixty millimolar NaOH solvent at a flow rate of 0.8 ml/min was used. Glycerol and trehalose standards were used for determining the position of the peaks and for measuring the levels of glycerol and trehalose in the samples.

4. EXPRESSION OF mRNAs AND PROTEINS DURING OVERWINTERING DIAPAUSE

4.1 DIAPAUSE ASSOCIATED PROTEINS

C. *fumiferana* produces large quantities of two proteins, diapause associated protein 1 (DAP1) and diapause associated protein 2 (DAP2) (Palli et al. 1998). The levels of these proteins start to increase soon after hatching and reach maximum levels by four days into the first instar. The high levels of these proteins are maintained throughout and until three days after the termination of diapause. The mRNAs coding for these proteins started to appear 24 hr after first-instar larvae emerged from the eggs and large quantities of mRNAs for these proteins were detected throughout the first-instar larval stage and in the second instars until they entered diapause (Fig. 2). Low levels of these mRNAs were detected in second-instar larvae that were in the prediapause, diapause and postdiapause stages. Both the mRNAs and proteins increase again in the sixth-instar larva as it prepares to pupate. Thus, these two abundant proteins are produced prior to two non-feeding stages, diapause and pupation. Purification of these two proteins followed by cloning of their full-length cDNAs revealed that these proteins are closely related to each other. The deduced amino acid sequences of DAP1&2 showed high similarity with the sequences of storage proteins identified in the last larval stages of lepidopteran insects such as *Trichoplusia ni* (Jones et al. 1993). Whether or not these proteins provide a source of amino acids for the developing second-instar larvae at the end of diapause or protect the larvae from desiccation during diapause has not been resolved. See Danks (2000) for a discussion on dehydration in dormant insects.

4.2 DEFENSIN-LIKE PROTEINS

We have used the differential display of mRNAs technique to isolate transcripts that are expressed during C. *fumiferana* overwintering diapause. One of the cDNAs that we have cloned during this process showed a very high similarity to insect defensin peptides cloned from insects belonging to the orders Diptera, Coleoptera, Odonata and Homoptera but not Lepidoptera (Hoffmann, 1995). The six cysteine residues that are involved in the disulphide linkages are conserved and are present at the appropriate positions in this C. *fumiferana* defensin-like protein. All the other amino acids that are conserved in insect defensins identified so far are also present in this protein, therefore we named this C. *fumiferana* defensin (Cfdefensin, Ladd, T.R. and Palli, S.R. unpublished results).

Low levels of Cfdefensin mRNAs were detected in the second-instar larvae that were held at 16ºC for two weeks in preparation for diapause (prediapause, Fig. 2). The Cfdefensin mRNA levels started to increase at five weeks after the second-instar larvae were transferred to 2ºC for diapause maintenance. The

Cfdefensin mRNA reached maximum levels by 10 weeks at 2°C; these maximum levels were maintained throughout their incubation at 2°C. The Cfdefensin mRNA levels decreased to low levels within two days after the larvae were transferred to 20°C for diapause termination (Gojtan, K. and Palli, S.R. unpublished results).

We are in the process of investigating the role of Cfdefensin during *C. fumiferana* overwintering diapause. Defensins are used by members of the orders Diptera, Coleoptera, Odonata and Homoptera to fight microbial infections. Defensins have not been reported from lepidopteran insects. Lepidopteran species are thought to use cecropin and similar proteins to fight microbial infections. It is conceivable that *C. fumiferana* larvae use these small peptides to fight microbial infections. Alternatively, these peptides may serve as protectants against freezing or desiccation. See the Walker et al. chapter in this book for a detailed discussion on antifreeze proteins.

4.3 GLUTATHIONE S-TRANSFERASE

During our attempts to isolate diapause associated proteins we identified a 23-kDa protein that was present at higher levels in diapausing second instar larvae than in feeding larvae. This 23-kDa protein was purified, and polyclonal antibodies were raised against this protein. The antibodies were used to isolate the full-length cDNA from the library constructed using RNA from second-instar diapausing larvae. The deduced amino acid sequence of this cDNA showed high similarity to glutathione S-transferase amino acid sequence and therefore, the cDNA was named *C. fumiferana* glutathione S-transferase (CfGST). Affinity-purification on a glutathione-sepharose 4B column and GST enzyme activity assay confirmed the identity of CfGST (Feng et al. 1999a).

Developmental Northern and Western blot analysis showed high levels of CfGST mRNA and protein in first-instar larvae and diapausing second-instar larvae. Both mRNA and protein levels gradually decreased after the termination of diapause, reaching lowest levels in the fifth instar. Both mRNA and protein increase once again beginning at 1.5 days after ecdysis to the sixth- instar larvae, reach maximum levels by the middle of the sixth- instar, and start decreasing at the time of head capsule slippage stage. Levels are undetectable by the time the larva is ready to molt. (Feng et al. 2000) Exposure to low temperatures did not induce an increase in CfGST mRNA or protein. But allowing the larvae to feed on balsam fir foliage induced expression of both CfGST mRNA and protein. The bacterial insecticide, *Bacillus thuringiensis* delta-endotoxin (Bt), the non-steroidal ecdysone analog tebufenozide, and the synthetic pyrethroid permethrin, induced the expression of *Cf*GST mRNA and protein in fifth-instar larvae (Feng et al. 2000).

Expression pattern of CfGST is similar to CfDAP1 and CfDAP2 proteins and both of them are expressed as the larvae prepare to diapause and pupate. One major difference in the expression profiles of CfGST and CfDAP1 and CfDAP2 is that the CfDAP1 and CfDAP2 mRNAs start decreasing when the second-instar larvae enter diapause and are not present in significant amounts during diapause, whereas CfGST mRNA continues to be present throughout diapause and decreases only after the termination of diapause when the larvae start feeding. The reason for *C. fumiferana* larvae producing large amounts of GST prior to and during diapause is not clear. Its production prior to two non-feeding stages implies similar functions as CfDAP1&2. In addition, since GSTs are known to detoxify alleochemicals it is conceivable that CfGST may be used to detoxify waste products accumulated during overwintering diapause.

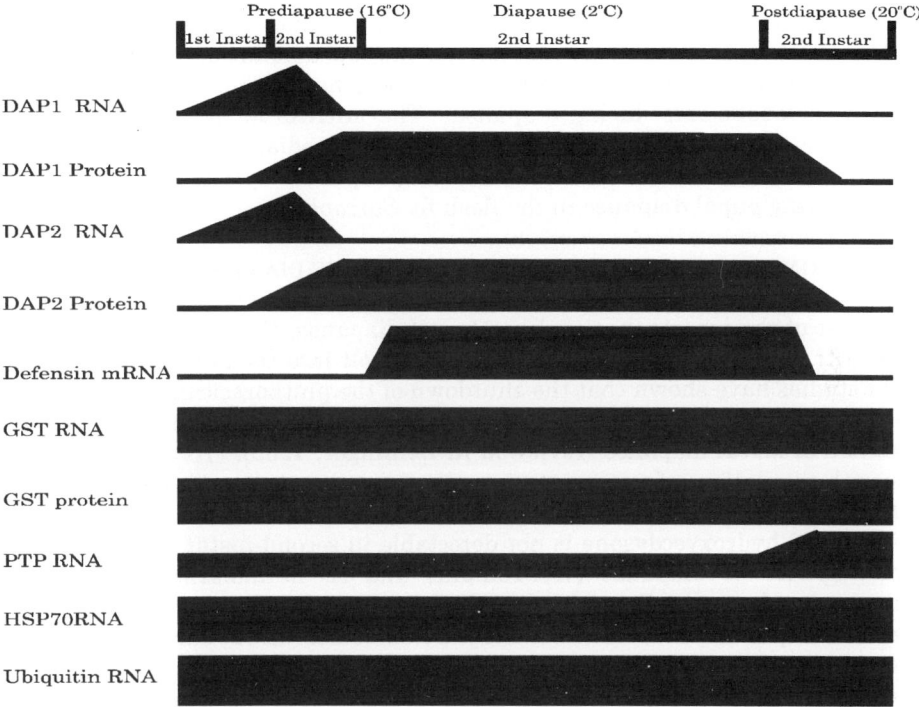

Figure 2. Schematic representation of expression profiles of some mRNAs and proteins during prediapause, diapause and postdiapause. After hatching, the first- instar larvae spend about five days spinning hibernacula in preparation for diapause. They moult into second instars inside the hibernacula. The second-instar larvae are kept at 16°C for 2 weeks (prediapause), prior to their transfer to 2°C for 20-30 weeks. At the end of diapause, the larvae are transferred to 20°C. The larvae crawl out of the hibernacula and start to feed in about four days after the transfer. The profiles are drawn based on Northern blots published by Palli et al. 1998; Feng et al. 1998; Feng et al. 2000; and unpublished results from our laboratory.

4.4 OTHER PROTEINS

We have also analysed the mRNA profiles of heat shock protein 70 (HSP70), polyubiquitin (ubiquitin) and mitochondrial phosphate transport protein (PTP) during the first instar and second instar in prediapause, diapause and postdiapause stages. High levels of PTP mRNA were detected in the first-instar larvae. The PTP mRNA levels decreased after the larvae moulted into the second-instar stage. These low levels were maintained through prediapause and postdiapause stages of the second instar. The PTP mRNA levels started to increase at the time of diapause termination and high levels of this RNA were detected throughout the feeding stages (Feng et al. 1998). Relatively low levels of PTP mRNA during diapause may reflect the low amount of phosphate being transferred to the mitochondria for use in ATP synthesis. This is in keeping with the slow metabolic rate during diapause. The mRNAs for both ubiquitin and HSP70 are present in high levels through prediapause, diapause and postdiapause (Fig. 2). Rinehart et al. (2000) reported upregulation of HSP70 mRNAs during pupal diapause in the flesh fly *Sarcophaga crassipalpis*.

5. ENDOCRINOLOGICAL CONTROL OF OVERWINTERING DIAPAUSE

In most of the insects that undergo larval diapause, the insects stop molting while their non-diapausing cohorts proceed to molt into the next stage. Several earlier studies have shown that the shutdown of the prothoracic glands, the main source of ecdysteroids, results in the failure to molt and is the major hormonal regulation of larval diapause (reviewed in Denlinger, 1985). To understand the hormonal regulation of overwintering diapause in *C. fumiferana*, we have analysed ecdysteroid levels in second-instar larvae that are in prediapause or diapause. 20-hydroxyecdysone is not detectable in second-instar larvae that are in prediapause or diapause (W. Tomkins and A. Retnakaran, unpublished results). As shown in Figure 3, mRNAs for both isoforms of *C. fumiferana* ecdysone receptor (CfEcRA and CfEcRB) are detectable in the larvae that are in diapause.   The mRNAs for *C. fumiferana* ultraspiracle (CfUSP), the heterodimeric partner for CfEcR, are also detectable in second-instar larvae that are in diapause. The mRNAs for ecdysone inducible transcription factors, *C. fumiferana* 75 (CHR75A, CHR75B) and *Choristoneura* hormone receptor 3 (CHR3), are not present in the second-instar larvae that are in diapause. They appear when second-instar larvae moult into the third instar. Therefore, it appears that all the machinery necessary for the ecdysone response is present in the second-instar larvae that are in diapause.   The developmental arrest is probably due to absence of ecdysone. Application of the stable ecdysteroid, RH-5992 to second-instar larvae that are in diapause leads to termination of diapause and progress in development as evidenced by appearance of CHR75 and CHR3 mRNAs (M. Primavera and A. Retnakaran, unpublished results). This observation further supports the hypothesis that lack of ecdysteroids is the main reason for the second instar larvae continuing to be in this stage during the overwintering diapause.

We recently cloned *C. fumiferana* juvenile hormone esterase (CfJHE) cDNA and studied the developmental expression of its mRNA throughout the life history *of C. fumiferana* (Feng et al. 1999b). A peak of JHE mRNA was detected on the day of moulting to every larval stage except into the second instar that are preparing to enter diapause. CfJHE mRNA was also not detected during diapause. It is possible that the main function of JHE on the day of molting is to reduce JH levels on subsequent days that allows synthesis and secretion of ecdysteroids for the next moult. In case of second-instar larvae that are preparing to enter diapause, JHE is not secreted resulting in higher JH levels, which in turn leads to lower ecdysteroids and developmental arrest or diapause. Further work is needed to verify this hypothesis.

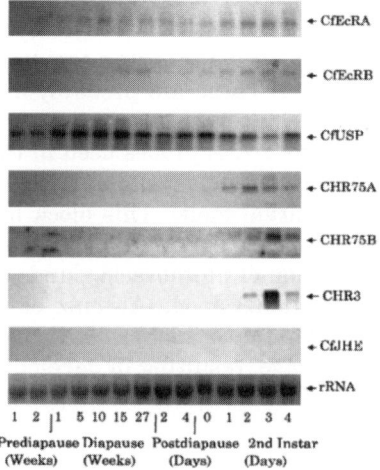

Figure 3. mRNA expression profiles of *Choristoneura fumiferana* ecdysone receptor A (CfEcRA, Perera et al. 1999), *C. fumiferana* ecdysone receptor B (CfEcRB, Kothapalli et al. 1995), *C. fumiferana* ultraspiracle (CfUSP, Perera et al. 1998), *C. fumiferana* hormone receptor 75A (CHR75A, Palli et al. 1997), *C. fumiferana* hormone receptor 75B (CHR75B, Perera, S.C and Palli, S.R. unpublished), *C. fumiferana* hormone receptor 3 (CHR3, Palli et al. 1996) and *C. fumiferana* juvenile hormone esterase (CfJHE, Feng et al. 1999b) mRNAs during second-instar larval stage. The second instar larvae were collected at 1 and 2 weeks after they were placed at 16°C (prediapause), at 1, 5, 10, 15, 25 weeks after they were moved to 2°C (diapause) and on 2 and 4 days after they were moved to 20°C for termination of diapause (postdiapause). The postdiapause second-instar larvae were collected on the day when they stated feeding and on 1, 2, 3, and 4 days after that. The total RNA was isolated and analysed on Northern blots using cDNA probes as described in publications cited above for each gene.

6. CONCLUSIONS AND FUTURE DIRECTIONS

We have described the simultaneous occurrence of the obligatory diapause and the overwintering process in the second-instar larvae of the spruce budworm, *C. fumiferana*. However, these two physiological events can be distinguished from each other on the basis of some of the biochemical changes that occur in these larvae during the obligatory overwintering diapause. Although all our studies are based on laboratory reared insects, and one needs to be cautious in interpreting these results, they nevertheless indicate trends that may hold true in field populations. Glycerol production and synthesis of defensin-like protein are clearly overwintering related events and these happen only after the larvae are exposed to cold treatment for certain lengths of time. If the larvae are not exposed to the cold treatment, they fail to synthesize glycerol or defensin-like protein. On the other hand, increases in trehalose levels and synthesis of CfDAP1 and CfDAP2, as well as increased synthesis of CfGST, occur as soon as the larvae hatch and start preparing for diapause. They also start spinning hibernacula at this time. These are probably diapause-related events. Experiments are in progress to identify the diapause or overwintering state of field populations using some of the cDNA probes used in this study.

Similar to other insects that undergo larval diapause, *C. fumiferana* seems to enter diapause by blocking the larval molt. This block in larval molt is mainly accomplished by preventing the synthesis of ecdysteroid necessary for the molt. The mechanism behind inactivation of prothoracic glands to decrease ecdysteroid synthesis and/or secretion is not understood. Absence of a JHE peak on the day of the second larval molt indicates that a higher JH titre may be responsible for lower ecdysteroid levels at this stage, resulting in developmental arrest. Further studies are required to elucidate the hormonal control of this obligatory overwintering diapause.

During the last few years we were able to identify and isolate cDNAs for some of the major players involved in the obligatory overwintering diapause in *C. fumiferana*. The next major challenge is to find out more about the functions of some of these proteins as well as to understand the hormonal control of diapause in this insect. Extending our studies from the laboratory reared insects to field populations is another area for investigation in the near future.

7. ACKNOWLEDGEMENTS

The supply of *C. fumiferana* larvae by the staff of the insect rearing facility at the Great Lakes Forestry Centre is gratefully acknowledged. The authors thank Mrs. Karen Jamieson for her editorial assistance. This research was supported in part by the National Biotechnology Strategy Fund and the Science and Technology Opportunities Fund of the Canadian Forest Service to the Biotechnology group at the Great Lakes Forestry Centre.

REFERENCES

Beck, S.D. 1968. Insect Photoperiodism. Academic Press. New York. 288 pp.

Danks, H.V. 2000. Dehydration in dormant insects. Journal of Insect Physiology 46, 837-852.

Denlinger, D.L. 1985. Hormonal control of diapause. In: Kerkut, G.A. and Gilbert, L.I. (Eds) Comprehensive Insect Physiology, Biochemistry and Pharmacology, Vol. 8. Pergamon Press, Oxford, pp. 354-412.

Feng, Q.-L., Davey, K.G., Pang, A.S.D., Primavera, M., Ladd, T.R., Zheng, S.-C., Sohi, S.S., Retnakaran, A., Palli, S.R. 1999a. Glutathione S-transferase from the spruce budworm, *Choristoneura fumiferana:* identification, characterization, localization, cDNA cloning, and expression. Insect Biochemistry and Molecular Biology 29, 779-793.

Feng, Q. L, Davey, K. G., Pang, A. S. D., Ladd, T. R., Zheng, S.-C., Retnakaran, A., and Palli, S. R. 2000. Glutathione S-transferase from the spruce budworm, *Choristoneura fumiferana*: Developmental expression and induction by various stresses. Journal of Insect Physiology (in press).

Feng, Q., Ladd, T.R., Tomkins, W.L., Sundaram, M., Sohi, S.S., Retnakaran, A., Davey, K.G., Palli, S.R. 1999b. Spruce budworm (*Choristoneura fumiferana)* juvenile hormone esterase: hormonal regulation, developmental expression and cDNA cloning. Molecular Cellular Endocrinology 148, 95-108.

Feng, Q., Palli, S.R., Ladd T.R., Retnakaran A., Davey, K.G. 1998. Identification and developmental expression of the mitochondrial phosphate transport protein gene from the spruce budworm, *Choristoneura fumiferana*. Insect Biochemistry and Molecular Biology 28, 791-799.

Grisdale, D. 1970. An improved laboratory method for rearing large numbers of spruce budworm, *Choristoneura fumiferana* (Lepidoptera:Tortricidae). The Canadian Entomologist 102, 1111-1117.

Han, E. N. Bauce, E. 1993. Physiological changes and cold hardiness of spruce budworm larvae, *Choristoneura fumiferana* (Clem.), during pre-diapause and diapause development under laboratory conditions. The Canadian Entomologist 125, 1043-1053.

Harvey, G.T. 1957. The occurrence and nature of diapause-free development in the spruce budworm, *Choristoneura fumiferana* (Clem.). Canadian Journal of Zoology 35, 549-572.

Hoffmann, J.A. 1995. Innate immunity in insects. Current Opinions in Immunology 7, 4-10.

Jones, G., Manczak, M., Horn, M. 1993. Hormonal regulation and properties of a new group of basic hemolymph proteins expressed during insect metamorphosis. The Journal of Biological Chemistry 268, 1284-1291.

Kothapalli, R., Palli, S. R., Ladd, T. R., Sohi, S. S., Cress, D., Dhadialla, T. S., Tzertzinis, G., Retnakaran, A. 1995. Cloning and developmental expression of the ecdysone receptor gene from the spruce budworm, *Choristoneura fumiferana*. Developmental Genetics 17, 319-330.

Palli, S.R., Ladd, T.R., Ricci, A.R., Primavera, M., Mungrue, I.N., Pang, A.S.D., Retnakaran, A., 1998. Synthesis of the same two proteins prior to larval diapause and pupation in the spruce budworm, *Choristoneura fumiferana*. Journal of Insect Physiology 44, 509-524.

Palli, S. R., Ladd, T. R., Ricci, A. R., Sohi, S. S., Retnakaran, A. 1997 Cloning and developmental expression of *Choristoneura* Hormone Receptor 75: A homologue of the *Drosophila* E75A gene. Developmental Genetics 20, 36-46.

Palli, S.R., Ladd, T.R., Sohi, S.S., Cook, B.J., Retnakaran, A. 1996. Cloning and developmental expression of *Choristoneura* hormone receptor 3, an ecdysone-inducible gene and a member of the steroid hormone receptor superfamily. Insect Biochemistry and Molecular Biology 26, 485-499.

Perera, S.C., Ladd, T.R., Dhadialla, T.S., Krell, P.J., Sohi, S.S., Retnakaran, A., Palli, S. R. 1999. Studies on two ecdysone receptor isoforms of the spruce budworm, *Choristoneura fumiferana*. Molecular Cellular Endocrinology 152, 73-84.

Perera, S.C., Palli, S. R., Ladd, T. R., Krell P. J., Retnakaran, A. 1998. The ultraspiracle of the spruce budworm, *Choristoneura fumiferana*: Cloning cDNA and developmental expression of mRNA. Developmental Genetics 22, 169-179.

Prebble, M.L. 1975. Spruce budworm, *Choristoneura fumiferana* Clem. Introduction. pp. 76 - 84. *In* Prebble, M.L. (Ed.), Aerial control of forest insects in Canada: a review of control projects employing chemical and biological insecticides. Department of the Environment, Government of Canada, Ottawa, Canada.

Retnakaran, A., Tomkins, W.L., Primavera M.J., Palli, S.R. 1999. Feeding behavior of the first instar *Choristoneura fumiferana* and *C. pinus pinus* (Lepidoptera: Tortricidae). The Canadian Entomologist 131, 79-84.

Rinehart, J.P., Yocum, G.P., Denlinger, D.L. 2000. Developmental upregulation of inducible hsp70 transcripts, but not the cognate from, during pupal diapause in the flesh fly, *Sarcophaga crassipalpis*. Insect Biochemistry and Molecular Biology 30, 515-521.

Storey, K.B., Storey, J.M. 1991. Biochemistry of cryoprotectants. In: Lee, R.E., Denlinger, D.L. (Eds) Insects at Low Temperature. Chapman and Hall, New York, NY, pp. 64-93.

Insect Timing: Circadian Rhythmicity to Seasonality
D.L. Denlinger, J. Giebultowicz and D.S. Saunders (Editors)
© 2001 Elsevier Science B.V. All rights reserved.

145

Insights for future studies on embryonic diapause promoted by molecular analyses of diapause hormone and its action in *Bombyx mori*

O. Yamashita, K. Shiomi*, Y. Ishida, N. Katagiri and T. Niimi

Graduate School of Bioagricultural Sciences, Nagoya University, Chikusa, Nagoya, Japan 464-8601

A brief historical review on the studies of embryonic diapause of the silkworm, *Bombyx mori*, is presented to recognize the current topics and to identify important future projects. Recent molecular and cellular studies provide basic information on the organization, expression and regulation of the diapause hormone gene, and on the induction of trehalase gene expression in the target organ by the hormone. In addition, we provide new information on regulatory mechanisms of gene expression and of peptide secretion from the neuroendocrine organ, and discuss a new diapause-associated metabolism. Finally, we propose some new directions for future studies of insect diapause.

1. INTRODUCTION

Diapause occurs as an alternative developmental program in the life cycle and is accomplished by the dynamic change of developmental, behavioral and physiological events. Expression of the diapause program is triggered by a particular set of environmental signals such as photoperiod, temperature and food quality, which are transduced into endogenous chemical messengers, hormones, in the neuroendocrine organs. Finally, hormones bring about the phase change from development to diapause or *vice versa* through structural and functional changes in the target organ (cf. Yamashita, 1996).

The silkworm, *Bombyx mori*, is a typical insect entering diapause at an early embryonic stage before dermal differentiation is completed. In addition to scientific interest, due to the technological innovation in sericultural industry, much knowledge has been accumulated on diapause mechanisms and applied to long-term egg preservation and controlled hatching of larvae.

The research history of silkworm diapause can be divided into three periods according to the topics and methodologies used. The first period (1920's-1940's) which covers genetic and ecophysiological research, discovered that the diapause nature (voltinism) is primarily determined by genetic capacity, and the expression of diapause in the bivoltine strain is under the control of environmental signals, temperature (Watanabe, 1922) and photoperiod (Kogure, 1933), experienced during the embryonic stage of the maternal generation (Yamashita and Hasegawa, 1985).

---

*Present address, Faculty of Textile Science, Shinsyu University, Ueda, Nagano, Japan 386-8567

The second phase, beginning in the 1950's, introduced endocrinological and biochemical studies on diapause; the most outstanding topic was the finding of the suboesophageal ganglion (SG)-diapause hormone (DH) system (Fukuda, 1952; Hasegawa, 1952), which led to our chemical study of DH. Since the first extraction of DH by Hasegawa (1957), we devoted much effort to isolate and determine the chemical structure of DH by using more than twenty million silkworm heads without success for more than 30 years until 1991. In 1991, DH was successfully isolated from an aqueous extract of 55,000 SGs and was determined to be a 24 amino acid peptide amide (Imai et al., 1991). Another topic was the discovery of sorbitol and glycerol in diapausing eggs, which promoted the biochemical characterization of diapause (Chino, 1958). These findings led us to investigate the mode of biochemical action of DH using its crude extract. The results provided evidence that DH acts on developing ovaries to enhance trehalase activity, thereby promoting hyperglycogenism in the eggs, a prerequisite metabolic event required for the establishment of diapause physiology (Yamashita and Hasegawa, 1985; Yamashita, 1996).

Now, since 1991, we are in the third period of diapause study in the silkworm. Based upon the chemical information on DH, much progress has been achieved on molecular analyses of the DH gene from the viewpoints of structure, expression and regulatory mechanisms (Sato et al., 1993; Xu et al., 1995a, 1995b). One of the most interesting facts is that the gene encodes a polyprotein precursor which yields DH, pheromone biosynthesis activating neuropeptide (PBAN) and three other neuropeptides, all of which shared the common C-terminal pentapeptide sequence, FXPRLamide. We thus named it the DH-PBAN gene (Xu et al., 1995a). The DH-PBAN gene is widely distributed in several lepidopteran insects, even in species which never enter embryonic diapause (Ma et al., 1994; Choi et al., 1998; Jacquin-Joly et al., 1998). Progress has also been made on the molecular research of DH action. DH has been demonstrated to induce directly trehalase gene expression in developing ovaries to bring about hyperglycogenism in mature eggs (Su et al., 1994). These molecular studies on the DH-PBAN gene and DH action have provided fundamental information that explains a series of biochemical and physiological events characteristic of diapause.

Through these molecular studies, we recognize several new biochemical and physiological questions, including the reverse molecular biology of diapause. This approach has just started, thus the fragmentary results obtained so far are not conclusive, but suggest a new angle for understanding diapause.

This article outlines briefly current topics on silkworm diapause, especially on the regulation of DH-PBAN gene expression analyzed by a transgenic *Drosophila* system, the regulation of DH peptide secretion, and diapause metabolism in eggs that are deficient in glycogen and sorbitol.

2. REGULATION OF DH-PBAN GENE EXPRESSION DETERMINED BY TRANSGENIC *DROSOPHILA*

Cloning and structural identification of the cDNA and the gene encoding DH and other FXPRLamide peptides permitted the undertaking of a series of molecular biological experiments on gene expression. *In situ* hybridization localized the gene expression to 12 neurosecretory cells in the SG. RT-PCR and northern blots of mRNA prepared throughout embryonic and post-embryonic development have demonstrated that the expression of the gene is under the control of the developmental program and environmental stimuli (Sato et al.,1994; Xu et al., 1995b). Thus, DH-PBAN gene expression is a prerequisite event for manifesting the physiological function of each peptide at a specific stage.

To analyze the regulatory mechanism of DH-PBAN gene expression, we first attempted to identify the *cis*-regulatory region by producing transgenic *Drosophila*. The transgenic animal provides a bioassay system for detecting any elements and factors including the *trans*-acting factors and the environmental signals for regulation of gene expression. First, we constructed the P-element vector which consisted of the 7 kb upstream region of the DH-PBAN gene-*hsp27* minimal promoter-*lacZ* reporter gene, and introduced it into *Drosophila* (*yw* strain) by using the P-element-mediated transformation method (Ishida et al., 1999). In the preliminary experiment, transformed flies were successfully generated, in which the reporter gene was expressed in a limited number of cells in the peripheral and central nervous systems (Ishida et al., 1999). Based on these results, we established 70 *Drosophila* lines that were transformed with 12 different constructs carrying different upstream regions of the DH-PBAN gene fused with the *lacZ* reporter gene. The reporter gene was constantly expressed in the six cells of the abdominal neuromere when transformed with the construct covering a 6 kb upstream region. A progressive deletion analysis finally indicated that a 650 bp fragment from –4,883 to –4,233 was sufficient and essential to direct DH-PBAN gene expression in these cells (Fig. 1). These six positive cells were characterized as being the peptidergic neurons producing the FXPRLamide peptide (Ishida et al., 2000). Thus, this region appears to include the *cis*-regulatory element for DH-PBAN gene expression, even in *Bombyx* neurosecretory cells.

3. REGULATION OF DH SECRETION FROM THE SG

As early as 1952, pioneer experiments on the endocrinological function of the SG suggested the possibility that the brain was involved in regulation of SG activity for induction of diapause eggs (Fukuda, 1952). On the other hand, the implantation of the corpus cardiacum(CC)-corpus allatum(CA) complex into day 4 non-diapause type pupae (non-diapause egg producers) induced diapause eggs (Takeda and Ogura, 1976), as did the

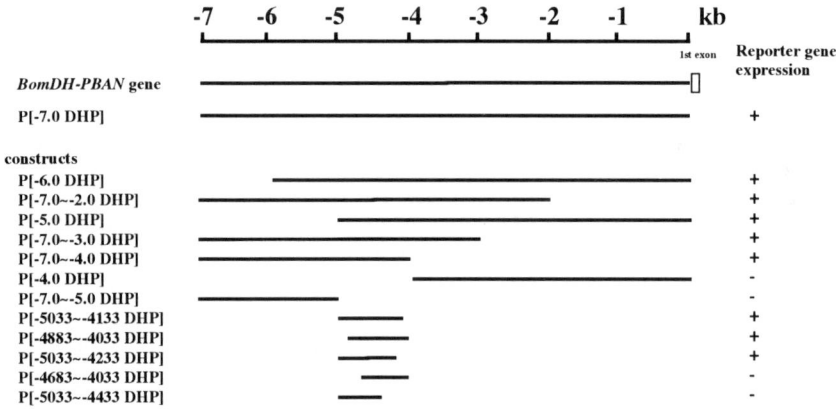

Figure 1. Deletion analysis of *cis*-regulatory region of 5'-flanking sequence of the DH-PBAN gene by the transformed *Drosophila*. Right column indicates *lacZ* reporter gene expression in the six neurosecretory cells of the abdominal neuromere. +, positive staining due to gene experssion; -, negative at a background level.

injection of DH, indicating that DH is temporarily stored in and released into hemolymph from these organs. By contrast, the SG needed to be implanted at earlier stages to become active in diapause induction (Yamashita and Hasegawa, 1985), indicating that the SG can directly release DH after becoming established in the host. These biological experiments suggested the sites of synthesis and release of DH, and the possible regulation mechanism without any direct evidence.

Our recent molecular studies have clearly demonstrated that the DH-PBAN gene is expressed in only 12 neurosecretory cells of the SG at a specific developmental stage (Sato et al., 1994; Xu et al., 1995b). However, each neuropeptide encoded by the gene is required to be released at different times of development for expression of its unique physiological function. DH is secreted at a middle stage (Yamashita and Hasegawa, 1985), and PBAN just before adult eclosion (Ichikawa, 1996). Such a stage-dependent secretion of each neuropeptide is likely to be achieved by different control mechanisms directed toward each neuropeptide. Now, we have initiated an investigation of the regulation mechanism of DH secretion, by which the DH titer in the hemolymph is decided.

To define the secretion pathway with the aid of immunochemical methodology, we first prepared two antibodies: one specifically recognized the C-terminal pentapeptide sequence of DH (aDH#1) and the other was a monoclonal antibody recognizing the N-terminal decapeptide of DH. The immunocytochemical study using aDH#1 has shown that FXPRLamide peptides are mainly localized in 12 somata of the SG projecting neuritis to the retrocerebral complex, including the CC-CA complex. The immunoreactivity decreased markedly in a pair of SLb somata of the SG in the diapause type as adult development progressed, whereas no decrease in immunoreactivity occurred in the non-diapause type (Sato et al., 1998). These results have demonstrated that DH is transported from the SLb somata of the SG to the CC-CA complex through the circumoesophageal connective and nervus corporis cardiaci-3 (Ichikawa et al., 1995), and the release of DH is stimulated in the diapause type of SG by an unknown mechanism.

Immunotitration was successively performed using the sandwich ELISA method, by which method DH could be detected at levels as low as 1 fmole per well. DH content was estimated to be ~200 fmol/SG on the day of pupation, and was at the same level in the SG from diapause and non-diapause types of pupae. In the diapause type of SG, DH content decreased precipitously at the middle stage when the diapause nature was determined, and reached trace levels at the last stage of pupal-adult development. In contrast, in the non-diapause type of SG, an initial level of DH was maintained until the end of the middle stage and declined toward adult eclosion. Like DH, total FXPRL amide peptides followed a similar developmental fate, an early decrease in the diapause type of SG and a delayed decrease in the non-diapause type of SG.

The difference between synthetic activity and releasing activity of DH in the SG should be reflected in the amount of DH reserved in the SG. The synthetic activity of DH was not directly measured in the present work, but the relative activity could be estimated from changes in DH-PBAN mRNA content measured by RT-PCR. The results indicate that the active synthesis of DH appeared during early and late stages, and the synthetic activity at the early stage was 2 to 3 times higher in the diapause type of SG than in the non-diapause type (Xu et al., 1995b). These facts suggest that the diapause type of SG is active in both synthesis and release of DH compared to the non-diapause type of SG. An old experiment had shown that the biological activity of SG extract was 3 to 5-fold higher in the diapause type of SG than in the non-diapause type of SG (Yamashita and Hasegawa, 1985). Consequently, it is concluded that there are two steps in the regulation of DH secretion from SG: at the level of gene expression and hormone release.

4. NEW INSIGHTS ON DIAPAUSE METABOLISM

Nearly a half-century has passed since Chino (1958) and Wyatt and Meyer (1959) discovered glycerol and sorbitol in diapausing insects. Until now, the high accumulation of polyols and sugars was generally accepted as the biochemical syndrome common to diapause physiology and was considered to be a biological adaptation against unfavorable conditions. Quite recently, Horie et al. (2000) have clearly demonstrated in the silkworm that sorbitol completely suppresses embryogenesis in an *in vitro* culture and proposed a unique physiological function for sorbitol as a growth regulator (an arrester).

We have conducted biochemical research on carbohydrate metabolism related to diapause of the silkworm since the 1960's and have proposed metabolic shifts that correlated to induction, onset and termination of diapause, which were evoked by the induction or enhancement of key enzymes. In this process, DH plays a key regulatory role to prepare the biochemical condition leading to diapause in eggs (Yamashita, 1996; Yamashita and Hasegawa, 1985). DH acts to enhance trehalase activity in developing ovaries, by which action a hyperglycogenism is brought about in mature eggs. The glycogen is carried over to laid eggs and becomes the sole resource for sorbitol production, thus the amounts of sorbitol in diapausing eggs are closely dependent on the glycogen content in mature eggs. Indeed, in the non-diapause type of eggs, no accumulation of sorbitol occurred due to less accumulation of glycogen reserves in mature eggs. If trehalase were intimately involved in the metabolic pathway executing the diapause program, the experimental inhibition of trehalase activity in ovaries would be excepted to convert diapause eggs to non-diapause eggs.

Recently, we have conducted a pharmacological experiment using a newly found trehalase inhibitor, trehazolin, to confirm our "DH-trehalase-diapause theory". Injection of trehazolin into young pupae did not affect vitellogenesis and choriogenesis but dramatically reduced glycogen accumulation to a trace level in mature eggs. Surprisingly, injection of trehazolin never affected DH action and permitted diapause egg production in despite of a considerably reduced glycogen content in the eggs (Katagiri et al., 1998). Thus, our finding provides a new method for production of eggs containing less amounts of glycogen and offers a novel system for analyzing the diapause-associated metabolism.

Trehazolin injection into young pupae interfered with oviposition and larval hatching behavior but had no serious effect on embryogenesis. We collected diapause eggs laid by moths treated with trehazolin at the pupal stage and analyzed glycogen, sorbitol, glycerol and trehalose at different stages after oviposition. The eggs laid by moths injected with trehazolin were characterized by trace levels of glycogen and sorbitol and a high level of trehalose, levels that were completely different from the biochemical attributes of control diapause eggs. Reduced glycogen storage should directly correlate with reduced sorbitol synthesis, so that eggs could not use sorbitol as the diapause-associated agent. The interesting observation was a high accumulation of trehalose in these eggs. At present, it remains unknown how trehalose is metabolically produced, which regulation occurs and what function is expressed by trehalose. Recently, many reports have indicated the unique biochemical functions of trehalose on stabilization and preservation of proteins, biomembranes, cells and organisms under extreme environments (Eroglu et al., 2000). If such a trehalose function is operating in the silkworm egg, the sorbitol-deficient eggs apparently have developed another metabolic pathway as a biochemical adaptation for diapause. The diapause of trehalose accumulated eggs can be terminated in the same way as natural diapause eggs by chilling at 5 °C as long as 70 days and then transferring them to 25°C. Consequently, we can conclude that the silkworm has evolved at least two programs for biochemical establishment of diapause.

150

Using differential display we have searched for new genes whose expression is induced or stimulated by DH (Fig. 2). In addition to the trehalase gene, three cDNA fragments were isolated as candidates responsible for DH action and were sequenced to characterize the genes; one gene (ID1) is a homolog to cytochrome P450, and the other two genes (ID3 and UP3) are not identical in sequence to each other, but are similar to retroviral Env protein. These three genes were induced as early as 30 min after DH injection and attained a maximum level at 1 h. The time course of gene expression was different from that of trehalase gene expression, thus that expression of these new genes is under a different control. Although the physiological functions of these genes are not known, it is true that DH manifests multiple functions through controlling different gene expression.

5. CONCLUDING REMARKS

The recent advance in molecular studies of silkworm diapause enables us to present a scenario for diapause using molecular terms. In this scenario, the key topics are the organization of the DH-PBAN gene, expression of the gene and its regulation, peptidogenesis and secretion from the neuroendocrine organ, the dynamics in circulation, the reception and expression of the hormonal signal in the target organ, and the regulation of target reactions leading to the metabolic change. To refine this draft of the diapause story in the silkworm, more detailed investigations are, of course, required at the molecular and cellular levels. Very little is known about the regulation mechanisms of gene expression by environmental temperature, the regulation mechanism of DH release and titer changes in circulation, and the receptor and the intracellular signal transduction in the target organs. These areas remain critical projects for the future. These questions will require considerable effort to be resolved, but several preliminary experiments in these areas have laid the foundation to address these problems.

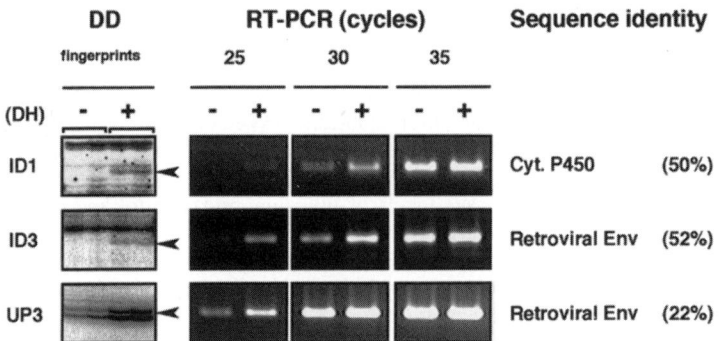

Figure 2. Identification and RT-PCR analysis of new genes of which expression are stimulated by DH. Differential display (DD) and RT-PCR analysis were performed using RNAs prepared from developing ovaries of day 5.5 pharate adults with (+) or without (-) DH injection. Deduced amino acid sequence of ID1 showed similarity to cytochrome P450 protein of *D. melanogaster* (Cyp305a1) (50% identical), and those of ID3 and UP3 showed similarities to Env protein of *T. ni* retrotransposon TED (52% and 22% identical).

To further advance diapause studies we need to propose new projects from different angles. The ideas may be derived from the identification of new genes of unknown function, the generation of transformed insects with new genes, the chemical modification of genes and peptides for new functions, and so on. Another approach will be to reexamine the central dogma that has been held as general knowledge for a long time. For example, how DH, PBAN and myotropins crossreact to elicit diapause induction, pheromone production and myostimulation of muscles can not be understood from the theory of one hormone for one function. Likewise, how silkworm eggs can undergo diapause without accumulation of glycogen-sorbitol is not apparent. By considering the present situation and future possibilities we should initiate new challenges for the further development of diapause studies, with the hope of opening a new page of insect science.

ACKNOWLEDGEMENTS

Work from the authors' laboratories was supported by a Grant-In–Aid (Nos. 03404008, 05556008 and 05760047) from the Ministry of Education, Science, Culture and Sports of Japan, and a Grant-In-Aid (BMP) from the Ministry of Agriculture, Forestry and Fisheries of Japan. The authors thank Professor David L. Denlinger of Ohio State University, Ohio, USA for critical reading of the manuscript.

REFERENCES

Chino H. 1958. Carbohydrate metabolism in the diapause egg of the silkworm, *Bombyx mori*. II. Conversion of glycogen into sorbitol and glycerol during diapause. Journal of Insect Physiology 2, 1-12.

Choi M. Y., Tanaka M., Kataoka H., Boo K. S. and Tatsuki S. 1998. Isolation and identification of the cDNA encoding the pheromone biosynthesis activating neuropeptide and additional neuropeptide in the oriental tabacco budworm, *Helicoverpa assulta* (Lepidoptera: Noctuidae). Insect Biochemistry and Molecular Biology 28, 759-766.

Eroglu A., Russo M. J., Bieganski R., Fowler A., Cheley S., Bayley H. and Toner M. 2000. Intracellular trehalose improves the survival of cryopreserved mammalian cells. Nature Biotechnogy 18, 163-167.

Fukuda S. 1952. Function of the pupal brain and suboesophageal ganglion in the production of non-diapause and diapause eggs in the silkworm. Annotationes Zoologicae Japoneneses 25, 149-155.

Hasegawa K. 1952. Studies on the voltinism in the silkworm, *Bombyx mori* L. with special reference to the organs concerning determination of voltinism. Journal of the Faculty of Agriculture, Tottori University 1, 83-124.

Hasegawa K. 1957. The diapause hormone of the silkworm, *Bombyx mori*. Nature 179, 1300-1301.

Horie Y., Kanda T. and Mochida Y. 2000. Sorbitol as an arrester of embryonic development in diapausing eggs of the silkworm, *Bombyx mori*. Journal of Insect Physiology 46, 1009-1016.

Ichikawa T., Hasegawa K., Shimizu I., Katsuno K., Kataoka H. and Suzuki A. 1995. Structure of neurosecretory cells with immunoreactive diapause hormone and pheromone biosynthesis activating neuropeptide in the silkworm *Bombyx mori*. Zoological Science 12, 703-712.

Ichikawa T., Shiota T., Shimizu I. and Kataoka H. 1996. Functional differentiation of neurosecretory cells with immunoreactive diapause hormone and pheromone biosynthesis activating neuropeptide of the moth, *Bombyx mori*. Zoological Science 13,

21-25.

Imai K., Konno T., Nakazawa Y., Komiya T., Isobe M., Koga K., Goto T., Yaginuma T., Sakakibara K., Hasegawa K. and Yamashita O. 1991. Isolation and structure of diapause hormone of the silkworm, *Bombyx mori*. Proceedings of Japan Academy Series B67, 98-101.

Ishida Y., Niimi T. and Yamashita O. 1999. The stage-and cell-specific expression of the *Bombyx mori* diapause hormone-pheromone biosynthesis activating neuropeptide (*BomDH-PBAN*) gene in the transformed *Drosophila*. Journal of Sericultural Science of Japan 68, 417-427.

Ishida Y., Niimi T. and Yamashita O. 2000. The *cis*-regulatory region responsible for the *BomDH-PBAN* gene expression in FXPRLamide peptide producing neurosecretory cells of the transformed *Drosophila*. Journal of Sericultural Science of Japan 69, 111-119.

Jacquin-Joly E., Burnet M., Francois M.-C., Ammer D., Meillour P. N.-L. and Descoins C. 1998. cDNA cloning and sequence determination of the pheromone biosynthesis activating neuropeptide of *Mamestra bassicae*: a new member of the PBAN family. Insect Biochemistry and Molecular Biology 28, 251-258.

Katagiri N., Ando O. and Yamashita O. 1998. Reduction of glycogen in eggs of the silkworm, *Bombyx mori*, by use of a trehalase inhibitor, trehazolin, and diapause induction in glycogen-reduced eggs. Journal of Insect Physiology 44, 1205-1212.

Kogure M. 1933. The influence of light and temperature on certain characters of the silkworm, *Bombyx mori*. Journal of the Department of Agriculture, Kyushu Imperial University 4, 1-93.

Ma P. W. K., Knipple D. C. and Roelofs W. L. 1994. Structural organization of the *Helicoverpa zea* gene encoding the precursor protein for pheromone biosynthesis activating neuropeptide and other neuropeptides. Proceeding of the National Academy of Science of the United States of America 91, 6506-6510.

Sato Y., Ikeda M. and Yamashita O. 1994. Neurosecretory cells expressing the gene for common precurosor for diapause hormone and pheromone biosynthesis activating neuropeptide in the suboesophageal ganglion of the silkworm, *Bombyx mori*. General and Comparative Endocrinology 96, 27-36.

Sato Y., Oguchi M., Menjo N., Imai K., Saito H., Ikeda M., Isobe M. and Yamashita O. 1993. Precursor polyprotein for multiple neuropeptides secreted from the suboesophageal ganglion of the silkworm, *Bombyx mori*: Characterization of the cDNA encoding diapause hormone precursor and identification of additional peptides. Proceeding of the National Academy of Science of the United States of America 90, 3251-3255.

Sato Y., Shiomi K., Saito H., Imai K. and Yamashita O. 1998. Phe-X-Pro-Arg-Leu-NH$_2$ peptide producing cells in the central nervous system of the silkworm, *Bombyx mori*. Journal of Insect Physiology 44, 333-342.

Su Z.-H., Ikeda M., Sato Y., Saito H., Imai K., Isobe M. and Yamashita O. 1994. Molecular characterization of ovary trehalase of the silkworm, *Bombyx mori* and its transcriptional activation by diapause hormone. Biochimica et Biophysica Acta 1218, 366-374.

Takeda S. and Ogura N. 1976. Induction of egg diapause by implantation of corpora cardiaca and corpora allata in *Bombyx mori*. Journal of Insect Physiology 22, 941-944.

Watanabe K. 1924. Studies on the voltinism of the silkworm, *Bombyx mori*. Bulletin of Sericultural Experiment Station, Tokyo 6, 411-455.

Wyatt G. R. and Meyer W. L. 1959. The chemistry of insect hemolymph. III. Glycerol. Journal of General Physiology 42, 1005-1011.

Xu W.-H., Sato Y., Ikeda M. and Yamashita O. 1995a. Molecular characterization of the gene encoding the precursor protein of diapause hormone and pheromone biosynthesis activating neuropeptide (DH-PBAN) of the silkworm, *Bombyx mori* and its distribution in some insects. Biochimica et Biophysica Acta 1261, 83-89.

Xu W.-H., Sato Y., Ikeda M. and Yamashita O. 1995b. Stage-dependent and temperature-controlled expression of the gene encoding the Precursor protein of diapause hormone and pheromone biosynthesis activating neuropeptide in the silkworm, *Bombyx mori*. Journal of Biological Chemistry 270, 3804-3808.

Yamashita O. 1996. Diapause hormone of the silkworm, *Bombyx mori*: Structure, gene expression and function. Journal of Insect Physiology 42, 669-679.

Yamashita O. and Hasegawa K. 1985. Embryonic diapause. In *Comprehensive Insect Physiology, Biochemistry and pharmacology* (Eds Kerkut G. A. and Gilbert L. I.) Vol. 1, pp. 407-434. Pergamon Press, Oxford.

Insect Timing: Circadian Rhythmicity to Seasonality
D.L. Denlinger, J. Giebultowicz and D.S. Saunders (Editors)
© 2001 Elsevier Science B.V. All rights reserved.

Stress proteins: a role in insect diapause?

David L. Denlinger[a], Joseph P. Rinehart[a] and George D. Yocum[b]

[a]Department of Entomology, Ohio State University, 1735 Neil Avenue, Columbus, OH 43210-1220, USA

[b]Red River Valley Agricultural Research Center, USDA-ARS, BRL, 1605 Albrecht Boulevard, Fargo, ND 58105-5674, USA

Genes encoding certain stress proteins (Hsp23 and 70) are highly upregulated during diapause, while others are either unaffected (Hsc70) or are downregulated (Hsp90). This disynchrony of expression is in marked contrast to the uniform upregulation of the stress protein genes in response to other stresses such as heat shock or cold shock. The diapause upregulation of the genes may be linked to a cryoprotective function of the proteins or a possible role in shutting down the cell cycle. The involvement of stress proteins in the dormancies of other animals and plants suggests a conserved mechanism contributing to the arrest of development.

1. INTRODUCTION

Heat shock proteins are best known as a highly conserved group of stress proteins that are quickly upregulated in response to environmental stress. Their synthesis was first noted in response to heat shock, hence the name "heat shock proteins", but "stress proteins" is perhaps a more appropriate term because we now know that their synthesis also can be elicited by a wide range of additional stresses including cold shock, desiccation, anoxia, ethanol, heavy metals, and other chemicals that cause denaturation of proteins. In response to stress, synthesis of the normal suite of proteins is suppressed, and instead the organism may switch almost exclusively to the synthesis of stress proteins.

The response is apparently ancient. Stress proteins with similar sequences are present in organisms as diverse as bacteria, yeast, plants and humans. The proteins are grouped into several families based on molecular mass. Three size categories are well documented in insects. The smallest of these are the small heat shock proteins, a cluster of proteins with molecular masses commonly in the range of 20-30kDa. The Hsp70 group, proteins with a molecular mass of approximately 70kDA, are the best known of the stress proteins and it is this group that usually shows the most dramatic response to stress. A group of higher molecular mass, the 90 kDa family, is the third common class of stress proteins in insects. The huge literature on the stress proteins and their functions is summarized in several recent reviews (Heikkila 1993, Parsell and Lindquist 1993, Feder and Hoffmann 1999).

Our interest in the association of stress proteins with diapause was prompted by the discovery that one of the diapause-upregulated clones we isolated from brains of the flesh fly, *Sarcophaga crassipalpis*, was a small heat shock protein (Flannagan *et al*, 1998). When the fly pupa enters diapause, even in the absence of temperature stress, the gene encoding a 23kDa small heat shock protein is quickly upregulated and remains upregulated throughout the duration of diapause (Yocum *et al.* 1998). We then asked whether the same is true for other stress proteins. Our experiments showed that the gene encoding Hsp70 is also upregulated by diapause (Rinehart *et al.* 2000), but this is not so for Hsp90 (Rinehart and Denlinger 2000). Of special interest is the prolonged expression of these genes during diapause. In other instances, the stress proteins are quickly upregulated in response to an environmental stress, but just as quickly disappear once the stress has been removed. But, diapause represents a very unusual case in which expression of the genes encoding the stress proteins persists for many months. Thus, expression of the stress protein genes during diapause is unusual in two ways: the upregulation usually noted in response to environmental stress is, in this case, exhibited developmentally when the pupa enters diapause, and the expression persists for a long time.

In this review we summarize the association of stress protein gene expression with pupal diapause in flesh flies and discuss the possible implications of the upregulation of certain of these proteins during this time. In addition, we link these findings with reports from other species showing the expression of stress proteins during periods of developmental arrest.

2. SMALL HEAT SHOCK PROTEINS

Of all the heat shock protein families, that of the small heat shock proteins (smHsps) is the most diverse. The smHsps belong to a family of proteins that includes the vertebrate eye structural protein, alpha-crystallin. Isoforms of alpha-crystallin and smHsps share similar structures and functions (MacRae 2000, Merck *et al.* 1993, Wistow 1985). smHsps range in size from 12 to 40 kDa, and number from 1 to 30 depending on the species. At the nucleic acid level the smHSPs are not highly conserved, but the two unifying features of the smHsps are a shared alpha-crystallin domain and the formation of similar higher-order structures. Within the alpha-crystallin domain of the four smHsps of *Drosophila melanogaster* there is approximately 68% identity with bovine alpha-crystallin (Ingolin and Craig 1982). The alpha-crystallin domain of the flesh fly *Sarcophaga crassipalpis* has 68% identity to *Gallus gallus* alpha-crystallin. Comparing the smHSPs of the two flies, the highest degree of identity is found within their alpha-crystallin domains (Yocum *et al.* 1998). *In vivo* the smHsps commonly form higher-order structures up to 800 kDa in size. The single known exception is *Caenorhabditis elegans*, in which Hsp12.6 does not form such a complex (Leroux *et al.* 1997).

The structure and phosphorylation status of the smHsps is critical for their function, as demonstrated in studies of actin polymerization (Benndorf *et al.* 1994; Lavoie *et al.* 1993, Miron *et al.* 1988 and 1991, Piotrowicz and Levin 1997). Phosphorylation causes a disassociation of the large multimer complexes of smHsps into monomers and dimers (Lambert *et al.* 1999). It is only the non-phosphorylated monomeric form of the smHSPs that will bind to the barbed end of actin, blocking its polymerization with other actin monomers (Benndorf *et al.* 1994). Neither the phosphorylated monomers nor multimeric forms of Hsp25 show an inhibitory effect on actin polymerization. In Chinese hamster cells engineered to over-express the human *hsp27* gene, phosphorylation of Hsp27 stimulates an increase in concentration of filamentous actin, causing changes in the cell surface associated with an active cytoskeleton (Lavoie *et al.* 1993). For a comprehensive review of the interaction of the heat shock protein families with the cytoskeleton see Ling and MacRae (1997a).

The chaperone activity that smHsps share with other heat shock protein families is dependent upon the smHsps existing as higher-order structures (Leroux *et al.* 1997). In *C. elegans* Hsp16-2 binds proteins in the early stages of denaturation before they aggregate with other proteins. Once aggregations are formed, Hsp16-2 cannot bind (Leroux *et al.* 1997). Although murine Hsp25, human Hsp27, and *C. elegans* Hsp16-2 can block formation of citrate synthase aggregations, only Hsp25 and 27 can refold the denatured protein (Jakob *et al.* 1993, Leroux *et al.* 1997). These results may be due to subtle experimental variations or may indicate that Hsp16-2 requires another chaperone to refold the denatured proteins.

Mouse Hsp25 inhibits elastase, a serine protease (Merck *et al.* 1993). One can envision a need to inhibit proteolysis during stress to protect proteins that have been denatured but can still be refolded. In mouse cells engineered to overexpress Hsp28, phosphorylation of Hsp28 leads to an inhibition of DNA synthesis (Sibanuma *et al.* 1992). Another property of some smHsps is their ability to covalently crosslink with other proteins through a transglutaminase-mediated reaction (Groenen *et al.* 1992, Merck *et al.* 1993). Merck *et al.* (1993) speculate that these crosslinked proteins may be involved in cellular organization.

Increased expression of the smHsps correlates positively with stress tolerance. Treating *D. melanogaster* S3 cells with 20-hydroxyecdysone induces expression of the smHsps with a parallel increase in thermotolerance (Ireland and Berger 1982, Berger and Woodward 1983). Genetically manipulating rodent cell lines to overexpress smHsps increases tolerance to normally lethal temperatures (Landry *et al.* 1989, Lavoie *et al.* 1993). The smHsps may also protect organisms against low temperature stress. smHsps are expressed during recovery from cold shock in the flesh fly, *S. crassipalpis* (Joplin *et al.* 1990a) and in the gypsy moth *Lymantria dispar* (Yocum *et al.* 1991). Though not perfect, there is a positive correlation between the synthesis of Hsp70 and 27 in human diploid IMR-90 fibroblasts and increased cold tolerance (Russotti *et al.* 1996).

During diapause in the flesh fly, *S. crassipalpis, hsp23* is upregulated immediately upon entry into diapause (Yocum *et al.* 1998), even in the absence of thermal stress (Fig. 1). Expression persists throughout diapause but declines within a few hours after diapause is terminated. Neither heat shock nor cold shock administered during diapause can further elevate the level of expression. Thus, *hsp23* is clearly developmentally upregulated during diapause. While the expression of many genes is shut down during diapause (Joplin *et al.* 1990b), *hsp23* belongs to a relatively small group of genes that are upregulated during this time (Flannagan *et al.* 1998). This fly does have additional smHsps, as well as a transcribed pseudogene of Hsp23 (unpublished observation, accession number AF156162). But, whether the other smHsps are also upregulated during diapause is still unknown.

The expression of *hsp23* in different tissues is not uniform during diapause. Highest levels of expression are noted in the brain, both the epidermal cells and gut show more modest levels of expression, and very little expression can be noted in the fat body (Fig. 2). Does this mean that the brain is the most vulnerable to injury? Or, as the center presiding over the hormonal regulation of development, is the shut-down of this tissue the most critical for assuring a developmental arrest?

Although the flesh fly experiments were the first to show *hsp* upregulation during insect diapause, earlier work with *Drosophila* cells demonstrated a nice correlation between cell cycle arrest and the expression of smHsps (Berger and Woodward1983). *Drosophila* S3 cells treated with 20-hydroxyecdysone enter a G2 cell cycle arrest, and concurrently the arrested cells express smHsps. A cell cycle arrest in cultured cells is clearly different from a whole organism's developmental arrest, yet a cell cycle arrest is a conspicuous feature of diapause as well (Tammariello and Denlinger 1998). We thus propose that one possible role for the smHsps during diapause is to facilitate and/or assure that the cells remain arrested.

A second possibility is that the smHsps may be contributing to cryoprotection during the overwintering diapause. They are present as conspicuous components of the response of nondiapausing flesh flies to both heat shock and cold shock (Joplin *et al.* 1990a). Although the diapausing pupae do not further upregulate these genes in response to thermal stress, the fact that expression is already high suggests that the activation of this gene family is a vital component of the protective mechanism operating during diapause. At this point we can still not be certain whether the high level of *hsp23* in flesh flies means that the protein is also abundant during diapause or whether translation occurs only when the protein is needed.

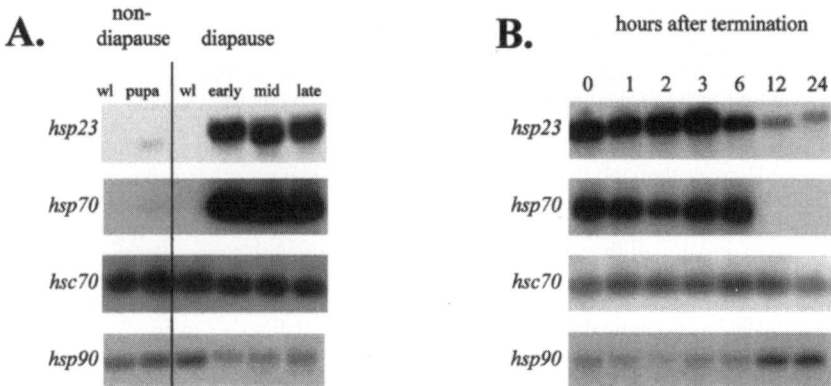

Figure 1. Northern blots showing the expression patterns for transcripts of four heat shock proteins in the flesh fly *Sarcophaga crassipalpis* in relation to pupal diapause and its termination. As shown in (A), transcripts of *hsp23* and *hsp70* are highly upregulated throughout diapause, while expression of the cognate gene *hsc70* is unaffected by diapause, and *hsp90* is downregulated during diapause. After diapause is terminated by application of hexane (B), *hsp23* and *hsp70* are downregulated, *hsc70* remains unaffected, and *hsp90* is again upregulated. "wl" refers to the wandering third instar larval stage. From Yocum *et al.* (1998), Rinehart *et al.* (2000) and Rinehart and Denlinger (2000).

No additional data on the involvement of smHsps is yet available for diapause in other species of insects, but studies on the dormancies in several other animals and plants suggest involvement of the smHsps. Most notable are the studies with p26, a 26 kDa heat shock protein, associated with diapause in the brine shrimp, *Artemia franciscana*. Individuals programmed to diapause encyst as gastrulae, enter diapause and desiccate. Under anoxic conditions, the brine shrimp bring their metabolism to a reversible standstill, and can survive 4-6 years in this condition (Clegg 1997, Clegg *et al.* 1999, Hand 1998). *A. franciscana* faces the challenge of maintaining structural integrity with no flow of energy. This challenge is met at least in part by the synthesis and accumulation of the disaccharide trehalose and p26. Trehalose stabilizes macromolecules and membranes (reviewed in Clegg *et al.* 1999). p26 is expressed only in diapausing embryos and in stages shortly before and after diapause. It can not be induced in other developmental stages by heat shock (Jackson and Clegg 1986, Liang and McRae 1999). What role could p26 serve in diapausing *A. franciscana*? Like other smHSPs, p26 may function as a molecular chaperone (Liang *et al.* 1997b), preventing proteins from denaturing as *A. franciscana* enters diapause and desiccates and later refolding these proteins when diapause ends. smHSPs also inhibit proteolysis (Merek *et al.* 1993). The protein profile of diapausing *A. franciscana* appears to be very stable even after 6 years under anoxic conditions (Clegg *et al.* 1999). The expression of p26 negatively correlates with DNA synthesis and cell division at both the initiation and termination of diapause, leading Clegg *et al.* (1999) to propose that p26 may also be involved in inhibiting DNA synthesis and cell division.

The smHsps are also associated with various forms of plant dormancy. They are expressed as normal components of seed development. Expression in seeds of peas

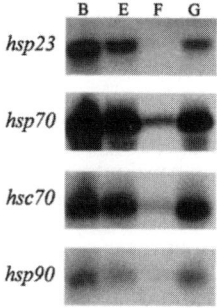

Figure 2. Northern blots showing the expression of four heat shock protein transcripts in different tissues of the flesh fly during pupal diapause. In all cases, highest expression is noted in the brain (B), followed by the epidermis (E) and midgut (G). Lowest levels of expression are noted in the fat body (F).

(DeRocher and Verling 1994), sunflowers (Coca *et al.* 1994) and *Arabidopsis* (Wehmeyer *et al.* 1996) starts at mid-maturation and increases as the seed dehydrates. *Arabidopsis* has four cytosolic smHsps, but a single protein, Hsp17.4, accounts for the majority of the smHsp expression during this time (Wehmeyer *et al.* 1996). Selection for reduced seed dormancy in *Arabidopsis*, however, does not result in lower levels of smHsps, suggesting that expression of the smHsps is not an absolute prerequisite for seed dormancy (Wehmeyer *et al.* 1996). smHsps are also associated with the overwintering dormancy of certain plants. In overwintering oak acorns, *hsp27* is expressed in dormant buds and in the quiescent centers of the root and shoot apex (Pla *et al.* 1998), but since these acorns were field collected during winter it is not possible to tell whether *hsp27* expression is normally a part of the overwintering program or if expression was induced by low temperatures. Similarly, a 20 kDa smHsp accumulates in cortical parenchyma cells of mulberry trees during cold acclimation during winter (Ukaji *et al.* 1999). Again, it is not clear whether this accumulation of the smHsp in mulberry is a direct response to dormancy or to the low temperatures encountered during the winter months.

The pathology of key human pathogens is dependent upon the microorganism being able to enter dormancy. *Mycobacterium tuberculosis* can remain dormant in human lungs for 30- 40 years. There is normally a drop in cell viability during the stationary growth stage due to autolysis. When *M. tuberculosis* transitions from log growth to the stationary growth phase, Hsp16 emerges as one of the major proteins (Yuan *et al.* 1996). *M. smegmatis* manipulated to overexpress Hsp16 was protected from autolysis after entering the stationary growth stage, suggesting a protective role for Hsp16. When *M. tuberculosis* and *M. bovis* are placed in a reduced oxygen environment, as may occur within granulomas in the lungs, they enter a stationary growth stage characterized by thickened cell walls. Immunological studies demonstrate that Hsp16 concentrates at the cell membrane, the cell wall skeleton, and within clusters throughout the cell (Cunningham and Spreadbury 1998). These results suggest a role for Hsp16 in the protection of cell structures during dormancy.

Bacteria can also form spores to escape environmental stress, and as with the stationary stage of the *Mycobacterium*, the smHsps are involved with this form of dormancy. The cotM gene from *Bacillus subtilis* encodes a 14 kDa smHsp, the expression of which is developmentally restricted to sporulation (Henriques *et al.* 1997). Strains of *B. subtilis* containing cotM mutants form outer coats with reduced amounts of outer coat proteins. As the spores age, the outer coat proteins progressively decrease, indicating structural instability of the outer coat. Henriques *et al.* (1997) suggest that Hsp14 serves as a substrate for transglutaminase-mediated crosslinking reactions with the outer coat proteins, a reaction thought to be critical in the formation of the spore's protective outer coat. A 21 kDa smHsp is also made by the bacterium *Stigmatella aurantiaca* during sporulation (Heidelbach *et al.* 1993).

Transferring the yeast *Saccharomyces cerevisiae* from a normal growth medium to a nitrogen-deficient medium causes a developmental switch from normal growth to spore formation. The cells reach the tetranucleate stage in 8 to 10 hours and sporulation is complete by 24 hours. Within 2 hours after the transfer to a nitrogen-deficient medium *hsp*26 transcripts are expressed, and the protein is detectable by the 10th hour. Both the *hsp*26 transcript and protein continue to be expressed at high levels following the completion of sporulation (Kurtz *et al.* 1986).

Several studies with mammalian cells link the expression of smHsps with cell cycle arrests. When *Ehrlich ascites* tumor cells are engineered to express *hsp*25 under a metallothionein promotor the cells show decreased growth rates with increased exposure to cadmium (Knauf *et al.* 1992). Incubating mouse osteoblastic MC3T3-E1 cell with various growth inhibitors induces phosphorylation of Hsp28 and inhibits cell growth (Shibanuma *et al.* 1992). These inhibitors only induce Hsp28 phosphorylation during the late G1 phase of the cell cycle, when the cells are normally sensitive to these inhibitors. In ras-transformations the inhibitors fail to induce Hsp28 phosphorylation and DNA synthesis is not blocked. These observations led Shibanuam *et al.* (1992) to conclude that phosphorylation of Hsp28 is linked to inhibition of DNA synthesis. A role for phosphorylation of smHsps in inhibiting cell growth may not be universal. Using various mammalian cells transformed with *hsp*27 and hypophosphorylatable or C-terminal deletion mutants, Arata *et al.* (1997) showed that phosphorylation and aggregation of Hsp27 is involved in inhibiting cell growth only in certain cell types.

3. HEAT SHOCK 70 FAMILY

While the small heat shock protein family is the most diverse, the heat shock protein 70 family is one of the least diverse but the most widely studied. Members of the Hsp70 family can be divided into two groups based on their expression pattern: those that are not expressed during normal conditions but are rapidly induced during stress (Hsp70), and those that are constitutively expressed during non-stress conditions, with little or no responsiveness to cellular injury (Hsc70).

The most well studied are those induced by stress. Originally described in *D. melanogaster,* the inducible Hsp70 proteins have been found in a wide variety of eukaryotic and prokaryotic organisms. As their name implies, they were first identified as heat shock proteins because they were rapidly and dramatically upregulated by high temperature. Transcripts of *hsp70* appear within 15 minutes of exposure to high temperature stress (Rinehart *et al.* 2000), and protein concentrations are quickly increased to 1000 times their pre-stress levels. Since their discovery, Hsp70 proteins have been found to be upregulated in response to a wide variety of stresses (Lindquist 1986, Feder 1996). In the flesh fly, Hsp70 is the dominant HSP induced by both heat shock and cold shock (Joplin and Denlinger 1990).

The sequence of Hsp70 is highly conserved, with nearly 50% amino acid identity among all eukaryotes (Parsell and Lindquist 1993). The highest identity is evident within the functional regions of the protein. These include an n-terminal region that exhibits adenosine binding activity, which appears to be involved in control of Hsp70 protein binding via ATP, and a central region involved in protein binding.

Hsp70 proteins have several functions during stress as well as non-stress conditions. During cellular stress, Hsp70 binds to misfolded or denatured proteins, acting as a chaperone, preventing the formation of harmful aggregation of denatured proteins, and allowing them to properly refold, leading to a restoration of function (Lindquist 1986). The specificity of Hsp70 binding to abherent proteins appears to lie in its binding affinity to hydrophobic peptides (Fourie *et al.* 1994), which would remain unexposed to binding under normal cellular conditions.

A classic demonstration of the abilities of Hsp70 to restore specific proteins after thermal damage was conducted using luciferase as a reported protein in *E.coli*. When *E. coli* cells are heat shocked, luciferase activity is halted until the cells are returned to ambient temperatures. An *in vitro* study confirmed that DnaK (bacterial Hsp70) plays an essential role in the restoration of luciferase activity (Schroder *et al.* 1993).

Hsp70 exhibits non-stress functions as well. It can function to hasten the degradation of abnormal proteins either as a result of stress or during normal cellular functions. In mammalian models, Hsp70 has been shown to achieve this by signaling for the lysosomal degradation of the proteins which are bound to it (Terlecky *et al.* 1992). Additionally, the developmental regulation of Hsp70 has led to the discovery of its role in meiosis. Disruption of Hsp70 during germ cell development leads to meiotic failure (Dix *et al.* 1995).

The second group of the heat shock protein 70 family are those that are constitutively expressed. The cognate heat shock 70 proteins (Hsc70) are expressed at substantial levels under normal conditions. These proteins are not responsive to heat shock, but have been shown to respond to cold temperatures in plants (Anderson *et. al.* 1994, Neven *et. al.* 1992) and in the flesh fly (Rinehart *et al.* 2000).

At the onset of stress, Hsc70 appears to act as a molecular thermometer (Craig and Gross 1991), releasing heat shock protein transcription factors to allow for the upregulation of other heat shock proteins. Hsc70 proteins have non-stress functions as well. During mitotic cell division, Hsc70 protein is moved into the nucleus at the onset of S phase (Gevers *et al.* 1997), where it associates with mitosis related structures. Blocking Hsc70 with antibody leads to inhibition of mitosis (Sconzo *et al.* 1999).

We have isolated clones of both *hsp70* and *hsc70* from the flesh fly to assess their expression patterns during periods of stress and during diapause. The clone of *hsp70* was developed by rt-PCR, while the clone of *hsc70* was isolated during a library screen. Both clones are partial, and include the distal regions of their respective reading frames. This region contains the greatest degree of divergence between *hsp70* and *hsc70*, making the clones ideally suited for individual analysis of the expression patterns of each gene.

Similar to *hsp23*, *hsp70* is upregulated when flesh fly pupae enter diapause, even in the absence of stress (Fig.1). This upregulation is evident only in diapausing pupae, and not in wandering larvae that are destined to enter diapause. The increased expression continues throughout diapause, and is quickly downregulated upon diapause termination. Unlike nondiapausing individuals, diapausing pupae are not able to further elevate *hsp70* expression in response to thermal injury. In contrast, *hsc70* expression was not affected by the induction of diapause. However, unlike *hsp70* and *hsp23*, *hsc70* remains responsive to cold shock during diapause (Rinehart *et al.* 2000).

Like *hsp23,* the expression of *hsp70* and *hsc70* is not uniform in all tissues during diapause (Fig. 2). Again, highest levels were found in the brain, lower levels in the gut and epithelium, and lowest levels in the fat body. Similar relative levels of Hsp70 were documented for the brain, gut and fat body of *D. melanogaster* immediately after heat shock (Krebs and Feder 1997b). In *D. melanogaster* the levels of Hsp70 correlated with the amount of cell death detected using vital dyes: the brain exhibited very little cell death, while the midgut showed the highest amount of cell death. Whether this interpretation can be applied to diapausing flesh flies remains to be determined. One could also argue that the high levels present in the brain reflect the vulnerability of the nervous system and the critical role played by this tissue in regulation of diapause. A shut-down of the brain is an essential feature of diapause, and an abundance of Hsps in this tissue may help assure a cessation of the cell cycle at this critical time.

To date, little information on *hsp70* and diapause exists from other species. A recent study on the adult diapause of *D. triauraria* suggests that *hsp70* is not upregulated in this species as a function of diapause but remains responsive to thermal stress (Goto *et al.* 1998). Additionally, previous data from the gypsy moth indicates that *hsp70* is not upregulated upon the initiation of diapause, but only after cold temperatures have been realized. We are currently investigating a possible role for *hsp70* in the diapause of other species. Some non-insect models of dormancy appear to involve elevated *hsp70* expression. During hibernation of the ground squirrel *Spermophilus tridecemlineatus* mammalian *hsp70* is not upregulated, although *grp75*, its mitochondria counterpart, is upregulated at the transcriptional level (Carey *et al.* 1999). However, hibernating bats (*Myotis lucifugus*) show upregulation of neither *hsp70* nor *grp78* (Storey, 1997). During the larval dauer stage of developmental arrest in the nematode *C. elegans*, *hsp70* transcripts are upregulated (Cherkasova *et al.* 2000), although previous researchers found no such upregulation (Dalley and Golomb 1992). In addition, mRNAs of high molecular weight heat shock proteins have been found to accumulate in the dormant conidiospores of the fungus *Neurospora crassa* (Pleovsky-Vig and Brambl 1985), although more recent studies indicate that *hsc70*, rather than the inducible form is involved (Rensing et al., 1998). *Hsp70* is also reported to be involved in the sporulation of the fungi *Blastocladiella emersonni* (Bonato, 1987) and during early spore germination of *Leptosphaeria maculans* (Patterson and Kapoor, 1995).

While the upregulation of *hsp70* during diapause in *S. crassipalpis* appears to be a unique observation thus far in insects, the developmental upregulation of *hsp70* is not. Upregulation of *hsp70* has been documented during spermatogenesis in *D. melanogaster* (Michaud 1997) and in mammalian models (Eddy 1999), as well as during oogenesis (Billoud *et al.* 1993) and embryogenesis (Luft and Dix 1999).

The function of Hsp70 during diapause remains unclear. One possibility is that Hsp70, perhaps in conjunction with the smHsps, is involved directly in the cell cycle (Tammariello and Denlinger, 1998, Nakagaki, 1991) and developmental arrest characteristic of diapause. Studies in *D. melanogaster* have suggested a link between *hsp70* expression and cell cycle arrest as well. *D. melanogaster* larvae that contain extra copies of the *hsp70* gene are characterized in part by decreased growth and slowed development (Krebs and Feder 1997a). In vitro studies also support this link: a *Drosophila* cell line that continuously expresses high levels of *hsp70* shows a reduction in cellular growth (Feder *et al.* 1992). Such a link is also evident in some non-insect models. Additionally, programmed cell death in *Caenorhabditis elegans* has been linked to hsp expression (Madi *et al.* 1997).

As we suggest for Hsp23, another potential role for Hsp70 during diapause is that it functions as part of the increased stress resistance characteristic of diapausing individuals. Data from diapausing pharate larvae of *Lymantria dispar* support this idea. When this insect enters its obligate diapause as a pharate larva, it does not harbor the increased tolerance to cold normally characteristic of diapausing insects in temperate regions. Additionally, Hsp70 is not upregulated upon entry to diapause. However, when diapausing individuals are subjected to cold temperatures, Hsp70 proteins are upregulated, and the pharate larvae become cold tolerant (Denlinger *et al.* 1992).

4. HEAT SHOCK 90 FAMILY

The third group of heat shock proteins that we have investigated in the flesh fly is the heat shock protein 90 (Hsp90) family, which includes proteins in the range of 80kDa to over 100kDa. In *D. melanogaster* and most other insects, Hsp90 proteins are approximately 83kDa (Blackman and Meselson 1986). A unique feature of proteins in this family is that they are constitutively expressed at substantial levels during non-stress conditions (Arbona 1993), with considerable upregulation observed at the onset of stress (Yiangou *et al.* 1997). In mammals, this appears to be achieved by two different genes:

one that is constitutively expressed under non-stress conditions and is downregulated during stress, and a second gene that is not constitutively expressed, but is upregulated during periods of stress. In insects, this does not appear to be the case: a single gene is responsible for both constitutive- and stress-inducible expression (Gething 1997).

Similar to the 70kDa heat shock proteins, the sequence of Hsp90 is highly conserved among known eukaryotes, with amino acid identities as high as 70% within its two highly conserved regions (Gupta 1995). Most of the functional domains of the protein lie within a c-terminus conserved region that includes a peptide binding domain, a steroid receptor binding domain, and a region that facilitates dimerization with other Hsp90 proteins (Minami et al. 1994).

The upregulation of Hsp90 in response to heat shock is not as robust as for other heat shock proteins. In most cases, a 10-15 fold increase in protein levels can be expected when comparing heat-stressed organisms to unstressed controls (Arbona 1993). However, this lack of dramatic upregulation is not indicative of a minor role for this protein during heat shock and other stresses. Hsp90 appears to have a specific function during recovery from stress, being especially important in the reactivation of stress-inactivated proteins (Nathan et al. 1997). Hsp90 appears to work in conjunction with other Hsps. Evidence exists for a combination of Hsp90 with Hsp70 in exerting protective effects in some instances, with the protein Hop playing a role in bringing the two proteins into contact with one another (Chen and Smith 1998). Unlike Hsp70, and as suggested by the lack of an adenosine binding domain, the binding of Hsp90 to aberrant protein is an ATP-independent process (Jakob et al. 1996).

Equally important to the stress functions of Hsp90 is the role it plays in signal transduction in unstressed cells, specifically in combination with steroid receptors (Ylikomi et al. 1998, Kimmins and MacRae 2000). In insects, Hsp90 has been found to be associated with the ecdysone receptor/ ultraspiracle complex, and its presence is essential for proper function (Arbeitman 2000).

The upregulation of other families of Hsps during diapause, and the role of Hsp90 in other models of dormancy led us to investigate the expression of hsp90 during flesh fly diapause. Towards this end, we isolated a clone of hsp90 from S. crassipalpis using degenerate primers in rt-PCR. Contrary to our expectations hsp90 was not upregulated, and was in fact downregulated during diapause (Rinehart and Denlinger 2000). The three fold downregulation occurred as the flies entered the pupal stage and expression remained low throughout diapause (Fig. 1). Upon termination of diapause, hsp90 transcripts were restored to prediapause levels within 12 hours. Much like hsc70, hsp90 remained responsive to stress incurred during diapause. While disynchrony of the Hsps response to stress is rarely noted, upregulation of select Hsp subsets at specific times during development is well documented. For example, oogenesis in D. melanogaster (Zimmerman et al. 1983) and phases of yeast development (Kurst et al. 1986) involve selective upregulation of Hsps.

Interestingly, even though hsp90 is downregulated during diapause, its relative tissue distribution is the same as for those hsps that are upregulated or remain unchanged during diapause (Fig. 2). While the higher levels in the brain might again be interpreted as a higher level of protection for this very important tissue, an alternative hypothesis might be that its prevalence in the brain is linked to its pivatal role in hormone reception and signal transduction.

The mechanism by which hsp90 is downregulated during diapause and is promptly restored upon diapause termination is likely to involve ecdysteroid signaling. Evidence exists for the control of hsp90 expression by 20-hydroxyecdysone, with the presence of the steroid leading to upregulation of hsp90, and the removal of 20-hydroxyecdysone causing gene downregulation (Thomas and Lengyel 1986). Pupal diapause in flesh flies has long been associated with the lack of ecdysteroids, with diapause termination brought about by their return (Zdarek and Denlinger 1975). Additional evidence in support of this theory is that hsp90 levels are restored to prediapause levels at 12 hours post termination, which correlates with the timepoint of ecdysteroid restoration (Walker and Denlinger 1980).

Although *hsp90* is not upregulated in flesh fly diapause, it does play a substantial role in other models of dormancy. In the yeast *Saccharomyces cervisiae*, the sporulation process involves *hsp90* upregulation, although transcripts are downregulated at the time of spore formation and do not appear to be a significant component of mature spores (Kurtz *et al.*1986). The most notable expression of *hsp90* during dormancy is in the dauer larvae of the nematode *C. elegans*, where transcripts are upregulated upon entry into the dauer stage and remain elevated until diapause is terminated (Cherkasova *et al.* 2000). This upregulation appears to involve the stabilization of *hsp90* transcripts throughout the dauer stage (Dalley and Golomb 1992). Although the exact function of Hsp90 during the dauer stage remains unclear, it appears to play a significant regulatory role. The dauer regulatory protein daf-21 has recently been identified as Hsp90 (Birnby *et al.*, 2000). *Daf-21*, in combination with *daf-11*, a gene encoding guanylyl cyclase, are the two known dauer constitutive genes (also known as *Daf-c* genes). These were previously known as temperature sensitive genes mutations that could lead to dauer larva formation by temperature alone, even if other environmental conditions, such as the presence of food and phermones, were not conducive to dauer formation.

5. RESPONSE TO LOW TEMPERATURES

There is now evidence from nondiapausing stages of several species, including the fruit fly *Drosophila melanogaster* (Burton *et al.* 1988) and other species of *Drosophila* (Goto and Kimura 1998), the flesh fly *Sarcophaga crassipalpis* (Joplin *et al.* 1990), and the biting midge *Culicoides variipennis sonoresis* (Nunnamaker *et al.*1996), that a cold shock can elicit synthesis of stress proteins. In these examples, the stress proteins are not made during the cold shock but immediately afterwards, upon the insect's return to a non-stressful temperature.

Is the same true for diapausing insects? We examined this question by testing the effect of cold shock on diapausing pupae of the flesh fly *S. crassipalpis*. As discussed above, *hsp23* and *hsp70* are already highly upregulated upon entry into diapause. Neither cold shock nor heat shock could elicit further upregulation (Yocum *et al.*1998, Rinehart *et al.* 2000). The high level of expression already operating in diapause could not be further elevated.

In contrast, *hsc70* and *hsp90* , the two stress genes characterized by low expression during diapause in *S. crassipalpis* were readily upregulated by cold shock during diapause. While *hsp90* responded to both heat shock and cold shock during diapause (Rinehart and Denlinger 2000), *hsc70* was upregulated only by cold shock (Rinhart *et al.* 2000). These results thus reflect disynchrony of the stress protein response during diapause and in response to low temperatures. Certain genes, *hsp23* and *hsp70,* are expressed as a function of the diapause program and show no further temperature regulation, while others, *hsc70* and *hsp90* are upregulated in diapausing pupae only when the insect is subjected to a cold shock. In *Drosophila triauraria*, a species with a reprodutive diapause, *hsp70* is not developmentally upregulated during diapause, but adults in diapause will respond to heat shock or cold shock by accumulating *hsp70* mRNA (Goto *et al.* 1998). And, in diapausing larvae of the goldenrod gallfly *Eurosta solidaginis,* heat shock proteins are made in response to heat shock but not cold shock (Lee *et al.* 1995). The gypsy moth *Lymantria dispar* diapauses at the end of embryonic development but before hatching from the egg. In this species the capacity to synthesize stress proteins in response to both high and low temperature stress (Yocum *et al.* 1991) is greatly enhanced once the embryos have experienced a period of chilling (Denlinger *et al.* 1992). Interestingly, the increased ability of the gypsy moth to synthesis stress proteins elicited by chilling also coincides with an increase in cold tolerance.

Thus, from these few examples it appears that there is considerable variation in the response among different species. Perhaps some of these differences will eventually be attributable to particular life stages. While the flesh flies diapause as pupae, the gall fly diapauses as a freeze-tolerant larva, the fruit flies diapause as adults, and the gypsy moth diapauses as a late embryo (pharate first instar larva).

7. FUTURE DIRECTIONS

The striking upregulation of *hsp23* and *hsp70* during pupal diapause has thus far been noted only in flesh flies. Clearly this same type of upregulation is not observed in the adult reproductive diapause of *Drosophila,* but whether this is a characteristic shared by many other species of insects in diapause remains unknown. Upregulation of stress proteins is evident in a number of non-insect species during arrested development, but additional comparative data is essential for determining how general this feature may be among other insects.

We have proposed two possible functions for the expression of stress proteins during diapause. First, the presence of stress proteins at this time may offer a cryoprotective role to the insect during diapause. During the low temperatures and desiccating environment of winter, the stress proteins could serve to preserve the integrity of critical proteins needed for cell functions during this stressful time. Secondly, the presence of stress proteins may serve to assure that development is halted. Numerous reports indicate a halt or slow-down of development when stress proteins are present. Their presence during diapause could contribute to a mechanism that effectively shuts down development at this time. Unfortunately, most of the data available is correlative. The correlations are a valuable first step, but experiments that go beyond the correlations are needed for understanding function. Manipulations of the stress protein titers and other alterations of this type using chemicals or knock-out techniques will be needed to decipher functions.

Calculations of turn-over rates of the mRNAs and the stress proteins are essential for understanding the dynamics of this response. Does the message simply persist much longer in the diapausing individuals? Are new stress proteins being made continuously or are they only made at critical times, perhaps in response to a direct stress? Is the half life of the proteins greatly extended during diapause?

To understand function it will also be valuable to determine the tissue and cellular distribution of the stress proteins. Our preliminary data indicates that the *hsp* transcripts are particularly high in the brain, whereas some other tissues such as the fat body show little activity. Does this mean that the brain is particularly vulnerable during diapause? The cellular distribution of the stress proteins during diapause and cold stress will be insightful. In response to heat shock, the Hsps are translocated from the cytoplasm to the nucleus. Is the same true for cold shock, and where within the cells are the proteins located during diapause? Even though we noted no further upregulation of *hsp23* and *hsp70* by low temperature stress during diapause, might stress result in higher rates of translation or intracellular translocation of the proteins?

The phosphorylation status of the stress proteins, especially that of the smHSPs, possibly contributes to their activity. But, at this point, no such information is known about these proteins in relation to diapause. Do the smHSPs during diapause form higher-order structures? This also remains unknown. Large multimeric complexes are formed in most species, but not for all of the smHSPs in *C. elegans.*

The persistant and high level of upregulation of the genes encoding stress proteins during diapause represents a novel situation. In most other cases, even cases of developmental regulation of stress proteins, the periods of upregulation are brief. In response to stress the genes are quickly upregulated and just as quickly downregulated when the stress is removed. Here, by contrast, upregulation persists for months at a time. The flesh flies we have worked on enter diapause in late August or September and remain in diapause until the following spring. This model thus offers a case where the stress response persists for an extremely long time and may suggest new roles for these widely-studied proteins.

ACKNOWLEDGEMENTS

This research was supported in part by grants from the USDA-NRI (98-35302-6659) and NSF (IBN-9728573).

REFERENCES

Anderson, J. V., LI, Q. B., Haskell, D. W., and Guy, C.L. 1994. Structural organization of the spinach endoplasmic reticulum-luminal 70-kilodalton heat-shock cognate gene and expression of 70-kilodalton heat-shock genes during cold-acclimation. Plant Physiology 104, 1359-1370.

Arata, S., Hamaguchi, S., Nose, K. 1997. Inhibition of colony formation of NIH 3T3 cells by the expression of the small molecular weight heat shock protein HSP27: involvement of its phosphorylation and aggregation at the C-terminal region. Journal of Cell Physiology 170, 19-26.

Arbeitman, M.N., Hogness, D.S. 2000. Molecular chaperones activate the Drosophila ecdysone receptor, an RXR heterodimer. Cell 101, 67-77.

Arbona, M., Defrutos, R., Tanguay, R. M. 1993. Transcriptional and translations study of the Drosophila-subobscura hsp83 gene in normal and heat-shock conditions. Genome 36, 694-700.

Benndorf, R., Hayefl, K., Ryazzntsev, S., Wieske, M., Behlke, J., Lutsch, G. 1994. Phosphorylation and supramolecular organization of murine small heat shock protein HSP25 abolish its actin polymerization-inhibiting activity. Journal of Biological Chemistry 269, 20780-20784.

Berger, E.M., Woodward, M.P. 1983. Small heat shock proteins in *Drosophila* confer thermal tolerance. Experimental Cell Research 147, 437-442.

Billoud, B., Rodriguezmartin, M.L., Berard, L. 1993. Consituitive expression of a somatic heat inducible hsp70 gene during amphibian oogenesis. Development 119, 921-932.

Birnby, D.A., Link, E.M., Vowels, J.J., Tian, H., Colacurcio, P.L., and Thomas, J.H. 2000. A transmembrane guanylyl cyclase (DAF-11) and Hsp90 (DAF-21) regulate a common set of chemosensory behaviors in *Caenorhabditis elegans*. Genetics 155, 85-104.

Blackman, R. K. and Meselson, M. 1986. Interspecific nucleotide sequence comparisons used to identify regulatory and structural features of the *Drosophila* hsp82 gene. Journal of Molecular Biology 188, 499-515.

Bonato, M. C. M., Silva, A. M., Gomes S. L., Maia J. C. C., Juliani M. H. 1987. Differential expression of heat-shock proteins and spontaneous synthesis of hsp70 during the life cycle of *Blastocladiella emersonii*. European Journal of Biochemistry 163, 211-220.

Burton, V., Mitchell, H.K., Young, P., Peterson, N.S. 1988. Heat shock protection against cold stress of *Drosophila melanogaster*. Molecular and Cellular Biology 8, 3550-3552.

Carey, H.V., Sills, N.S., and Gorham, D.A. 1999 Stress proteins in mammalian hibernation. American Zoologist. 39, 825-835.

Chen, S.Y. and Smith, D.F. 1998. Hop as an adaptor in the heat shock protein 70 (Hsp70) and Hsp90 chaperone machinery. Journal of Biological Chemistry 273, 35194-35200.

Cherkasova, V., Ayyadevara, S., Egilmez, N., Reis, R.S. 2000. Diverse *Caenorhabditis elegans* genes that are upregulated in dauer larvae also show elevated transcript levels in long-lived, aged, or starved adults. Journal of Molecular Biology. 300, 433-448.

Clegg, J.S., 1997. Embryos of *Artemia franciscana* survive four years of continuous anoxia: the case for complete metabolic rate depression. The Journal of Experimental Biology 200, 467-475.

Clegg, J.S., Drinkwater, L.E., Sorgeloos, P. 1996. The metabolic status of diapause embryos of *Artemia franciscana* (SFB). Physiological Zoology 69, 49-66.

Clegg, J.S., Willsie, J.K., Jackson, S.A., 1999. Adaptive significance of a small heat shock/alpha-crystallin protein (p26) in encysted embryos of the brine shrimp, *Artemia franciscana*. American Zoologist 39, 836-847.

Coca, M.A., Almonguera, C., Jordano, J. 1994. Expression of sunflower low-molecular-weight heat-shock proteins during embryogenesis and persistence after germination: localization and possible functional implication. Plant Molecular Biology 25, 479-492.

Craig, E.A. and Gross, C.A. 1991. Is hsp70 a cellular thermometer? Trends in Biochemical Sciences 16, 135-140.

Cunningham, A.F., Spreadbury, C.L. 1998. Mycobacterial stationary phase induced by low oxygen tension: cell wall thickening and localization of the 16-kilodalton alpha-crystallin homolog. Journal of Bacteriology 180, 801-808.

Dalley, B.K., Golomb, M. 1992. Gene expression in the *Caenorhabditis elegans* dauer larva: developmental regulation of hsp90 and other genes. Developmental Biology 151, 80-90.

Denlinger, D.L., Lee, R.E., Yocum, G.D., Kukal, O. 1992. Role of chilling in the acquisition of cold tolerance and the capacitation to express stress proteins in diapausing pharate larvae of the gypsy moth, *Lymantria dispar*. Archives of Insect Biochemistry and Physiology 21, 271-280.

DeRocher, A.E., Vierling, E. 1994. Developmental control of small heat shock proteins expression during pea seed maturation. Plant Journal 5, 93-102.

Dix, D.J., Bobseine, K.L., Allen, J.W., Collins, B.W., Mori, C., Nakamura, N., Poormanallen, P., Goulding, E.H., and Eddy, E.M. 1995. Targeted disruption of hsp70-2 leads to failed meiosis, germ-cell apoptosis, and male-infertility. Molecular Biology of the Cell 6, 1829-1829.

Eddy, E.M. 1999. Role of heat shock protein HSP70-2 in spermatogenesis. Reviews of Reproduction 4, 23-30.

Feder, M.E. 1996. Ecological and evolutionary physiology of stress proteins and the stress response: the *Drosophila melanogaster* model. In *Animals and Temperature: Phenotypic and Evolutionary Adaptations* (I.A. Johnston and A.F. Bennett, Eds), pp. 79-102. Cambridge University Press, Cambridge

Feder, M.E., Hofmann, G.E. 1999. Heat-shock proteins, molecular chaperones, and the stress response: evolutionary and ecological physiology. Annual Review of Physiology 61, 243-282.

Feder, J.H., Rossi, J.M., Solomon, J., Solomon, N., and Linquist, S. 1992. The consequences of expressing hsp70 in *Drosophila* cells at normal temperatures. Genes and Development. 6, 1402-1413.

Flannagan, R.D., Tammariello, S.P., Joplin, K.H., Cikra-Ireland, R.A., Yocum, G.D., Denlinger, D.L., 1998. Diapause-specific gene expression in pupae of the flesh fly *Sarcophaga crassipalpis*. Proceedings of the National Academy of Sciences, USA 95, 5616-5620.

Fourie, A.M., Sambrook, J.F., Gething M.J.H. 1994. Common and divergent peptide binding specificities of hsp70 molecular chaperones. Journal of Biological Chemistry 269, 30470-30478.

Gething, M. J. 1997. Guidebook to molecular chaperones and protein-folding catalysts. Oxford University Press, Oxford.

Gevers, M., Fracella, F., and Rensing, L. 1997. Nuclear translocation of constitutive heat shock protein 70 during S phase in synchronous macroplasmodia of *Physarum polycephalum*. FEMS Microbiology Letters 152, 89-94.

Goto, S.G., Kimura, M.T. 1998. Heat- and cold-shock responses and temperature adaptations in subtropical and temperate species of *Drosophila*. Journal of Insect Physiology 44, 1233-1239.

Goto, S.G., Yoshida, K.M., Kimura, M.T. 1998. Accumulation of Hsp70 mRNA under environmental stresses in diapausing and nondiapausing adults of *Drosophila triauraria*. Journal of Insect Physiology 44, 1009-1015.

Groenen, P.J.T.A., Bloemendal, H., de Jong, W.W., 1992. The carboxy-terminal lysine of alpha B-crystallin is an amino-substrate for tissue transglutaminase. European Journal of Biochemistry 205, 671-674.

Gupta, R. S. 1995. Phylogenetic analysis of the 90kD heat-shock family of protein sequences and an examination of the relationship among animals, plants, and fungi species. Molecular Biology and Evolution 12, 1063-1073.

Hand, S.C. 1998. Quiescence in *Artemia franciscana* embryos: reversible arrest of metabolism and gene expression at low oxygen levels. Journal of Experimental Biology 201, 1233-1242.

Heidelbach, M., Skladny, H., Schairer, H.U. 1993. Heat shock and development induce synthesis of a low-molecular-weight stress-response protein in the myxobacterium *Stigmatella aurantiaca*. Journal of Bacteriology 175, 7479-7482.

Heikkila, J.J. 1993. Heat shock gene expression and development. I. An overview of fungal, plant, and poikilothermic animal developmental systems. Developmental Genetics 14, 1-5.

Henriques, A.O., Beall, B.W., Morgan, C.P., Jr. 1997. CotM of *Bacillus subtilis*, a member of the alpha-crystallin family of stress proteins, is induced during developmental and participates in spore outer coat formation. Journal of Bacteriology 179, 1887-1897.

Ireland, R.C., Berger, E.M. 1982. Synthesis of the low molecular weight heat shock proteins stimulated by 20-hydroxyecdysone in a cultured *Drosophila* cell line. Proceedings of the National Academy of Sciences, USA 79, 855-859.

Ingolin, T.D., Craig, E.A. 1982. Four small *Drosophila* heat shock proteins are related to each other and to mammalian alpha-crystallin. Proceedings of the National Academy of Sciences, USA 79, 2360-2364.

Jackson, S.A., Clegg, J.S. 1996. Ontogeny of low molecular weight stress protein p26 during early development of the brine shrimp, *Artemia franciscana*. Development Growth & Differentiation 38, 153-160.

Jakob, U., Gaestel, M., Engel, K., Buchner, J. 1993. Small heat shock proteins are molecular chaperone. Journal of Biological Chemistry 268, 1517-1520.

Jakob, U., Scheibel, T., Bose, S., Reinstein, J., and Buchner, J. 1996. Assessment of the ATP binding properties of Hsp90. Journal of Biological Chemistry 271, 10035-10041.

Joplin, K. H., and Denlinger, D. L. 1990. Development and tissue specific control of the heat shock induced 70kDa related proteins in the flesh fly *Sarcophaga crassipalpis*. Journal of Insect Physiology 36, 239-249.

Joplin, K.H., Yocum, G.D., Denlinger, D.L., 1990a. Cold shock elicits expression of heat shock proteins in the flesh fly, *Sarcophaga crassipalpis*. Journal of Insect Physiology 36, 825-834.

Joplin, K.H., Yocum, G.D., Denlinger, D.L. 1990b. Diapause specific proteins expressed by the brain during the pupal diapause of the flesh fly, *Sarcophaga crassipalpis*. Journal of Insect Physiology 36, 775-783.

Kimmins, S. and MacRae, T.H. 2000. Maturation of steroid receptors: an example of functional cooperation among molecular chaperones and their associated proteins. Cell Stress and Chaperones 5, 76-86.

Knauf, U., Bielka, H., Gaestel, M. 1992. Over-expression of the small heat-shock protein, hsp25, inhibits growth of *Ehrlich ascites* tumor cells. FEBS Letters 309, 297-302.

Krebs, R.A. and Feder, M.E. 1997a. Deleterious consequences of Hsp70 overexpression in *Drosophila melanogaster* larvae. Cell Stress and Chaperones 2, 60- 71.

Krebs, R.A. and Feder, M.E. 1997b. Tissue-specific variation in Hsp70 expression and thermal damage in *Drosophila melanogaster* larvae. Journal of Experimental Biology 200, 2007-2015.

Kurtz, S., Rossi, J., Petko, L., Lindquist, S. 1986. An ancient developmental induction: Heat-shock proteins induced in sporulation and oogenesis. Science 231, 1154-1157.

Lambert, H., Charette, S.J., Bernier, A.F., Guimond, A., Landry, J. 1999. HSP27 multimerization mediated by phosophorylation-sensitive intermolecular interactions at the amino terminus. Journal of Biological Chemistry 274, 9378-9385.

Landry, J., Chretien, P., Lambert, H., Hickey, E., Weber, L.A. 1989. Heat shock resistance conferred by expression of the human HSP27 gene in rodent cells. Journal of Cell Biology 109, 7-15.

Lavoie, J.N., Gingras-Breton, G., Tanguay, R.M., Landry, J. 1993. Induction of Chinese hamster HSP27 gene expression in mouse cells confer resistance to heat shock. Journal of Biological Chemistry 268, 3420-3429.

Lavoie, J.N., Hickey, E., Weber, L.A., Landry, J. 1993. Modulation of actin microfilament dynamics and fluid phase pinocytosis by phosphorylation of heat shock protein 27. Journal of Biological Chemistry 268, 24210-24214.

Lee, R.E., Jr., Dommel, R.A., Joplin, K.H., Denlinger, D.L. 1995. Cryobiology of the freeze-tolerant gall fly *Eurosta solidaginis*: overwintering energetics and heat shock proteins. Climate Research 5, 61-67.

Leroux, M.R., Ma, B.J., Batelier, G., Melki, R., Candido, P.M. 1997. Unique structural features of a novel class of small heat shock proteins. Journal of Biological Chemistry 272, 12847-12853.

Liang, P., Amons, R., MacRae, T.H., Clegg, J.S. 1997b. Purification, structure and in vitro molecular-chaperone activity of *Artemia* p26, a small heat-shock/alpha-crystallin protein. European Journal of Biochemistry 243, 225-232.

Liang, P., MacRae, T.H. 1997a. Molecular chaperones and the cytoskeleton. Journal of Cell Science 110, 1431-1440.

Liang, P., MacRae, T.H. 1999. The synthesis of a small heat shock/alpha-crystallin protein in *Artemia* and its relationship to stress tolerance during development. Developmental Biology 207, 445-456.

Lindquist, S. 1986. The heat shock response. Annual Review of Biochemistry 55, 1151-1191.

Luft, J.C. and Dix, D.J. 1999. Hsp70 expression and function during embryogenesis. Cell Stress and Chaperones 4, 162-170.

Madi, A., Punyiczki, M, and Fesus, L. 1997. Lessons to learn from the cell death and heat shock genes of *Caenorhabditis elegans*. Acta Biologica Hungarica 48, 303-318.

Merek, K.B., Groenen, P.J.T., Voorter, C.E.M., de Haard-Hoekman, W.A., Horwitz, J., Bloemendal, H., de Jong, W.W., 1993. Structural and functional similarities of bovine alpha-crystallin and mouse small heat-shock protein. A family of chaperones. Journal of Biological Chemistry 268, 1046-1052.

MacRae, T.H. 2000. Structure and function of small heat shock/alpha-crystallin proteins: established concepts and emerging ideas. Cellular and Molecular Life Sciences (in press).

Michaud, S., Marin, R., Westwood, J. T., and Tanguay, R. M. 1997. Cell-specific expression and heat-shock induction of hsps during spermatogenesis in *Drosophila melanogaster*. Journal of Cell Science 110, 1989-1997.

Minami, Y., Kimura, Y., Kawasaki, H., Suzuki, K., Yahara, I. 1994. The carboxy-terminal region of mammalian hsp90 is required for its dimerization and function in-vivo. Molecular and Cellular Biology 14, 1459-1464.

Miron, T., Wilchek, M., Geiger, B. 1988. Characterization of an inhibitor of actin polymerization in vinculin-rich fraction turkey gizzard smooth muscle. European Journal of Biochemistry 178, 543-553.

170

Miron, T., Vancompernolle, K., Vandekerckhove, J., Wilchek, M., Lutsch, G. 1991. A 25-KD inhibitor of actin polymerization is a low molecular mass heat shock protein. Journal of Cell Biololgy 114, 255-261.

Nakagaki M, Takei R, Nagashima E, Yaginuma T. 1991. Cell-cycles in embryos of the silkworm, *Bombyx-mori*- G2- arrest at diapause stage. Rouxs Archives of Developmental Biology 200, 223-229.

Nathan, D.F., Vos, M.H., and Lindquist, S. 1997. In vivo functions of the *Saccharomyces cerevisiae* Hsp90 chaperone. Proceedings of the National Academy of Sciences, USA. 94, 12949-12956.

Neven, L.G., Haskell, D.W., Guy, C.L., Denslow, N. Klein, P.A., Green, L.G., Silverman A. 1992. Association of 70-kilodalton heat-shock cognate proteins with acclimation to cold. Plant Physiology 99, 1362-1369.

Nunamaker, R.A., Dean, V.C., Murphy, K.E., Lockwood, J.A. 1996. Stress proteins elicited by cold shock in the biting midge, *Culicoides variipennis sonoresis* Wirth and Jones. Comparative Biochemistry and Physiology B - Biochemistry and Molecular Biology 113, 73-77.

Parsell, D.A. and Lindquist, S. 1993. The function of heat-shock proteins in stress tolerance-degradation and reactivation of damaged proteins. Annual Review of Genetics 27, 437-496.

Patterson, N.A. and Kapoor, M. 1995. Developmetnally-regulated expression of heat-shock genes in *Leptosphaeria maculans*. Canadian Journal of Microbiology 41, 499-507.

Piotrowicz, R.S., Levin, E.G., 1997. Basolateral membrane-associated 27-kDa heat shock protein and microfilament polymerization. Journal of Biological Chemistry 272, 25920-25927.

Pla, M., Huguet, G., Verdaguer, D., Puigderrajols, P., Llompart, B., Nadal, A., Molinas, M. 1998. Stress proteins co-expressed in suberized and lignified cells and in apical meristems. Plant Science 139, 49-57.

Plesofsky-Vig, N., Brambl, R. 1985. The heat shock response of fungi. Experimental Mycology 9, 187-194.

Pyza, E., Mak, P., Kramarz, P. Laskowski, R. 1997. Heat shock proteins (HSP70) as biomarkers in ecotoxicological studies. Ecotoxicology and Environmental Safety 38, 244-251.

Rensing, L, Monnerjahn, C., Meyer, U. 1998. Differential stress gene expression during the development of *Neurospora crassa* and other fungi. FEMS Microbiology Letters 168, 159-166.

Rinehart, J.P., Yocum, G.D., Denlinger, D.L. 2000. Developmental upregulation of inducible hsp70 transcripts, but not the cognate form, during pupal diapause in the flesh fly, *Sarcophaga crassipalpis*. Insect Biochemistry and Molecular Biology 30, 515-521.

Rinehart, J.P., Denlinger, D.L. 2000. Heat shock protein 90 is downregulated during pupal diapause in the flesh fly, *Sarcophaga crassipalpis*, but remains responsive to thermal stress. Insect Molecular Biology 9, 641-645.

Russotti, G., Brieve, T.A., Toner, M., Yarmush, M.L. 1996. Induction of tolerance to hypothermia by previous heat shock using human fibroblasts in culture. Cryobiology 33, 567-580.

Thomas, S. R. and Lengyel, J. A. 1986. Ecdysteroid-regulated heat-shock gene expression during *Drosophila melanogaster* development. Developmental Biology 115, 434-438.

Schroder, H., Langer, T., Hartl, F.U., Bukau, B. 1993. DNAK, DNAJ AND GRPE form a cellular chpaerone machinery capable of repairingheat-induced protein damage. EMBO Journal 12, 4137-4144.

Sconzo, G., Palla, F., Agueli, C., Spinelli, G., Giudice, G., Cascino, D., Geraci, F. 1999. Constitutive hsp70 is essential to mitosis during early cleavage of *Paracentrotus lividus* embryos: The blockage of constitutive hsp70 impairs mitosis. Biochemical and Biophysical Research Communications 260, 143-149.

Shibanuma, M, Kuroki, T., Nose, K. 1992. Cell-cycle dependent phosphorylation of HSP28 by TGF beta 1 and H2O2 in mouse osteoblastic cells (MC3T3-E1), but not in their ras-transformants. Biochemical Biophysical Research Communiation 187, 1418-1428.

Storey, K.B. 1997. Metabolic regulation in mammalian hibernation: enzyme and protein adaptations. Comparative Biochemistry and Physiology A- Physiology 118, 1115-1124.

Tammariello, S.P., Denlinger, D.L. 1998. G0/G1 cell cycle arrest in the brain of *Sarcophaga crassipalpis* during pupal diapause and the expression pattern of the cell cycle regulator, proliferating cell nuclear antigen. Insect Biochemistry and Molecular Biolology 28, 83-89.

Terlecky, S.R., Chiang, H.L., Olson, T.S., and Dice, J.F. 1992. Protein and peptide binding and stimulation of in vitro lysosomal proteolysis by the 73-kDa heat-shock protein. Journal of Biological Chemistry 267, 9202-9209.

Ukaji, N., Kuwabara, C., Takezawa, D., Arakawa, K., Yoshida, S., Fujikawa, S. 1999. Accumulation of small heat-shock protein homologs in the endoplasmic reticulum of cortical parenchyma cells in mulberry in association with seasonal cold acclimation. Plant Physiology 120, 481-489.

Walker, G. P. and Denlinger, D. L. 1980. Juvenile hormone and moulting hormone titres in diapause- and non-diapause destined flesh flies. Journal of Insect Physiology 26, 661-664.

Wehmeyer, N., Hernandez, L.D., Finkelstein, R.R., Vierling, E. 1996. Synthesis of small heat-shock proteins is part of the developmental program of late seed maturation. Plant Physiology 112, 747-757.

Wistow, G. 1985. Domain structure and evolution in alpha-crystallins and small heat shock proteins. FEBS Letters 181, 1-6.

Yiangou, M., Tsapogas, P., Nikolaidis, N., Scouras, Z. G. 1997. Heat shock gene expression during recovery after transient cold shock in *Drosophila auraria* (Diptera : Drosophilidae). Cytobios 92, 91-98.

Ylikomi, T., Wurtz, J.M., Syvala, H., Passinen, S., Pekki, A., Haverinen, M., Blauer, M., Tuohimaa, P., and Gronemeyer, H. 1998. Reappraisal of the role of heat shock proteins as regulators of steroid receptor activity. Critical Reviews in Biochemistry and Molecular Biology 33, 437-466.

Yocum, G.D., Joplin, K.H., Denlinger, D. L. 1991. Expression of heat shock proteins in response to high and low temperature extremes in diapausing pharate larvae of the gypsy moth, *Lymantria dispar*. Archives of Insect Biochemistry and Physiology 18, 239-249.

Yocum, G.D., Joplin, K.H., Denlinger, D. L. 1998. Upregulation of a 23 kDa small heat shock protein transcript during pupal diapause in the flesh fly, *Sarcophaga crassipalpis*. Insect Biochemistry and Molecular Biolology 28, 677-682.

Yuan, Y., Crane, D.D., Barry, C.E., Jr. 1996. Stationary phase-associated protein expression in *Mycobacterium tuberculosis*: function of the mycobacterial ± -crystallin homolog. Journal of Bacteriology 178, 4484-4492.

Zdarek, J. and Denlinger, D. L. 1975. Actions of ecdysteroids, juvenoids, and non-hormonal agents on termination of pupal diapause in the flesh fly. Journal of Insect Physiology 21, 1193-1202.

Zimmerman, J. L., Petri, W., and Meselson, M. 1983. Accumulation of a specific subset of *D. melanogaster* heat shock mRNAs in normal development without heat shock. Cell 32, 1161-1170.

Insect Timing: Circadian Rhythmicity to Seasonality
D.L. Denlinger, J. Giebultowicz and D.S. Saunders (Editors)

Regulation of the cell cycle during diapause

Steven P. Tammariello

Department of Biological Sciences, State University of New York at Binghamton, Binghamton, NY, 13902, USA

Insect diapause provides an engaging model for exploring the intricate interplay between signal transduction, gene expression, protein processing and behavior. At the level of the whole organism, development is halted for several months until environmental or genetic cues terminate diapause and trigger progressive development. In at least three insect species the developmental stasis observed during diapause appears to be linked to a cell cycle arrest. Cell proliferation is arrested at diapause initiation and this arrest persists until the conclusion of diapause. This chapter will be separated into three topics: (i) a review of the recent literature in eukaryotic cell cycle regulation, (ii) cell cycle control during insect diapause, and (iii) cell cycle regulation during dormancy periods in non-insect eukaryotes.

1. THE EUKARYOTIC CELL CYCLE

1.1 Introduction to proliferation

Within the last decade a wealth of data has been generated regarding the regulation of the eukaryotic cell cycle (reviewed in Sherr, 2000). For the purpose of this chapter, I will concentrate on the best known enzymes that control cell proliferation. The cell cycle is a highly controlled and ordered process that requires the precise interaction of several proteins. This ordered cycling ensures that cells exiting mitosis will undergo DNA replication before proceeding to another mitotic division.

The eukaryotic cell cycle is normally divided into four distinct phases, G1, S, G2, and M. The S phase is the period of DNA replication, the M phase corresponds to Mitotic division, G1 is the gap phase between mitosis and DNA replication, and G2 is the gap phase between DNA replication and mitosis (Fig. 1). Cells can exit the cell cycle in a stage (normally called G0) during periods of terminal differentiation, senescence, apoptosis, or quiescence, however only quiescent cells are capable of re-entering the cell cycle and resuming proliferation (Cook et al., 2000).

The cell cycle is driven by a series of timed events that involve protein signaling cascades, proteolysis, and differential gene expression. These pathways have been well characterized, and the cellular machinery necessary for cellular proliferation has been identified (reviewed in Sherr, 2000). These cell cycle proteins are highly conserved at the molecular level between organisms as evolutionarily distinct as yeast and humans (reviewed by Jacob, 1995).

Cell proliferation is controlled by the interaction of specific protein kinases and their cyclin partners. This family of protein kinases, known as cyclin-dependent kinases (cdks), complex with a variety of other cell cycle regulatory proteins to maintain the correct order of the cell cycle (reviewed by Nasmyth, 1996). Several cdks genes have been isolated and cloned and their primary partners (called cyclins) have been identified and characterized. The cdk acts as a functional kinase, while its cyclin partner recognizes the substrate to be phosphorylated (reviewed in Sherr, 1994). Complexes necessary for successful proliferation include Cdc2-Cyclin B, Cdk4-Cyclin D, Cdk2-Cyclin A, and Cdk2-Cyclin E (Fig. 1), although this represents only a partial list of functional cyclin/cdk complexes. The temporal gene expression and activity of the main cyclins and cdks throughout the cell cycle has been defined, and a basic model of cellular proliferation has been proposed (Fig. 1). Cyclins D and E are primarily involved in G1 to S phase transition, while Cyclins A and B are involved in ordering the S and G2 stages, respectively.

1.2 G1/S phase progression

Progression from G1 to S is highly regulated by a cellular "checkpoint" due to the importance of fidelity preceding DNA replication. Cells can be arrested in response to a myriad of environmental cues either by decreasing the expression of proliferation genes such as cdks and cyclins, or by the expression of cyclin-dependent kinase inhibitors (CKIs) that bind directly to cyclin/cdk complexes inhibiting kinase activity (See section 1.3).

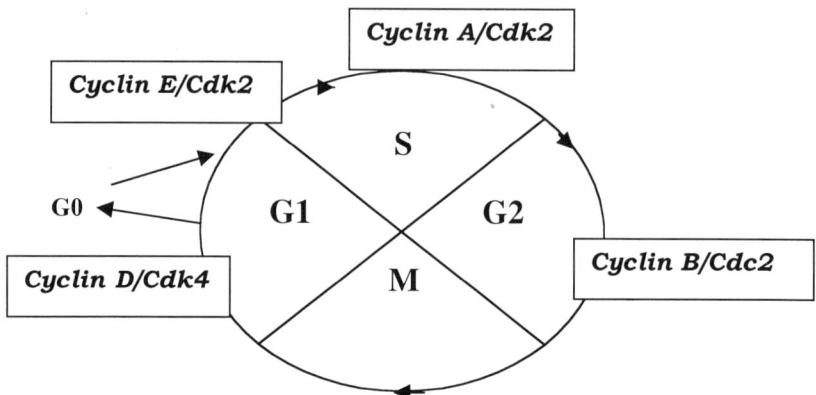

Figure 1. Eukaryotic cell cycle progression by activity of complexes composed of cyclins and cyclin-dependent kinases. Cells can exit the cell cycle in G0 during periods of quiescence, senescence, apoptosis, and terminal differentiation.

Although both Cyclin D and Cyclin E are necessary for G1/S phase transition in most eukaryotic organisms, the specific activation of S phase during embryogenesis in the fruit fly, *Drosophila melanogaster* is linked to the expression of cyclin E (Knobloch et al., 1994). Ectopic expression of cyclin E during a G1 arrest following the exit of the cell cycle in *D. melanogaster* embryogenesis results in progression of the cells into the S phase, and cells from Cyclin E-deficient flies are terminally arrested in G1 (Richardson et al., 1995). Ectopic expression of Cyclin D at this stage does not promote entrance into S phase, and thus is not considered to be sufficient for G1/S phase transition.

Another protein that is necessary for G1 phase progression is the Proliferating Cell Nuclear Antigen (PCNA). This protein is an essential factor for simian virus 40 (SV40) DNA replication *in vitro* and is involved in the elongation stages of DNA replication (Prelich et al., 1987), as well as in cellular chromosomal DNA replication *in vivo* (Jaskulski et al., 1988; Waseem et al., 1992). More specifically, PCNA is a processivity factor of DNA polymerase ∂, an enzyme necessary for DNA replication. Fukuda et al. (1995) proposed the model that PCNA forms a homotrimeric ring that acts as a molecular "clamp" during DNA synthesis, interacting directly with both the DNA strand and the DNA polymerase. In addition to its essential role in replication, PCNA is required for nucleotide excision repair of DNA (Zeng et al., 1994) and also may participate in cell cycle control as demonstrated by an interaction with multiple cdk-cyclin complexes (Xiong et al., 1993). Therefore, PCNA is multifunctional through interaction with specific partners, and all of the functions seem essential for cell proliferation.

1.3 Cyclin-dependent kinase inhibitors

Recent research efforts have focused on isolating and characterizing proteins that can arrest the cell cycle. One such protein is the tumor suppressor p53 that is expressed in response to DNA damage by chemical mutagens or irradiation (reviewed in North and Hainaut, 2000). While the role of p53 as a direct cdk inhibitor has not yet been established, it is known to function as a transcription factor that can mediate the expression of *waf1/cip1 (p21)*. The *waf1/cip1* gene encodes a protein that inhibits cdk-cyclin kinase activity by direct interaction with cdk/cyclin complexes (El Diery et al., 1993; Harper et al., 1993). Currently, at least four members of the p21 family of cyclin dependent kinase inhibitors (CKIs) have been isolated and characterized in eukaryotes (Ravitz and Wenner, 1997).

In vitro studies suggest p21 is a universal cyclin/cdk complex inhibitor, but *in vivo* p21 appears to be a G1-specific cell cycle inhibitor (Xiong et al., 1993). Further, p21 was originally thought to be expressed only under the direct control of *p53* following DNA damage or during apoptosis. Subsequent studies have shown that *p21* can be expressed independently of *p53* in cells undergoing senescence, quiescence, and differentiation (Dulic et al., 1993; El-Diery et al., 1993; Steinman et al., 1994).

Northern blot analysis has identified p53- and p21-like molecules in cell lines from the fall armyworm, *Spodoptera frugiperda* and the fruit fly, *Drosophila melanogaster* in response to DNA-damaging agents (Bae et al., 1995). The *Drosophila* p21 homolog, named *dacapo*, has been cloned and it encodes an inhibitor of cyclin E/cdk2 complexes that is required for epidermal cell cycle arrest during *Drosophila* embryogenesis (Lane et al., 1996).

1.4 G2/M checkpoint

Another cell cycle checkpoint occurs at the G2/M phase transition. This checkpoint ensures that cells complete DNA replication prior to progressing into mitosis. This checkpoint is regulated by the Mitosis Promoting Factor (MPF), which is a complex of cdc2 and cyclin B (Gautier et al., 1990). This complex is activated by removal of a specific phosphate group from the MPF by the Cdc-25 phosphatase (O'Connor, 1997). In *Drosophila*, mitosis entry of most cells can be triggered by brief bursts of the transcription of string (*stg*), the *Drosophila* homolog of Cdc-25 phosphatase. Expression of *stg* ultimately leads to CyclinB-Cdc2 activation and subsequent entry into mitosis (Lehman et al., 1999).

2. CELL CYCLE ARREST DURING INSECT DIAPAUSE

Many insects undergo a period of developmental stasis called diapause in order to circumvent periods of adverse environmental conditions. The life stage at which an insect enters diapause is species specific, but one consistent characteristic is a developmental arrest. Once favorable environmental conditions are restored, insects terminate diapause and resume development. Although a profusion of information has been published on the physiology of insect diapause, very little is known about the molecular mechanism(s) underlying this period of dormancy. In this section, I will focus attention on the seminal papers that link cell cycle arrest to insect diapause in three different insect species (Table 1).

2.1 Flesh fly diapause

The flesh fly, *Sarcophaga crassipalpis,* exhibits a facultative pupal diapause that is induced by short photoperiod coupled with low temperatures (Denlinger 1972; Denlinger, 1985). Four days of short daylength (less than 13.5 hours) must be perceived by the fly during the photosensitive stage, which occurs during late embryonic and early larval life, to initiate the diapause program in *S. crassipalpis* (Denlinger, 1972). Flesh flies reared under short day conditions develop normally until two days after pupation when diapause is initiated and development ceases. Denlinger and colleagues have thoroughly defined the physiology of this model as a period that represents a truly alternate developmental pathway with diapause-specific transcripts and proteins (Flannagan et al, 1998; Joplin et al., 1990).

We first looked for evidence of cell cycle arrest during diapause by performing flow cytometry on flesh fly brains to assess the cell cycle status throughout the diapause program. Brain cells from diapausing pupae are arrested in the G0/G1 phase of the cell cycle, however, this arrest is not observed in brain cells from prediapausing larvae or continuously developing pupae that were reared under long day conditions (Fig. 2B; Tammariello and Denlinger, 1998). Only 60% of the cell population from the pre-diapause brain are in the G0/G1 phase, but the proportion increases to greater than 95% in brain cells from diapausing pupae 10 days after pupariation (Fig. 2B). Concomitantly, the S-phase fraction decreases to <2%, compared to 20% in wandering larvae. This 10-fold decrease in the S-phase fraction between prediapause larvae and 10 day diapausing pupae

Table 1
Insect species with defined cell cycle arrests during diapause.

Insect Species	Diapause Stage	Cell Cycle Arrest	Protein Involved
S. crassipalis	Pupal	G1	PCNA
M. sexta	Pupal	G2	?
B. mori	Embryonic	G2	Cdc2, Cdc25

implies a population of G0/G1-arrested cells. Results from 20 day diapausing pupal brains also indicate a G0/G1 arrest (Fig. 2B); by this time the G0/G1 fraction increases to >97%, and the S-phase fraction drops to less than 1% of the population. Two days after using hexane to terminate diapause, cells begin to break from the G0/G1 arrest. The fraction of G0/G1 cells decreases to less than 80%, which is coupled with an increase in the cells in S phase to approximately 12% of the population. Three days after hexane application, a greater percentage of cells have re-entered the cell cycle suggesting that the cell population is no longer arrested (Fig. 2B). The re-entry of the brain cells into the cell cycle directly corresponds to known developmental markers such as the migration of imaginal discs and the development of the adult eye.

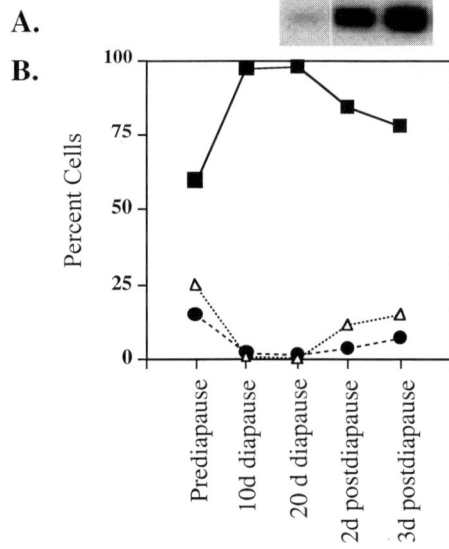

Figure 2. (A.) The expression of *pcna* using Northern blot hybridization at multiple timepoints (Diapause, 2d postdiapause and 3d postdiapause). Notice that increased expression of *pcna* correlates with cell cycle status of the flesh fly brain. (B.) Percentage of cells arrested in Go/G1 (■), S (Δ), and G2/M (●) phases of the cell cycle before, during and after diapause calculated from flow cytometric analysis performed on whole fly brains (Adapted from Tammariello and Denlinger, 1998).

Subsequent analysis using 5-bromodeoxyuridine (BUdR) verified that brain cells from diapausing flies are arrested prior to S phase of the cell cycle. Brain cells from flies in diapause show very little BUdR incorporation while brain cells from flies that had broken diapause exhibited high BUdR incorporation, suggesting that the cells were progressing into the S phase of the cell cycle (data not shown).

Northern blot analysis and *in situ* hybridization identified that one G1/S phase gene, *pcna*, is differentially expressed in brain cells from diapausing and non-diapausing pupae (Tammariello and Denlinger, 1998). Expression of *pcna* is significantly downregulated during diapause, and transcript levels remain low until diapause termination. By 12 hours after diapause termination *pcna* levels increase approximately 15-fold, and this expression pattern correlates with the cell cycle data from flow cytometric analysis as increased *pcna* levels promote proliferation (Fig. 2A). Three other genes that are known to be important in the G1/S phase transition, *cyclin E, p21 (wafl/cipl), and p53*, do not appear to be differentially expressed between diapausing and non-diapausing pupae (Tammariello and Denlinger, 1998).

2.2 Cell cycle status during diapause in Lepidopteran insects

The tobacco hornworm, *Manduca sexta*, undergoes a pupal diapause in response to short daylength experienced during all five larval instars (Bell, 1975). The development of the *M. sexta* eye has been studied extensively, and it serves as a superb model to observe cell cycling during diapause. Champlin and Truman (1998) performed a series of intricate experiments to link ecdysteroid release and cell cycle status during diapause in *M. sexta*. They were able to demonstrate that exposing pupal brains to ecdysteroid levels under a critical threshold arrests cells of the optic lobe anlagen (OA) in the G2 phase of the cell cycle (Champlin and Truman, 1998). This arrest is reversible and proliferation resumes with the addition of suprathreshold levels of 20-hydroxyecdysone.

During uninterrupted (long day) development cells in the OA continuously cycle throughout the pupal stage, however these cells enter a G2 cell cycle arrest when the larvae are reared under short day conditions. This arrest persists until diapause termination at which time cells resume proliferation. Champlin and Truman (1998) used 5-bromodeoxyuridine (BUdR) incorporation to study cell cycle progression in diapausing and nondiapausing pupae. Through pupal day two, high levels of BUdR incorporation, indicative of cell cycling, are visualized in brains from both long day-reared and short day-reared hornworms. By pupal day three, however, BUdR incorporation disappears in short day animals while incorporation in long day animals remains robust. BUdR label continues to be low in the OA of diapausing *M. sexta* throughout diapause, however by 18 hours after diapause termination, BUdR incorporation is restored to the level of long day-reared hornworms.

To verify the G2 arrest in the OA during diapause, Champlin and Truman (1998) used the mitotic inhibitor colcemid in a series of washout experiments. Data from these experiments suggests that OA cells had to go through mitosis before entering S phase after diapause termination, supporting the data that suggested the OA cells were arrested in G2 phase of the cell cycle.

A G2 cell cycle arrest has also been reported during the embryonic diapause of the silkworm, *Bombyx mori* (Nakagaki et al., 1991). Flow cytometric analysis of whole embryos verified the G2 arrest, and subsequent experiments have been initiated to characterize the molecular events associated with this arrest. Yaginuma and colleagues have cloned several genes from *B. mori* that regulate the G2/M transition, including two cdc2-related Ser/Thr kinases, and they have commenced expression studies of these genes to define the role of these genes during *B. mori* diapause (Iwasaki et al., 1997).

3. CELL CYCLE ARREST IN DORMANCY PERIODS OF NON-INSECT SPECIES

3.1 Nematode dormancy is regulated by a G1 cell cycle arrest

Under adverse conditions, such as starvation, the nematode *Caenorhabditis elegans* enters a reversible developmental arrest called the dauer stage or dauer diapause (reviewed in Thomas, 1993). Dauer diapause represents an alternative third larval stage adapted for dispersal and long-term survival. Following such arrest, which may exceed three times the usual life-span of *C. elegans*, worms resume development to form reproductive adults of normal longevity. The dauer stage is under the control of the dauer-formation (*daf*) family of genes, and mutations of these genes can extend life-span two- to fourfold, even in adults that mature without entering the dauer stage (reviewed in Vanfleteren and Braeckman, 1999). Recent evidence suggests that during the dauer stage, cells arrest in the G1 stage of the cell cycle until the termination of the dauer stage when cells resume proliferating (Hong et al., 1998). However, in this model the G1 arrest is regulated by the expression of a cyclin-dependent kinase inhibitor (CKI-1) that appears to be a member of the p21 family of CKIs (Hong et al., 1998).

A similar developmental arrest has been described in the parasitic nematode *Toxocara canis*, whose life cycle contains periods of developmental stasis that are linked to a cell cycle arrest (Loukas and Maizels, 1998). In this model, the cell cycle arrest appears to be linked to the differential expression of the *prohibitin* transcript that encodes a protein known to be intimately involved in the inhibition of cell proliferation and molting in *C. elegans* and *D. melanogaster*.

This phenomenon is not limited to insects and nematodes. The annual killfish, *Austrofundulus limnaeus* also relies on a diapause-like developmental state to circumvent periods when the ephemeral ponds in which they live dry out on a seasonal basis (Podrabsky and Hand, 1999). This normally results in complete mortality of the adults and juveniles, however the populations persist because killfish embryos enter a dormancy state in the sediment of the pond (Podrabsky and Hand, 1999). This period of developmental arrest is characterized by the cessation of somite proliferation until the termination of the dormancy period, athough the molecular control of this period has not yet been defined.

3.2 Apical dominance in the pea plant

Perhaps the most similar molecular model of developmental arrest to that observed in the flesh fly occurs in the dormancy-to-growth transition in the pea plant, *Pisum sativum*. Development and growth is limited to the terminal bud, and there is little or no outgrowth of the axillary buds that lie underneath. However, when the terminal bud is damaged, the axillary buds break from the arrest and begin to proliferate in the same manner as the terminal bud (reviewed in Phillips, 1975). Northern blot analysis confirmed that most of the cells in the dormant axillary buds arrest in the G1 phase of the cell cycle (Shimizu and Mori, 1998a). When the transcription of several cell cycle regulatory genes was measured, *pcna* was found to be expressed at very low levels during the dormancy period in axillary buds (Shimizu and Mori, 1998a). Once the terminal bud is decapitated, *pcna* mRNA levels increase leading to a high level of proliferation and growth in the bud. However, PCNA protein levels remain constant between dormant and growing axillary buds (Shimizu and Mori, 1998b). Why *pcna* transcript is differentially expressed but PCNA protein levels remain constant is not yet understood, but scientists in the field are focusing on proteolysis cycles between dormant and growing buds to establish the turnover rate of PCNA in this model.

4. CONCLUSIONS

From the studies outlined earlier in this chapter, it becomes clear that cell cycle regulation plays an important role during insect dormancy. In at least three insect species cell proliferation is arrested during diapause, however the cell cycle phase of the arrest appears to be species-specific. Brain cells from the flesh fly, *Sarcophaga crassipalpis,* enter a G1 cell cycle arrest during diapause, while the optic lobe cells of the tobacco hornworm, *Manduca sexta*, and embryonic cells from the silkworm, *Bombyx mori*, arrest in the G2 phase of the cell cycle. Truman and colleagues have performed a set of meticulous experiments linking hormonal activity with cell cycle status throughout the diapause program. We are currently examining the early signal transduction events that precede hormone release at diapause termination to establish a potential link between protein signaling cascades and cell cycle re-entry. Several groups, including ours, have also launched efforts to clone cell cycle regulatory genes to study their expression patterns during diapause, however these experiments are still in their infancy. One gene in particular, the proliferating cell nuclear antigen *(pcna)* has been found to be down regulated throughout the pupal diapause of the flesh fly, *Sarcophaga crassipalpis.* Other genes currently under investigation include the G2/M transition genes *cdc2* and *cdc-25* phosphatase and the G1/S transition genes *dacapo* and *cyclin E.* Defining the mechanism of cell cycle control during diapause would not only be of potential benefit for insect control but could ultimately yield significant and novel results in the fields of eukaryotic signal transduction, differential gene expression, and protein processing.

REFERENCES

Bae, I., Smith, M.L. and Fornace, A.J. 1995. Induction of p53-, mdm2-, and waf1/cip1-like molecules in insect cells by DNA damaging agents. Experimental Cell Research 217, 541-545.

Bell, R.A., Rasul, C.G. and Joachim, F.A. 1975. Photoperiodic induction of the pupal diapause in the tobacco hornworm, *Manduca sexta.* Journal of Insect Physiology 21, 1471-1480.

Champlin, D.T. and Truman, J.W. 1998. Ecdysteroid control of cell proliferation during optic lobe neurogenesis in the moth *Manduca sexta.* Development 125, 269-277.

Cook, S.J., Balmanno, K., Garner, A., Millar, T., Taverner, C., and Todd, D. 2000. Regulation of cell cycle re-entry by growth, survival and stress signalling pathways. Biochemical Society Tranactions 28, 233-240.

Denlinger, D.L. 1972. Induction and termination of pupal diapause in *Sarcophaga* flesh flies. Biological Bulletin 142, 11-24.

Denlinger, D.L. 1985. Hormonal control of diapause. In *Comprehensive Insect Physiology, Biochemistry and Pharmacology*, (Eds. Kerkut, G.A. and Gilbert, L. I.) Vol 1, pp. 353-412. Pergamon Press, New York.

Dulic, V., Drullinger, L.F., Lees, E., Reed, S., and Stein, G.H. 1993. Altered regulation of G1 cyclins in senescent human diploid fibroblasts:accumulation of inactive cyclinE-cdk2 and cyclin D1-cdk2 complexes. Cell Biology 90, 11034-11038.

El-Deiry, W.S., Tokino, T., Velculescu, V.E., Levy, D.B., Parsons, R., Trent, J.M., Lin, D., Mercer, W.E., Kinzler, K.W. and Vogelstein, B. 1993. Waf1, a potential mediator of p53 tumor suppression. Cell 75, 817-825.

Flannagan, R.D., Tammariello, S.P, Joplin, K.H., Cikra-Ireland, R.A., Yocum, G.D. and Denlinger, D.L. 1998. Diapause specific gene expression in pupae of the flesh fly, *Sarcophaga crassipalpis.* Proceedings of the National Academy of Sciences USA 95, 5616-5620.

Fukuda, K., Morioka, H., Imajou, S., Ikeda, S., Ohtsuka, E. and Tsurimoto, T. 1995. Structure-function relationship of the eukaryotic DNA replication factor, proliferating cell nuclear antigen. Journal of Biological Chemistry. 270, 22527-22534.

Gautier, J., Minshull, J., Lohka, M.,Glotzer, M., Hunt, T. and Maller, J. 1990. Cyclin is a component of maturation-promoting factor from *Xenopus.* Cell 60, 487-494.

Harper, J.W., Adami, G.R., Wei, N., Keyomarsi, K. and Elledge, S.J. 1993. The p21 Cdk-interacting protein CIP1 is a potent inhibitor of G1 cyclin-dependent kinases. Cell 75, 805-816.

Hong, Y., Roy, R., Ambros, V. 1998. Developmental regulation of a cyclin-dependent kinase inhibitor controls postembryonic cell cycle progression in *Caenorhabditis elegans.* Development 125, 3585-3597.

Iwasaki, H., Takahashi, M., Niimi, T., Yamashita, O., Yaginuma, T. 1997. Cloning of cDNAs encoding *Bombyx* homologues of Cdc2 and Cdc2-related kinase from eggs. Insect Molecular Biology 6, 131-141

Jacobs, T.W. 1995. Cell cycle control. Annual Review of Plant Molecular Biology 46, 317-339.

Jaskluski, D., DeRiel, J.K., Mercer, W.E., Calabretta, B. and Baserga, R. 1988. Inhibition of cellular proliferation by antisense oligodeoxynucleotides to PCNA cyclin. Science 240, 1544-1546.

Joplin, K.H., Yocum, G.D., and Denlinger, D.L. 1990. Diapause specific proteins expressed by the brain during the pupal diapause of the flesh fly, *Sarcophaga crassipalpis.* Journal of Insect Physiology 36, 775-783.

Knobloch, J.A., Sauer, K., Jones, L., Richardson, H.E., Saint, R. and Lehrer, C.F. 1994. Cyclin E controls S phase progression and its down-regulation during *Drosophila* embryogenesis is required for the arrest of cell proliferation. Cell 77, 107-120.

Lane, M.E., Sauer, K., Wallace, K., Jan, Y.N., Lehner, C.F., Vaessin, H. 1996. Dacapo, a cyclin-dependent kinase inhibitor, stops cell proliferation during *Drosophila* development. Cell 87, 1225-1235.

Lehman, D.A., Patterson, B., Johnston, L.A., Balzer, T., Britton,J.S., Saint, R., Edgar B.A. 1999. Cis-regulatory elements of the mitotic regulator, string/Cdc25. Development 126, 1793-1803.

Loukas, A., Maizels, R.M. 1998. Cloning and characterisation of a prohibitin gene from infective larvae of the parasitic nematode *Toxocara canis*. DNA Sequencing 9, 323-328.

Nakagaki, M., Takei, R., Nagashima, E., and Yaginuma, T. 1991. Cell cycles in embryos of the silkworm, *Bombyx mori*:: G2-arrest at diapause stage. Roux's Archives of Developmental Biology 200, 223-229.

Nasmyth, K. 1996. Viewpoint: Putting the cell cycle in order Science 274, 1672-1677.

North, S. and Hainaut, P. 2000. p53 and cell-cycle control: a finger in every pie. Pathological Biology 48, 255-270.

O'Connor, P.M. 1997. Mammalian G1 and G2 phase checkpoints. Cancer Survey 29, 151-182.

Phillips, I. D. 1975. Apical dominance. Annual Review of Plant Physiology 26, 341-367.

Podrabsky, J.E. and Hand, S.C. 1999. The bioenergetics of embryonic diapause in an annual killifish, *Austrofundulus limnaeus*. Journal of Experimental Biology 202, 2567-2580.

Prelich, G., Tan, C-K., Kostura, M., Matthews, M.B., So, A.G., Downey, K.M. and Stillman B. 1987. Functional identitiy of proliferating cell nuclear antigen and a DNA polymerase auxiliary protein. Nature 326, 517-520.

Ravitz, M.J. and Wenner, C.E. 1997. Cyclin-dependent kinase regulation during G1 phase and cell cycle regulation by TGF-beta. Advances in Cancer Research 71, 165-207.

Richardson, H.E., O'Keefe, L.V., Marty, T. and Saint, R. 1995. Ectopic cyclin E expression induces premature entry into S-phase and disrupts pattern formation in the *Drosophila* eye imaginal disc. Development 121, 3371-3379.

Sherr, C. J. 1996. Cancer cell cycles. Science 274, 1672-1677.

Sherr, C. J. 2000. The Pezcoller lecture: cancer cell cycles revisited. Cancer Research 60, 3689-3695.

Shimizu, S. and Mori, H. 1998a. Analysis of cycles of dormancy and growth in pea axillary buds based on mRNA accumulation patterns of cell cycle-related genes. Plant Cell Physiology 39, 255-262.

Shimizu, S. and Mori, H. 1998b. Changes in protein interactions of cell cycle-related genes during the dormancy-to-growth transition in pea axillary buds. Plant Cell Physiology 39, 1073-1079.

Steinman, R.A., Hoffman, B., Iro, A., Guillouf, C., Liebermann, D.A., and El-Houseini, M. E. 1994. Induction of p21 during differentiation. Oncogene 9, 3389-3396.

Tammariello, S.P.and Denlinger, D.L. 1998. G0/G1 cell cycle arrest in the brain of *Sarcophaga crassipalpis* during pupal diapause and the expression pattern of the cell cycle regulator, proliferating cell nuclear antigen. Insect Biochemistry and Molecular Biology 28, 83-89.

Thomas, J.H. 1993. Chemosensory regulation of development in *C. elegans*. Bioessays 15, 791-797.

Vanfleteren, J.R. and Braeckman, B.P. 1999. Mechanisms of life span determination in *Caenorhabditis elegans*. Neurobiology of Aging 20, 487-502.

Waseem, N.H., Labib, K., Nurse, P., Lane, D.P., 1992. Isolation and analysis of the fission yeast gene encoding polymerase delta accessory protein PCNA. EMBO Journal. 11, 5111-5120.

Xiong, Y., Zhang, H., and Beach, D. 1993. p21 is a universal inhibitor of cyclin kinases. Nature 366, 701-704.

Zeng, X.-R., Jiang, Y., Zhang, S.-J., Hao, H., Lee, Y.-W.-T. 1994. DNA polymerase delta is involved in the cellular response to UV damage in human cells. Journal of Biological Chemistry 269, 13748-13751.

Insect Timing: Circadian Rhythmicity to Seasonality
D.L. Denlinger, J. Giebultowicz and D.S. Saunders (Editors)
© 2001 Elsevier Science B.V.   All rights reserved.

Significance of specific factors produced throughout diapause in pharate first instar larvae and adults

Koichi Suzuki, Hiromasa Tanaka and Ying An

Department of Agro-Bioscience, Faculty of Agriculture, Iwate University, Moiroka 020-8550, Japan

Specific factors and molecules associated with diapause in pharate first instar larvae of the wild silkmoth and adults of the leaf beetle are discussed in the context of the regulatory basis for insect diapause and as a source of molecules of potential interest for biotechnology.

1. INTRODUCTION

In most insects of the temperate zone, diapause and hibernation are strategies for reproductive synchronization and survival during difficult seasons in their life cycle. In general, the role of insect diapause has been understood to be an arrest of development that frequently overcomes unsatisfactory seasons. Individuals in diapause accumulate cryoprotectants such as polyols and antifreeze proteins. In this article, we describe specific factors produced during diapause, but factors associated with coldhardiness are covered in other recent reviews (Storey, 1997; Sømme, 1999). To examine specific factors and molecules involved in diapause we have focused on the diapause syndrome in two species. We have especially sought new entomoresources that may be available for the development of insect biotechnology in the new century. This article describes experimental results from a Japanese wild silkmoth *Antheraea yamamai* (Lepidoptera: Saturniidae), a producer of wild silk and a huge experimental insect, and results from the leaf beetle *Gastrophysa atrocyanea* (Coleoptera: Chrysomelidae: Chrysomelinae), a species being used for biological control of the dock weed.

2. LIFE CYCLES OF THE WILD SILKMOTH AND LEAF BEETLE

Insect diapause can occur in any developmental stadium such as egg, larva, pupa, or adult. However, there are many species excluded from this simple classification. Umeya (1946, 1950) observed that egg diapause occurs in the blastoderm or reversed blastoderm embryo, late gastrular embryo, segmented embryo, embryo with appendages, and pharate first instar larva. In his studies, Chippendale (1977) considered diapause of pharate first instar larvae to be a kind of larval diapause. In the case of *A. yamamai*, pharate first instar larvae enter diapause within the chorion in late September and hibernate overwinter (Fig. 1A). In the temperate zone of Japan, larvae hatch in synchrony with the appearance of fresh leaves, and after completing larval development , this species enters a pupal arrest for one month in mid-summer. In early September adults emerge, mate and oviposit their eggs on fine branches of oak. Embryonic development is completed and pharate first instar larvae swallow yolk cells and arrest their development just before hatching. This feature has been observed in many lepidopteran insects, including the gypsy moth *Lymantria dispar* and *Parnassius glacialis*. However, there are some differences between these species because pharate first instar larvae of *L. dispar* enter diapause before swallowing the yolk cells (Fig. 1B), and in *P. glacialis* the pharate first instar larvae swallow the yolk cells in June but do not hatch until the following spring (Fig. 1C). Regardless of species differences, all are obligatory diapauses and are controlled genetically without input from environmental cues.

Figure 1. Diapause features of pharate first instar larvae of three species of Lepidoptera and the leaf beetle adult. A, *Antheraea yamamai*; B, *Lymantria dispar japonica*; C, *Parnassius glacialis*; D, *Gastrophysa atrocyanea*. A, B, and C are naked diapausing pharate first-instar larvae removed from their egg shells. Diapausing adults of *G. atrocyanea* have burrowed into the soil (August 6, 1997) (D).

The leaf beetle *G. atrocyanea* hibernates in the adult stage and mates in spring. Newly emerged adults burrow into the soil and enter diapause after feeding on dock weeds for only one week (Fig. 1D). From June until the following spring, the beetle remains in an obligatory diapause. The Colorado potato beetle *Leptinotarsa decemlineata*, which belongs to the same subfamily, also has an adult diapause, but it is facultative and controlled by photoperiod and other environmental signals. In the case of the leaf beetle, its food plant, the dock weed *Rumex obtrusifolius*, dies back in mid-summer, but fresh leaves appear in the fall and can survive under heavy snow.

3. DIAPAUSE TERMINATION USING ARTIFICIAL COMPOUNDS

Insect diapause can sometimes be broken artificially with chemical compounds. Development in the silkworm *Bombyx mori* is arrested late in the gastrular stage of embryonic development, and this diapause can be prevented or broken with hydrochloride, as reported by Italian scientists over 100 years ago. This method has been improved in Japan and is now used for mass rearing of commercial races. The mechanism behind this popular artificial method for hatching, however, has not yet been elucidated.

Diapause of some cricket eggs can be broken by solutions of urea and organic solvents, and embryonic diapause in *Atrachya menetriesi* can be terminated by dipping in mercuric chloride solution. Although we know many examples of this type, the mechanism of artificial hatching is not yet understood (Yamashita and Suzuki, 1991). In larval and pupal diapause, there are also many reports of artificial diapause termination. Some organic solvents (hexane, cyclohexane, diethyl ether, etc.) are able to break pupal diapause in the flesh fly *Sarcophaga crassipalpis* (Denlinger et al., 1980). Adult diapause is ascribed to a deficiency of juvenile hormone(s) (JHs) even if this deficiency is caused by environmental stimuli or a genetic program. Accordingly diapause in adults can be terminated by the application of JHs or juvenile hormone analogues (JHA). The leaf beetle diapauses in the soil for 10 months, but continuous applications of a JHA (methoprene) to diapausing adults will break the arrest earlier in females than in males (Ichimori et al., 1987). These chemical compounds had been used as tools to explain the control mechanisms. But, they have not yet provided good information about the mode of action of diapause termination or diapause mechanism.

Many anti-juvenile hormone agents (anti-JHA) have been synthesized as new insect growth regulators. Among them, Kuwano et al. (1985) have reported a large number of 1, 5-disubstituted imidazoles, that induce clear precocious metamorphosis in the silkworm. We focused on these chemicals because we anticipated that any kind of anti-JHA could break diapause in pharate first instar larvae of the wild silkmoth, since this diapause has larval features (Chippendale, 1977). We found that an imidazole compound, 1-benzyl-5-[(E)-2,6-dimethyl-1,5-heptadienyl] imidazole (KK-42) does successfully break diapause in pharate first instar larvae of the wild silkmoth. The hatched larvae were reared on fresh oak leaves or artificial diet, and beautiful, green cocoons were produced (Suzuki et al., 1989). Before analyzing the mechanism of this artificial hatching with the imidazole compound, we demonstrated that mass artificial hatching could be accomplished. We first prepared naked

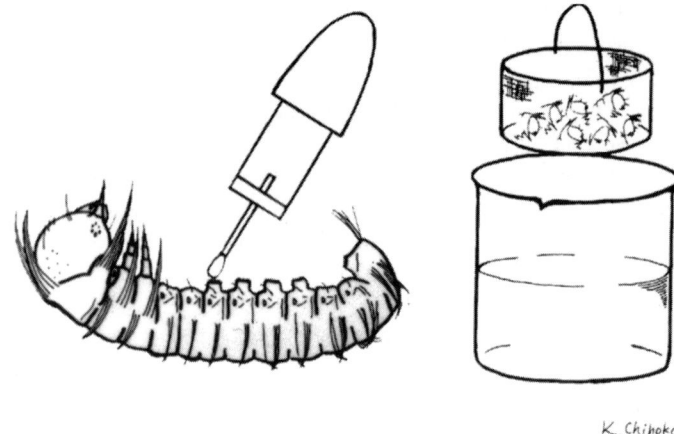

K. Chihoko

Figure 2. Application of imidazole compound to naked diapausing pharate first instar larva (left) and mass artificial hatching (right) in the wild silkmoth.

diapausing pharate first instar larvae by removing the egg shell. A few hundreds of naked individuals were dipped into 0.02% KK-42 in acetone solution for two seconds. KK-42 dissolved in acetone $(0.1\mu g/0.5\mu l)$ was topically applied to the ventral side of each diapausing individual (Fig. 2). This method was more flexible and attainable using naked diapausing pharate first instar larvae, rather than direct dipping of the eggs, because when intact eggs were used it was first necessary to chill them for 10 days at 5 °C just before or after application of the imidazole compound, and dipping of the eggs caused high mortality when a 2% concentration of KK-42 was used. This method is comparable to the techniques of artificial hatching with hydrochloride in *Bombyx* eggs. Two breedings per year could be achieved using this method with fresh oak leaves. If we combined artificial hatching with artificial diet, four breedings could be accomplished theoretically in one year. Recently, Nakamura et al. (1999) reported that KK-62, 1-benzyl-5-[2,6-dimethylheptyl] imidazole is even more suitable for diapause termination than KK-42.

This artificial hatching method using the imidazole compound is also effective in diapausing pharate first instar larvae of the gypsy moth. Superficially the diapause features of this species resemble those of the wild silkmoth, but pharate first instar larvae of the gypsy moth enter diapause before swallowing the yolk cells (Fig. 1B). The application of KK-42 can break this diapause although it is less effective than in the wild silkmoth (Suzuki et al., 1993). This effect on the gypsy moth has also been observed by other research groups

(Bell, 1996; Lee and Denlinger, 1997). In the gypsy moth, the genetic selection of a non-diapause strain is useful for rearing healthy individuals throughout the year for laboratory studies (Hoy, 1977). Artificial hatching with the imidazole compound is noteworthy for analyzing control mechanisms of this diapause but has not yielded results applicable to pest control.

As in diapausing pharate first instar larvae of the wild silkmoth, *P. glacialis* swallows yolk cells but enters diapause already in June as described above (Fig. 1C). We tested high concentrations of KK-42 solution on diapausing individuals but they never broke diapause. This indicates that the double features of a summer and winter diapause in *P. glacialis* are likely different from the winter-only diapause in the wild silkmoth.

4. FUNCTION OF IMIDAZOLE COMPOUNDS

A large number of 1-substituted imidazoles capable of inducing precocious metamorphosis and breaking diapause in pharate first instar larvae of the wild silkmoth have already been prepared by Kuwano et al. (1991) and Nakamura et al. (1998). One of these, KK-42, decreases the JH titer in the hemolymph of the silkworm (Akai and Mauchamp, 1989). This chemical, however, prevents ecdysteroid synthesis in prothoracic glands of the silkworm *in vitro* at a low dose (Yamashita et al., 1987). In the case of this anti-ecdysteroid effect, KK-42 administered to pupae inhibits pupal-adult development (Kadono-Okuda, 1987). Thus, imidazole compounds have been suggested to function as both anti-JHs and anti-ecdysteroids.

Since our findings of diapause termination in response to KK-42 (Suzuki et al., 1989), both possibilities of anti-JHs and anti-ecdysteroids have been investigated by examining the effects of diapause rescue with JHs and JHA, and by determining ecdysteroid titers in diapausing and diapause-terminated individuals. We reported that KK-42 does not function as an anti-JH and anti-ecdysteroid in diapause termination of pharate first instar larvae in the wild silkmoth (Suzuki et al., 1991). We also obtained the same evidence for diapause termination in pharate first instar larvae of the gypsy moth (Suzuki et al., 1993), and consequently we proposed that KK-42 has a hitherto unknown action. Lee and Denlinger (1997) and Lee et al. (1997), however, suggest that KK-42 functions as an anti-ecdysteroid in the case of the gypsy moth, and ecdysteroids are essential for the initiation and maintenance of this diapause.

If imidazole compounds exert an unknown action in diapause termination of the wild silkmoth, it is reasonable to speculate that there is a KK-42-binding protein(s) that acts on the target tissues. Fortunately, we obtained a resin-coupled imidazole compound with sepharose, and isolated two binding proteins from the cytosol fraction in diapausing pharate first instar larvae, using affinity chromatography (Fig. 3, unpublished data). Although we believe that these binding proteins will provide a clue to analyze the mechanism of artificial hatching with imidazole compounds, the candidate proteins also may be involved as a receptor associated with the events of chilling, a prerequisite for diapause termination in the wild silkmoth.

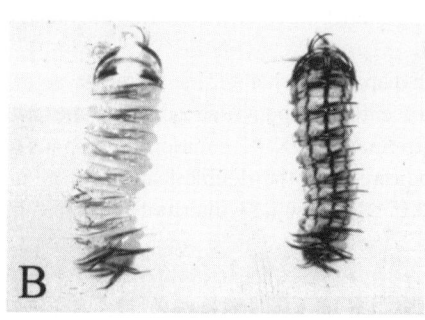

Figure 3. Imidazole compound-binding proteins and the site of action of imidazole compound. A, cytosolic KK-42 binding proteins (45- and 40- kDa) by silver staining; B, Larvae were lighted between the prothorax and mesothorax and the anterior region to the ligation was cut off. Left, acetone recipient, and right individual renewed a KK-42 solution and broke diapause.

Many hibernating plants and insects have regulation systems responsive to long periods of low temperature exposure. The completion of the processes leading to pupal diapause termination requires the presence of the brain, but the molecules responsive to low temperature have not been identified (Chippendale, 1983). In diapause eggs of the silkworm, the metabolism of polyols produced from glycogen is reversely altered by acclimation to low temperatures (Chino, 1958). At diapause termination following acclimation at 5°C, NAD-sorbitol dehydrogenase is activated and sorbitol is converted to glycogen (Yaginuma et al., 1990). This low temperature-sensitive mechanism is complex and involves a number of biochemical and physiological changes. Yaginuma's group reports that mRNA of NAD-sorbitol dehydrogenase increases in the yolk cells during exposure to 5 °C (Niimi et al., 1993), and they have cloned a specific gene, 'Samui', which is expressed by low temperatures in diapausing eggs (Moribe et al., 1999).

The low temperature events that lead to diapause termination may consist of slow and diverse reactions, but the diapause termination caused by chemicals is fast. Although both mechanisms are possibly related, there is still no coherent molecular explanation of this event.

In diapausing pharate first instar larvae of the wild silkmoth, the experiments combining chilling, ligation and KK-42 application indicate that acclimation at 5 °C over long periods and imidazole treatment may act via the same temperature-sensitive receptor localized in the thorax (Suzuki et al., 1994). A probe protein, such as a KK-42-binding protein(s) (Fig. 3), may be a prospective molecule involved in the regulation system responsive to low temperatures. We hence propose a model in which both the imidazole compound and chilling elicit artificial hatching through the same mechanism (Fig. 4).

5. FACTORS PRODUCED THROUGHOUT DIAPAUSE OF PHARATE FIRST INSTAR LARVAE AND ADULTS

Besides ecdysteroids, JHs, prothoracicotropic hormones, and diapause hormone leading to the induction, maintenance, and termination of diapause, a number of biochemical and physiological changes of carbohydrates, amino acids, lipids, and nucleic acids are reported. Progress has been made in many studies, but there is still no understanding how developmental arrest and low metabolism continue. Some storage proteins produced throughout diapause are useful for monitoring diapause but they produce very little regulating information (Ichimori et al., 1990).

Surprisingly, Horie et al. (2000) found that sorbitol produced at diapause initiation and diminished at diapause termination in the silkworm, functions to arrest embryonic

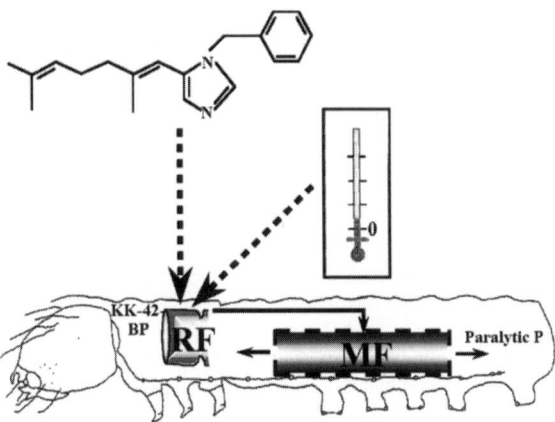

Figure 4. A new model of the control mechanism of diapause of pharate first instar larvae in the wild silkmoth. Imidazole compound (KK-42) and exposure to low temperatures for diapause termination may act via the same receptor localized in the thorax. RF, repressive factor; MF, maturation factor; KK-42-BP, imidazole compound-binding protein; Paralytic P, a member of the paralytic peptide family.

development *in vivo*, and if it is excluded from the culture medium, arrested embryos can initiate development. Consequently, they propose that sorbitol is an intrinsic arrester during diapause. This concentration that is inhibitory in the culture medium is the same level (0.2 M) as sorbitol produced throughout diapause in the silkworm eggs. In diapausing eggs of the silkworm, embryonic cells are arrested in the G_2 phase (Nakagaki et al., 1991). Although it is unclear whether the addition of sorbitol to the culture medium also halts development at the G_2 phase of the cell cycle, sorbitol is a prospetive molecule to evaluate for its role in the maintenance of diapause. Molecular work on diapause-specific genes also provides insight on control mechnism of pupal diapause. The flesh fly *S. crassipalpis* expresses heat shock protein 70 (*hsp* 70) throughout diapause ; the transcript is upregulated at diapause initiation (Rinehart et al., 2000). The role of this gene may be involved in the protection inherent to diapause, but other contributions remain to be analyzed.

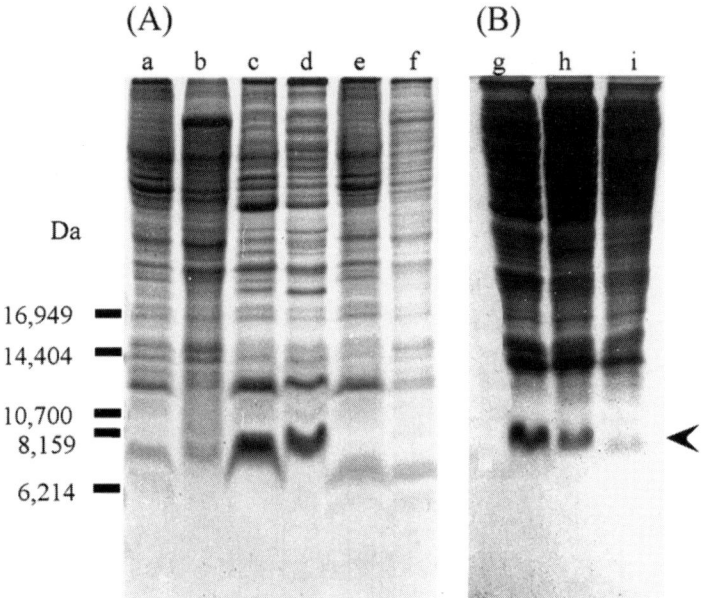

Figure 5. Changes of electrophoretic pattern in tricine SDS-PAGE. (A) Changes of electrophoretic pattern of adult samples (head portion and thorax-abdomen portion) throughout the periods of pre-diapause and diapause, and the overwintered stage. Head portion of pre-diapause (a), diapause (c), and overwintered (e) stages; Thorax-abdomen portion of pre-diapause (b), diapause (d), and overwintered (f) stages. Each lane (a-f) contained about 50 µg of total proteins. (B) Changes of electrophoretic pattern of diapause (g, treated by acetone), terminated diapause by JHA (h) and overwintered (i) adults. Numerals on the left side are the molecular weight of standard proteins. The arrow head indicates a specific band (7.9 kDa).

We found a specific peptide produced throughout diapause in the leaf beetle *G. atrocyanea* (Tanaka et al., 1998). This peptide appears only during diapause and diminishes when diapause is terminated (Fig. 5). This diapause-specific peptide has been isolated and the partial amino acid sequence is AVRIGPXDQVXPRIVPERHEXXRAHGRSGYAYXSGG. No homologous peptides or proteins are known. The molecular weight is approximately 7.9kDa with SDS-PAGE but 4466.3 Da by mass spectrometry. From the immuno-histochemistry and western blotting, it is evident that this peptide is synthesized in diapausing fat body and secreted into the hemolymph. When this antibody is injected into diapausing adults, diapause termination does not occur. These results suggest that the specific peptide has no role in maintaining diapause but may offer a significant defense function against microrganisms. The complete amino acid sequence, cDNA, and its function throughout diapause will be reported to help understand the control mechanism of adult diapause.

A decade ago we proposed a model that two factors are involved in the control mechanism of diapause in pharate first instar larvae of the wild silkmoth. Throughout diapause, a repressive factor (RF) is secreted from the mesothorax and a maturation factor (MF) is secreted from the 2nd of 5th abdominal segments (Suzuki et al., 1990). MF is a peptide-like material and induces melanized stripes in the integument characteristic of post-diapause development (Naya et al., 1994; Fig. 6). This insect has no non-diapause strain, thus for RF

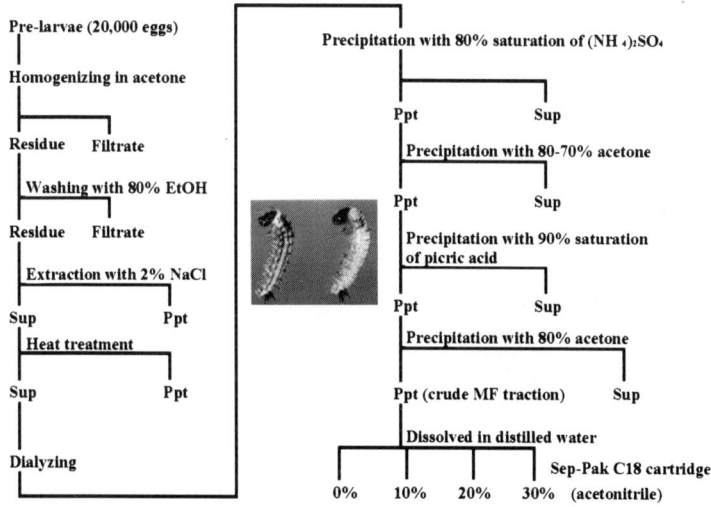

Figure 6. Procedure for partial purification of the maturation factor (MF) from pharate first instar larvae destined for diapause termination. The inset shows a photograph of bioassayed MF activity. Left, injected with MF fraction (0% acetonitrile); Right, injected with distilled water (control).

assay, pharate first instar larvae destined to diapause termination by ligation between the mesothorax and metathorax or by the application of an imidazole compound are used to bioassay RF activity. Thus we are using as an index of RF activity, the delay of diapause termination or the restoration of diapause. The active fractions are extracted from the mesothoracical region of the gut dissected from diapausing individuals or directly from diapausing whole bodies. The activity is heat-stable and partially isolated by reverse-phase HPLC (unpublished data, Fig. 7). These results suggest that RF also is peptide-like material and functions to maintain diapause. Additionally, we observed immunohistochemically that innervation of the anterior region of midgut, close to the foregut-midgut boundary, contains many FMRFamide-like bipolar neurons (An et al., 1998; Fig. 8). In the near future we expect to report a novel sequence, secretory organ, and function for the RF molecule.

6. PROSPECTS FOR APPLICATION OF INSECT DIAPAUSE MOLECULES

In this article, we propose an improved model for regulation of obligatory diapause in pharate first instar larvae of the wild silkmoth. Fig. 4 shows four potentially critical molecules : imidazole compound-binding protein, RF, MF, and Paralytic Peptide (Paralytic P). Paralytic P was isolated from larval hemolymph in the wild silkmoth, and its amino acid sequence indicates it is a novel member of the paralytic peptide family (Seino et al., 1998)

Figure 7. Diapause restoration by the repressive factor fraction in pharate first-instar larvae destined for diapause termination. By a single ligation of the intersegmental membrane between the mesothorax and metathorax, the thorax (metathorax)- abdomen compartments are produced. The posterior compartments show diapause termination by water injection (left) and diapause restoration by double injections of the gut extract (right). Scale bar: 1 mm.

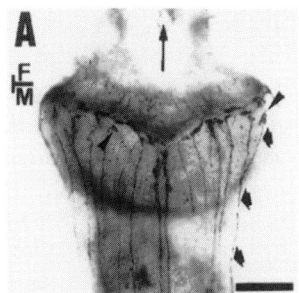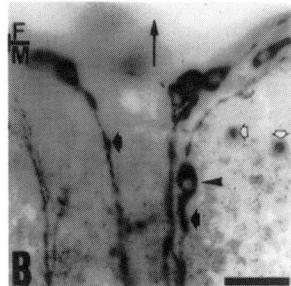

Figure 8. FMRFamide-like immunoreactivity on the surface of a whole-mounted gut of pharate first instar larvae of *A. yamamai*. A, the foregut-midgut boundary; B, expanded from the ventral side of A. F/M indicates the foregut-midgut boundary. Long arrows show the longitudinal body axis. Arrowheads indicate the FMRFamide-like immunoreactive bipolar neurons at the anterior part of the midgut near the foregut-midgut boundary. Black arrows indicate axons, and white arrows the FMRFamide-like immunoreactive endocrine cells in the midgut epithelium. Scale bars: A, 145 μm; B, 200 μm.

(recently, proposed as 'ENF' peptide family by Strand et al., 2000). Paralytic P is included in the figure, even though it is not yet clear whether the sequence from pharate first instar larvae is similar to that in larvae. Activity of Paralytic P appears during diapause, but it is in a family that has multiple biological activities. As described above, these molecules may not only be involved in the control mechanism of diapause of pharate first instar larvae but may play a role in diapause of other stages as well.

In many animals, as well as in insects, diapause or hibernation is an essential feature of the life cycle. From work on the nematode *Caenorhabditis elegans*, it is evident that a *daf*-2 gene of the insulin receptor family exerts diapause-like states and longevity (Kimura et al., 1997). In the chipmunk, a mammalian hibernator, novel blood proteins (20-, 25-, 27-, 55-kDa) start to diminish in concentration before and disappear during hibernation. These proteins reappear in the blood as hibernation ceases and remain during non-hibernation (Kondo and Kondo, 1992). These key genes and proteins may provide valuable tools for understanding molecular mechanism of diapause incidence, but we have not yet found a common regulatory basis. Diapause features and their mechanisms in organisms are surely diversified, thus it may be difficult to achieve an integrated understanding. However, it is possible that common regulatory molecules do exist. Molecules that function as arresters of development or agents that may exihibit interesting antibiotic properties offer attractive potential for biotechnology in the new century.

ACKNOWLEDGEMENTS

We wish to thank Professor D. L. Denlinger of the Department of Entomology, The Ohio State University, for critical reading of the manuscript. This study was supported by a grant from the Research for the Future Program of the Japan Society for the Promotion of Science (JSPS-RFTF 99L01203).

REFERENCES

Akai, H., Mauchamp, B., 1989. Suppressive effects of an imidazle derivative, KK-42 on JH levels in hemolymph of *Bombyx* larvae. The Journal of Sericultural Science of Japan, 58, 73-74.

An, Y., Nakajima, T., Suzuki, K., 1998. Immunohistochemical demonstration of mammalian- and FMRF amide-like peptides in the gut innervation and endocrine cells of the wild silkmoth, *Antheraea yamamai* (Lepidoptera: Saturniidae) during diapause and post-diapause of pharate first-instar larvae. European Journal of Entomology, 95, 185-196.

Bell, R. A., 1996. Manipulation of diapause in the gypsy moth, *Lymantria dispa* L., by application of KK-42 and precocious chilling of eggs. Journal of Insect Physiology, 42, 557-563.

Chino, H., 1958. Carbohydrate metabolism in diapause eggs of the silkworm, *Bombyx mori*. II. Conversion of glycogen into sorbitol and glycerol during diapause. Journal of Insect Physiology, 2, 1-12.

Chippendale, G. M., 1977. Hormonal regulation of larval diapause. Annual Review of Entomology, 22 , 121-138.

Chippendale, G. M., 1983. Larval and pupal diapause. In: Downer, R. G., Laufer, H. (Eds.), Endocrinology of Insects, Alan R. Liss, New York, pp. 343-356.

Denlinger, D. L., Campbell, J. J., Bradfield, J. Y., 1980. Stimulatory effect of organic solvents on initiating development in diapausing pupae of the flesh fly, *Sarcophaga crassipalpis*, and the tobacco hornworm, *Manduca sexta*. Physiological Entomology, 5 , 7-15.

Horie, Y., Kanda, T., Mochida, Y., 2000. Sorbitol as an arrester of embryonic development in diapausing eggs of the silkworm, *Bombyx mori*. Journal of Insect Physiology, 46, 1009-1016.

Hoy, M. A., 1977. Rapid response to selection for a nondiapausing gypsy moth. Science, 196, 1462-1463.

Ichimori, T., Suzuki, K., Kurihara, M., 1987. Sexual difference in the termination of adult diapause of the leaf beetle, *Gastrophysa atrocyanea* Motschulsky (Coleoptera: Chrysomelidae). Applied Entomology and Zoology, 22, 107-109.

Ichimori, T., Ohtomo, R., Suzuki, K., Kurihara, M., 1990. Specific protein related to adult diapause in the beetle, *Gastrophysa atrocyane*a. Journal of Insect Physiology, 36,   85-91.

Kadono-Okuda, K., Kuwano, E., Eto, M., Yamashita, O., 1987. Anti-ecdysteroid action of some imidazole derivatives on pupal-adult development of the silkworm, *Bombyx mori* (Lepidoptera: Bombycidae). Applied Entomology and Zoology, 22, 370-379.

Kimura, K. D., Tissenbaum, H. A., Liu, Y., Ruvkun, G., 1997. *daf-2*, an insulin receptor-like gene that regulates longevity and diapause in *Caenorhabditis elegans*. Science, 277, 942-946.

Kondo, N., Kondo, J., 1992. Identification of novel blood proteins specific for mammalian hibernation. The Journal of Biological Chemistry, 267, 473-478.

Kuwano, E., Takeya, R., Eto, M., 1985. Synthesis and anti-juvenile hormone activity of 1-substituted-5-[(*E*)-2,6-dimethyl-1,5-heptadienyl] imizadoles. Agricultural and Biological Chemistry, 49, 483-486.

Kuwano, E., Fujisawa, T., Suzuki, K., Eto, E., 1991. Termination of egg diapause by imidazoles in the silkmoth, *Antheraea yamamai*. Agricultural and Biological Chemistry, 55, 1185-1186.

Lee, K.-Y., Denlinger, D. L., 1997. A role for ecdysteroids in the induction and maintenance of pharate first instar diapause of the gypsy moth, *Lymantria dispar*. Journal of Insect Physiology, 43, 289-296.

Lee, K.-Y., Valaitis, A. P., Denlinger, D. L., 1997. Further evidence that diapause in the gypsy moth, *Lymantria dispa*, is regulated by ecdysteroids: a comparison of diapause and nondiapause strains. Journal of Insect Physiology, 43, 897-903.

Moribe, Y., Niimi, T., Yaginuma, T., 1999. A cold-inducible *Bombyx* gene, *Samui*, encoding a BAG-protein. Proceedings of the Japan Society for Comparative Endocrinology, No.14, p. 24.

Nakagaki, M., Takei, R., Nagashima, E., Yaginuma, T. 1991. Cell cycles in embryos of the silkworm, *Bombyx mori*: G_2-arrest at diapause stage. Roux's Archives of Developmental Biology, 200, 223-229.

Nakamura, A., Sagisaka, A., Suzuki, K., Kuwano, E., 1998. Precocious metamorphosis in *Bombyx mori* larvae and diapause termination in pharate first-instar larvae of *Antheraea yamamai* induced by 1-substituted imidazoles. Journal of Pesticide Science, 23, 117-122.

Nakamura, A., Kuwano, E., Suzuki, K., 1999. Effects of imidazole compounds on control of life cycle of the wild silkmoth, *Antheraea yamamai*. International Journal of Wild Silkmoth & Silk, 4, 65-72.

Naya, S., Sagisaka, A., Suzuki, K., 1994. Identification of a maturation factor inducing post-diapause development in pharate first-instar larvae of the wild silkmoth, *Antheraea yamamai*. International Journal of Wild Silkmoth & Silk, 1, 195-200.

Niimi, T., Yamashita, O. Yaginuma, T., 1993. A cold-inducible *Bombyx* gene encoding a protein similar to mammalian sorbitol dehydrogenase: yolk nuclei-dependent gene expression in diapause eggs. European Journal of Biochemistry, 213, 1125-1131.

Rinehart, J. P., Yocum, G. D., Denlinger, D. L., 2000. Developmental upregulation of inducible hsp 70 transcripts, but not the cognate form, during pupal diapause in the flesh fly, *Sacrophaga crassipalpis*. Insect Biochemistry and Molecular Biology, 30, 515-521.

Seino, A., Sato, Y., Yamashita, T., Sato, Y. Suzuki, K., 1998. Identification of a novel member of the paralytic peptide family in the silkmoth *Antheraea yamamai*. The Journal of Sericultural Science of Japan, 67, 473-478.

Sømme, L., 1999. The physiology of cold hardiness in terrestrial arthropods. European Journal of Entomology, 96, 1-10.

Storey, K. B., 1997. Organic solutes in freezing tolerance. Comparative and Biochemical Physiology, 17A, 319-326.

Strand, M. R., Hayakawa, Y., Clark, K. D., 2000. Plasmatocyte spreading peptide (PSP1) and growth blocking peptide (GBP) are multifunctional homologs. Journal of Insect Physiology, 46, 817-824.

Suzuki, K., Fujisawa,T., Kurihara, M., Abe, S., Kuwano, E., 1989. Artificial hatching in the silkworm, *Antheraea yamamai*: application of KK-42 and its analogs. In: Akai, H., Wu, Z. S. (Eds.), Wild Silkmoth '88, International Society for Wild Silkmoths, Tokyo, pp.79-84.

Suzuki, K., Minagawa, T., Kumagai, T., Naya, S., Endo, Y., Osanai, M., Kuwano, E., 1990. Control mechanism of diapause of the pharate first-instar larvae of the silkmoth *Antheraea yamamai*. Journal of Insect Physiology, 36, 855-860.

Suzuki, K., Naya, S., Kumagai, T., Minakawa,T., Fujisawa, T., Kuwano, E., 1991. The mode of action of KK-42 on diapause breakdown in pharate first instar larvae of the wild silkmoth, *Antheraea yamamai*. In: Akai, H., Kiuchi, M. (Eds.), Wild Silkmoths '89•'90, International Society for Wild Silkmoths, Tokyo, pp.73-79.

Suzuki, K., Nakamura, T., Yanbe, T., Kurihara, M., Kuwano, E., 1993. Termination of diapause in pharate first-instar larvae of the gypsy moth *Lymantria dispar japonica* by an imidazole derivative KK-42. Journal of Insect Physiology, 39, 107-110.

Suzuki, K., Kumagai, T., Naya, S., Kuwano, E., 1994. Effect of chilling on diapause termination of pharate first-instar larvae in the wild silkmoth, *Antheraea yamamai*. International Journal of Wild Silkmoth & Silk, 1, 60-64.

Tanaka, H., Sudo, C., An, Y., Yamashita, T., Sato, K. Kurihara, M., Suzuki, K., 1998. A specific peptide produced adult diapause of the leaf beetle, *Gastrophysa atrocyanea* Motschulsky (Coleoptera: Chrysomelidae). Applied Entomology and Zoology, 33, 535-543.

Umeya, Y., 1946. Embryonic hibernation and diapause in insects from the viewpoint of the hibernating-eggs of the silkworm. Bulltein of The Sericultural Experiment Station, 12, 393-481. In Japanese with English summary.

Umeya, Y., 1950. Studies on embryonic hibernation and diapause in insects. Proceedings of The Japan Academy, 26, 1-9.

Yaginuma, T., Kobayashi, M., Yamashita, O., 1990. Distinct effects of different low temperatures on the induction of NAD-sorbitol dehydrogenase activity in diapause eggs of the silkworm, *Bombyx mori*. Journal of Comparative Physiology B, 160, 277-285.

Yamashita, O., Kadono-Okuda, K., Kuwano, E., Eto, M., 1987. An imidazole compound as a potent anti-ecdysteroid in an insect. Agricultural and Biological Chemistry, 51, 2295-2297.

Yamashita, O., Suzuki, K., 1991. Roles of morphogenic hormone in embryonic diapause. In: Gupta, A. P. (Ed.), .Morphogenic Hormones in Arthropods 3, Rutger University Press, New Brunswick, New Jersey, pp. 82-128.

Surviving winter with antifreeze proteins: studies on budworms and beetles

Virginia K. Walker[a], Michael J. Kuiper[a], Michael G. Tyshenko[a], Daniel Doucet[a], Steffen P. Graether[b], Yih-Cherng Liou[b], Brian D. Sykes, Zongchao Jia[b], Peter L. Davies[a,b] and Laurie A. Graham[b]

Departments of Biology[a] and Biochemistry[b], Queen's University, Kingston, Ontario K7L 3N6 Canada

The rigors of cold climates have resulted in the evolution of unique, hyperactive antifreeze proteins that bind to microscopic ice crystals. Gene isolation and structure-function studies on these antifreeze proteins have led to an increased appreciation of the strategies employed by insects for winter survival.

1. OVERWINTERING STRATEGIES AND ANTIFREEZE PROTEINS

The onset of winter heralds a host of changes to the physiology of insects adapted for survival at seasonally low temperatures. Some insects, those that are freeze tolerant, may upregulate metabolic pathways for the production of cryoprotectants such as sugars or polyhydroxy alcohols, as well as amino acids (Duman et al., 1991; Storey and Storey, 1991; Danks et al., 1994). Freeze-tolerant insects also raise their supercooling point, with some producing ice nucleators to ensure freezing at high subzero temperatures, and as well they may accumulate other macromolecules such as enzymes for anaerobic glycolysis (Storey and Storey, 1983; Storey and Storey, 1986; Duman et al., 1991). Most studied overwintering insects, however, have an alternative strategy for winter survival and are freeze susceptible. To avoid freezing, and consequent death, these insects decrease their supercooling points. They may synthesize low molecular weight cryoprotectants such as polyols to lower the freezing point of their hemolymph. To avoid inoculating ice they may seal themselves in cocoons or hibernacula and eliminate materials that could act as ice nucleators. Additional strategies include an increase in the production of molecules which serve as energy sources during the long period of food deprivation, and the synthesis of stress proteins for protection against cold shock (Lee, 1991; Denlinger et al., 1991; Danks et al., 1994; Joanisse and Storey, 1996). Some freeze-susceptible insects may also synthesize thermal hysteresis proteins (THPs) which depress the hemolymph freezing point, relative to the melting point, by inhibiting the growth of ice until the non-equilibrium freezing point is reached. Although thermal hysteresis (TH) activity was originally described in insects (Ramsay, 1964; Grimstone et al. 1968), the subsequent discovery and extensive characterization of proteins with similar properties from fish, termed antifreeze proteins (AFPs), has prescribed a name change on the insect proteins.

All AFPs have TH activity, which can be determined by a number of techniques. In our laboratories, we use a nanoliter osmometer (Clifton Technical Physics, Hartford, NY, USA) according to established methods (Chakrabartty and Hew, 1991), and TH activity is plotted as a function of protein concentration. Ice crystal morphology as viewed through the microscope, is useful too, as an indication of ice-binding activity, particularly at low AFP concentrations. We also use an ice recrystallization inhibition assay (Knight *et al.*, 1984) that is based on the ability of AFPs to keep ice crystal size small after a snap freeze followed by warming to a temperature close to the melt. Other laboratories have used alternative techniques to determine TH activity (Hansen and Baust, 1988; Duman *et al.*, 1991). All are useful, but different techniques can lead to variations in reported values, making it difficult to directly compare one laboratory's results with another.

The high concentrations of AFPs in the serum of certain north Atlantic and Antarctic fish aided the characterization of the fish proteins and the cloning of the corresponding genes (reviews: Davies and Hew, 1990; Cheng and DeVries, 1991; Davies and Sykes, 1997; Ewart *et al.*, 1999). In contrast, hemolymph from overwintering insects is more difficult to obtain. As well, some insects show lower concentrations of AFP than are typically found in fish serum. As a consequence, almost 30 years of work in laboratories around the world elapsed between the observation of protein-mediated TH in insects and the first reports of successful purification and cloning of the corresponding cDNAs for these AFPs (Graham *et al.*, 1997; Tyshenko *et al.*, 1997; Duman *et al.*, 1998).

2. BUDWORMS AND BEETLES

Although TH activity had been reported in a variety of insects, the two freeze-susceptible species, spruce budworm (*Choristoneura fumiferana*) and mealworm beetles (*Tenebrio molitor*) represented particularly promising candidates for the purification of AFPs as both can be reared under laboratory conditions. Spruce budworm is a major pest of the boreal forest in North America and overwinters as 1 mm-long second-instar larvae in hybernacula at the tips of conifer branches, at temperatures of -30°C or lower. Conveniently, the budworm's obligatory diapause can be imitated in the laboratory by keeping hatched larvae on cheesecloth at 2-4°C, for 12-30 weeks. While in diapause, glycerol concentration increases 10-fold (Han and Bauce, 1993; 1995) and AFP activity is present (Hew *et al.*, 1983). In the forest, spruce budworm will terminate diapause in the spring and molt to the third-instar. Development then proceeds quickly with rapid molts to ensure that the reproductively active adult stage is reached by mid-summer.

The mealworm beetle, *T. molitor*, is a pest of stored grains in temperate climates and overwinters as quiescent larvae at indeterminate instars, requiring at least a year to complete the life cycle (Cotton and St. George, 1929). Under laboratory conditions, mealworms develop year round. Larvae constitutively produce AFP, but after several weeks at 4°C, elevated levels of TH activity are observed (Horwath and Duman, 1983; Graham *et al.*, submitted). In contrast to other overwintering larvae such as the spruce budworm, and the fire-colored beetle, *Dendroides canadensis*, there have been no reports of polyol accumulation correlated with decreased temperatures in *T. molitor* (Duman *et al.*, 1991). This suggests that although the temperatures to which mealworms are exposed in granaries may not be as extreme as those experienced by less domesticated insects, AFPs may nevertheless, contribute substantially to this insect's survival due to the lack of high concentrations of other cryoprotectants.

3. SPRUCE BUDWORM AFP

C. fumiferana second-instar larvae produce a group of 9-12 kDa proteins with TH activity (Tyshenko *et al.*, 1997; Doucet *et al.*, submitted). After purification of the AFP from crude homogenates, endoprotease digestion products were subjected to amino acid sequencing, allowing degenerate oligonucleotide primers to be designed for the isolation of a partial cDNA by polymerase chain reaction (PCR). Full-length clones were then recovered from a second-instar cDNA library and expressed to give recombinant protein in bacterial extracts. The protein was subsequently purified from the bacterial inclusion bodies and slowly refolded *in vitro*. It showed TH activity of 3.8°C at 2.4 mg/ml (Tyshenko *et al.*, 1997; Gauthier *et al.*, 1998; Fig. 1). The production of a functional recombinant AFP from the cDNA sequence confirmed the identity of the sequence as spruce budworm AFP (sbwAFP) cDNA and also showed that this protein could account for the high TH activity seen in overwintering larvae. The activity of sbwAFP was significantly higher than any AFP isolated from fish, plants or bacteria, and on a molar basis, was estimated to be 10-30 times more active than any known fish AFP (Fig. 1).

Insight into the hyperactivity of sbwAFP was not conspicuous in the amino acid sequence of this unique protein (Fig. 2). Similar sequences were not found in other proteins, including those of the characterized fish AFPs. Overall, sbwAFP is rather hydrophilic, in contrast to fish AFPs (Sönnichsen *et al.*, 1995; Jia *et al.*, 1996). However, threonine residues, which are also found in some fish AFPs, are abundant, immediately suggesting that this amino acid might be important for TH activity in sbwAFP. There are 8 cysteine residues in the 9 kDa, 90-amino-acid AFP, and all of these Cys are disulfide bonded (Gauthier *et al.*, 1998). There is a rapid loss of TH activity when these bonds are broken by reducing agents.

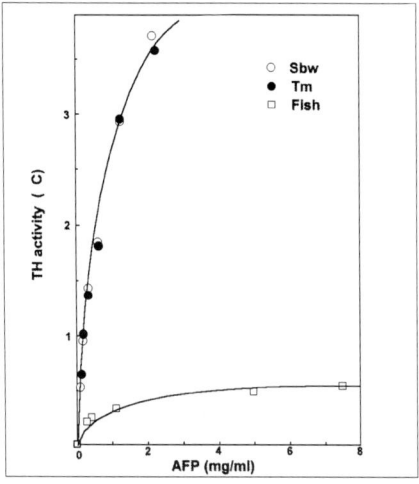

Figure 1. Thermal hysteresis activity as a function of concentration. Recombinant sbwAFP, TmAFP and fish AFP (Type III) are compared

Figure 2. Sequence of two sbwAFP isoforms, encoding a 9 and a 12 kDa protein (upper and lower sequence), showing the repetitive structure and imperfect T-X-T motifs.

In order to identify important residues for ice interaction, additional cDNAs encoding sbwAFP isoforms were isolated and sequenced (Doucet *et al.*, submitted). Since they diverged as much as 36% from the first sequence identified, alignment of "conserved" amino acids in each sequence revealed that there was a sometimes imperfect "Thr-X-Thr" motif (where X could be any residue) spaced approximately 15 residues apart (Fig. 2). Although sbwAFP is not as obviously repetitive as some other AFPs, particularly some fish and the *T. molitor* AFP, one could argue against sbwAFP's initial designation as "non-repetitive" (Cheng, 1998). The importance of some of the Thr residues in individual Thr-X-Thr motifs was demonstrated when they were substituted with Leu by *in vitro* mutagenesis. The alteration of individual residues resulted in up to 80-90% loss of TH activity (Graether *et al.*, 2000). This result may not be surprising since certain Thr residues in fish Type I and III AFPs (Chao *et al.*, 1997; Haymet *et al.*, 1998; Jia *et al.*, 1996) have a central role in ice-adsorption, through hydrogen bonding and/or hydrophobic interactions with water molecules in the ice surface.

The sbwAFP structure was solved using nuclear magnetic resonance (NMR) spectroscopy and showed that the protein was folded into a left-handed, β-helix, with a triangular cross section and ~15 residues to each loop (Graether *et al.*, 2000; Fig. 3). The flat face consisting of Thr-X-Thr motifs appears to mediate ice adsorption. This structure is stabilized by parallel β-sheets and a hydrophobic core, as well as by disulfide bonds. The structure appears to be unique (PDB™, Research Collaboratory for Structural Bioinformatics); certainly it is unlike any of the fish AFPs solved to date.

Ice crystals produced in the presence of sbwAFP are also unique. Ice crystals formed without AFPs are thin circular disks, whereas in the presence of fish AFPs they grow as bipyramid shapes (Fig. 4). With sbwAFP, ice growth not only requires a significantly lower temperature at a given protein concentration than is used with fish AFP solutions, but the microscopic ice

Figure 3. Ribbon illustrations of sbwAFP (left) and TmAFP (right). In each, threonine residues in the ice binding motifs are shown on the β-helix, adjacent to the prism plane of an ice crystal.

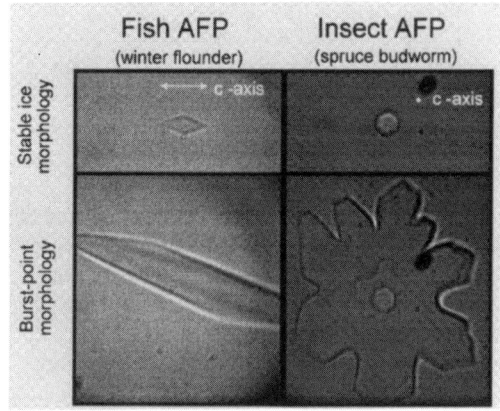

Figure 4. Ice crystal morphology in the presence of fish AFP (Type I) and sbwAFP. Ice crystals were viewed microscopically prior to (stable morphology) and after (burst-point morphology) the non-equilibrium freezing point was reached.

crystals resemble flat, hexagonal-shaped disks. When the temperature of sbwAFP solutions is lowered beyond the non-equilibrium freezing point and freezing of the entire solution commences, the ice grows in waves of smooth-edged layers, in contrast to the burst of sharp spicules, characteristic of the fish AFPs (Fig. 4).

4. SPRUCE BUDWORM AFP EXPRESSION

AFPs were purified from thousands of larvae from a budworm population that had recently originated from the wild and thus there were proteins, and corresponding cDNAs, that were very closely related, likely representing allelic differences. However, more important and significant diversity was found in isolated cDNAs with only 64-78% identity, including those with an internal duplication corresponding to a 30-31 amino acid insertion. These undoubtedly originate from distinct genes and encode distinct isoforms (Doucet *et al.*, submitted). Some cDNAs can also be distinguished by 3' untranslated regions of various lengths (~200, ~500 and ~1000 nucleotides). Southern analysis, using moderately stringent conditions and DNA isolated from single insects, has confirmed that there are ~15 sbwAFP genes (Doucet *et al.*, unpublished). The expression of these genes confers at least 4°C of TH activity to the larvae. Some fish AFP gene families are larger, but copy number often depends on their geographic location; Newfoundland ocean pout have ~150 AFP genes, but ocean pout caught in waters only several degrees in a more southerly latitude have ~40 AFP genes. Winter flounder from the coastal waters of the north Atlantic have ~30 AFP genes. Presumably, marine fish living in ice-laden waters have increased the gene copy number in order to increase amounts of gene product and therefore their TH activity (Hew *et al.*, 1988; Davies *et al.*, 1989; Chen *et al.*, 1997a). Although the number of fish AFP genes correlates with TH activity, the relationship between serum AFP concentrations and TH activity is not linear (Fig. 1). Therefore, notwithstanding the large numbers of AFP genes in some fish, TH activity is ~4-times less than that seen in the budworm, leading to speculation that, had the insect evolved a fish-type AFP, even gene amplifications leading to copy numbers two or three orders of magnitude higher would not have resulted in the observed budworm TH activity. Physiological adaptations including the expression of a small family of hyperactive sbwAFPs, as well as the production of glycerol, provides protection after the onset of seasonally cold weather. Thus the budworm does not require extensive AFP gene amplification that could be costly for such small, streamlined genomes of Lepidopterans and Dipterans (Tyshenko and Walker, 1997).

Expression of the sbwAFP genes appears to be developmentally regulated (Doucet *et al.*, unpublished). Transcripts corresponding to some of the cDNAs are present at egg hatch, but there is an accumulation of many transcripts during the second-instar diapause stage. Levels quickly drop after molting to third instar, which in wild populations, corresponds to the time of spring bud emergence. Synthesis of cryoprotectants in other species appears to be regulated by temperature and/or photoperiod (Horwath and Duman; 1983; Lee, 1991), and not necessarily by the hormonal regimen of diapause (Danks *et al.* 1994). Although AFP gene regulation by diapause signals would be attractive, it has been argued that diapause may have evolved in the tropics, in response to seasonal drought (Danks, 1987). Thus it will be interesting to examine the regulation of sbwAFP mRNA, perhaps even in a transgenic model (e.g. Walker *et al.*, 1995). Although it is not known what factors control sbwAFP gene expression, the isolation of genomic DNA corresponding to several of the cDNA clones would allow this investigation to be initiated.

5. BEETLE AFP

T. molitor larvae produce 8-12 kDa AFP proteins (TmAFP) with very high TH activity; depending upon the isoform, they may have up to 100 times the specific activity of fish AFPs (Graham *et al.*, 1997; Liou *et al.*, 1999; Fig. 1). Since larvae of advanced instars have TH activity, and at 2 cm, are considerably larger than the second-instar budworms, it was feasible to collect hemolymph for protein isolation, and thus initiate the purification with an enriched source of AFP. A leg was severed and hemolymph was collected from the wound in a glass microcapillary tube. Several hundred larvae were bled, which was sufficient for the purification of several micrograms of distinct TmAFP isoforms. In a manner similar to that described for sbwAFP, oligonucleotide primers based on the amino acid sequence allowed the PCR amplification of larval cDNA. The subsequent cDNA library hybridization yielded complete coding sequences (Graham *et al.*, 1997; Liou *et al.*, 1999). TmAFPs are rich in Cys and Thr residues, together representing ~40% of the amino acids. Recombinant TmAFP was recalcitrant to proper folding; a time consuming protocol involving slow oxidation, followed by high pressure liquid chromatography to purify the properly folded form, was required to optimise TH activity after bacterial expression (Liou *et al.*, 2000a). Somewhat reminiscent of the problems with bacterial expression of sbwAFP, the difficulty was again attributed to the necessity for proper disulfide bond formation between the many Cys residues.

The primary sequence of TmAFP is made up of 12-amino-acid repeats of Thr-Cys-Thr-X-Ser-X-X-Cys-X-X-Ala-X, unlike the sequence of any fish AFPs (Graham *et al.*, 1997; Fig. 5). Subsequently, an AFP from another beetle, *D. canadensis*, was shown to have 40-66% amino acid identity with TmAFP and have the same 12-amino-acid repeat, but with an occasional additional residue, and thus designated as a 13-residue repeat (Duman *et al.*, 1998; Li *et al.*, 1998a). The TmAFP structure was solved by X-ray crystallographic analysis (Liou *et al.*, 2000b). The protein folds as a right-handed β-helix; each repeat represents a single turn of the helix and the first three residues of each repeat, Thr-Cys-Thr, together form a flat surface with the Thr residues precisely aligned. This structural analysis substantially confirmed the model of the TmAFP three-dimensional fold that had been deduced from isoform sequences and the

```
           R1              R2              R3              R4              R5              R6
4-9   ACTGCGNCPNAV TCTNSQHCVKAT TCTGSTDCNTAV TCTNSKDCFEAQ TCTDSTNCYKAT ACTNSTGCPGH
1-3   .........H   ...D.KN....A .......K....R ............K ............ ............
2-14  .........    ..........N  ..........Q  ............N ............ .......S....
C-9   ..S.R....K   ........R.R  ........R.M  ............K ............ ....T.......
D-16  .........H   ...D......A  ..........R  ............A ............ .....H......
      TCT.S..C..A. TCT.S..C..A. TCT.S..C..A. TCT.S..C..A. TCT.S..C..A. TCT.S..C..A.
```

Figure 5. Sequences of the 12 residue repeats (R1-R6) found in 5 TmAFP isoforms. Residues in each repeat that match with the consensus are shaded grey.

disulfide bond constraints, in which each 12-amino-acid repeat was stacked to form a β-helix. An interior location for the intra-repeat, disulfide-bonded Cys suggested that the Thr side groups, belonging to the Thr-Cys-Thr motif, might project into the solvent and be ideal for ice interactions (Liou *et al.*, 1999; Fig. 3). The 13-residue repeat found in *D. canadensis* AFP (DAFP) also contains a Thr-Cys-Thr motif and Li *et al.* (1998b) showed that the two Cys residues within each repeat form a disulfide bond with each other. The structure of DAFP has not been reported, but it is likely that the two beetle AFPs share a similar three-dimensional fold.

Although the structure of TmAFP is completely unrelated to any fish AFP, it was apparent that when the sbwAFP structure was solved by NMR techniques, the two insect AFPs showed some remarkable similarities (Fig. 3; Table 1). Both fold as a single β-helix; the sbwAFP with a left-handed β-helix and the TmAFP with a right-handed β-helix of a slightly smaller perimeter, as there are ~3 fewer residues per turn. The sbwAFP ice binding motif, Thr-X-Thr, matches the TmAFP ice binding consensus and in both proteins the three residue sequences are arrayed to form a flat surface with the Thr residues showing a near perfect match to the ice lattice (Liou *et al.*, 2000b; Graether *et al.*, 2000).

Beetle AFPs are represented by large families of isoforms. More than 17 TmAFP cDNAs, encoding 9 protein isoforms, are known. The cDNAs show polymorphisms, either in the coding region, the polyadenylation site, or the 5' untranslated region (Graham *et al.*, 1997; Liou *et al.*, 1999). Isoforms have also been reported in *Dendroides* where more than a dozen different cDNAs have been sequenced (Duman *et al.*, 1998; Andorfer and Duman, 2000). As has been found in sbwAFP, some TmAFP coding region variants are longer. Differences in length corresponded to multiples of the 12-amino-acid repeat with the most common variant containing 6 of the repeats, but with other isoforms corresponding to 7 or 9 repeats (Graham *et al.*, 1997; Liou *et al.*, 1999). In addition to these variants, there is additional isoform diversity due to glycosylation variability. A given protein isoform can be unglycosylated or contain one of two types of N-linked glycan. The sugar groups are attached to an Asn residue at the carboxyl end of the AFP, apparently away from the putative Thr-Cys-Thr ice-binding site. Perhaps not surprisingly then, isoforms show TH activity, independent of their glycosylation status. It is unlikely that AFPs from other insects are extensively modified in this way, as there are few consensus N-glycosylation signals in the published DAFP sequences and none in sbwAFP.

Ice crystals formed in the presence of TmAFP are small, rounded hexagons, radically different from those formed in fish AFP solutions and unlike the flat hexagons of the sbwAFPs. When the temperature is lowered beyond the non-equilibrium freezing point, however, ice forms in smooth layers indistinguishable from the appearance of ice formed in

Table 1
A comparison of spruce budworm, beetle and fish AFPs[1]

	sbwAFP	TmAFP	Fish AFP (Type I)
Molecular weight	9-12 kDa	8-12 kDa	3-5 kDa
TH (1 mg/ml)	3°C	2.5-5°C	0.3°C
Ice crystal morphology	Flat hexagon	Rounded hexagon	Bipyramid
Burst at the non-equilibrium freezing point	Smooth layers	Smooth layers	Sharp spicules
Sequence	~15 residue repeats:	12 residue repeats:	11 residue repeats
	$(TXTX^S/_XXXXXXXXXCXX)_{5-7}$	$(TCTXSXXCXXAX)_{6-9}$	$(TXX^D/_NXXXXXXX)_{3-4}$
Ice binding motif (in bold italics)	*TXT*	*TCT*	TX_3AX_3A
Structure	β-helix	β-helix	α-helix
Gene copy	~15	30-50	30

[1]References for the insect AFPs originated in our laboratories and are cited in the text. Data on the fish Type I AFP (from winter flounder) can be found in Chao *et al.* (1996), Davies and Sykes (1997), Ewart *et al.* (1999) and Baardsnes *et al.* (1999).

the presence of sbwAFP (Fig. 4). Again, like all AFPs tested to date, including sbwAFP, TmAFP is effective at inhibiting ice recrystallization (Kuiper *et al.*, unpublished).

6. BEETLE AFP EXPRESSION

As discussed earlier, TmAFP isoforms are abundant, indicating that they are encoded by a moderately large gene family (Liou *et al.*, 1999). Southern analysis shows 30-50 linked gene copies. Expression of these genes is low throughout larval development, but increases moderately in the final instar, and is elevated to high levels during quiescence (Graham *et al.*, submitted), a delay in development that is a result of environmental stress such as desiccation, starvation or low temperatures. Thus the larvae, which are the overwintering stage, can be induced to accumulate higher levels of AFP transcripts and TH activity depending upon the conditions. This suggests that some aspect of the hormonal control of quiescence may be responsible for TmAFP regulation. In *D. canadensis*, there is also a seasonal increase in transcript and DAFP abundance, but it appears that not all isoforms are identically regulated (Andorfer and Duman, 2000). Although it would be of interest to examine regulatory regions in beetle genomic DNA, this work may be somewhat hampered, at least in *Tenebrio*, by the large copy number and the clustering of linked copies.

7. THE CHALLENGES AHEAD

Although progress has been made towards an understanding of the structure and expression of the insect AFPs, many more questions remain. It is exasperating that enzymologists may know very well how a substrate fits into the active site of their favourite enzyme, and although AFP's "ligand", ice, is made up of only a tetrahydral arrangement of water, we do not yet understand how these proteins adsorb to its surface. *In vitro* mutagenesis and the sequencing of additional isoforms may give clues to important residues for TH activity, molecular modelling and computer docking studies may provide inspiration for additional experimental work, and ice etching (Knight *et al.*, 1991; Graether *et al.*, 2000), as well as ice crystal morphologies, can provide information on which of the possible ice planes, each AFP adsorbs. Notwithstanding this data, however, the details of how these proteins actually bind to ice and function to protect insects, some freeze susceptible, and others freeze tolerant, is not known.

The insect AFPs are much more active relative to AFPs from fish, insects and bacteria (Ewart *et al.*, 1999). What makes them so? We suspect that, at least for sbwAFP, the protein may adsorb to more than one ice face, thereby inhibiting ice growth until a lower temperature is reached (Graether *et al.*, 2000), but we can only speculate why this would be advantageous. In overwintering freeze-susceptible budworm, the hybernaculum could offer protection from innoculative freezing (see Sakagami *et al.*, 1985), glycerol may colligatively lower both the freezing point and supercooling point, and sbwAFP could efficiently adsorb to microscopic ice crystals and lower the freezing point still further. As well, the AFP could also possibly stabilize a supercooled state of the hemolymph and inhibit heterogeneous ice nucleation sites (Duman *et al.*, 1991). Despite all these adaptations, however, it is not really known how these larvae survive temperatures that can dip to -30°C. As well, why would freeze-tolerant insect species sometimes produce AFPs? Presumably their ice nucleators will initiate freezing at elevated subzero temperatures, but AFPs could prevent ice from recrystallizing and thus reducing mechanical damage from large ice crystals (Knight and Duman, 1986). However, do these types of AFPs, by adsorbing to ice crystals in the hemolymph, also change the morphology of the growing ice so that it forms in waves of smooth-edged layers, perhaps again, making it less likely that there will be damage to the insect tissues?

The evolutionary origins of the insect antifreezes are also of interest. Beetles and budworms are unrelated and distinct insects, separated by ~150 million years of evolutionary history. Cold hardy orders are scattered throughout insect phylogeny and appear to have evolved independently (Danks, 1978; Danks, 1981). The ancestry of TmAFP and sbwAFP, as reflected in their primary sequence, is obviously different, but the models of their respective AFP structures are close enough to suggest that the beetle and the budworm AFPs may be an example of convergent evolution (Fig. 3; Table 1). There is some precedence for this phenomenon as similar fish antifreeze glycoproteins, with different genomic origins, have been reported in the evolutionarily and geographically separate northern cod and Antarctic notothenioids (Chen *et al.*, 1997b; Cheng, 1998). What sequence encoding the 12/13-residue and ~15-residue repeats were selected to be reiterated in the beetle and budworm ancestors, respectively? The periodicity of the resulting three-dimensional fold reflected the spacing on the ice crystal lattice and would have lead to the extant insect AFPs. Presumably natural selection favoured those insects that subsequently amplified these genes in the respective genomes, so that they became increasingly tolerant of low winter temperatures. It is possible that we will neither discover the genic origin of these putative ice adsorption motifs, Thr-Cys-Thr and Thr-X-Thr, nor answer all the other posed questions, but continuing research on these hyperactive AFPs promises to be both exciting and rewarding.

8. ACKNOWLEDGEMENTS

Warmest thanks go to Ms. S. Gauthier for technical assistance. Brian Sykes' home department (Biochemistry, University of Alberta) is acknowledged for hosting the NMR studies on sbwAFP. The work has been supported, for the most part, by the Natural Sciences and Engineering Research Council (Canada) and the Medical Research Council (Canada).

REFERENCES

Andorfer, C.A., Duman, J.G., 2000. Isolation and characterization of cDNA clones encoding antifreeze proteins of the pyrochroid beetle *Dendroides canadensis*. Journal of Insect Physiology 46, 365-372.

Baardsnes, J., Kondejewski, L.H., Hodges, R.S., Chao, H., Kay, C., Davies, P.L., 1999. New ice-binding face for type I antifreeze protein. FEBS Letters 463, 87-91.

Chakrabartty, A., Hew, C.L., 1991. The effect of enhanced alpha-helicity on the activity of a winter flounder polypeptide. European Journal of Biochemistry 202, 1057-1063.

Chao, H., Hodges, R.S., Kay, C.M., Gauthier S.Y., Davies, P.L., 1996. A natural variant of type I antifreeze protein with four ice-binding repeats is a particularly potent antifreeze. Protein Science 5, 1150-1156.

Chao, H., Houston, M.E., Hodges, R.S., Kay, C.M., Sykes, B.D., Lowen, M.C., Davies, P.L., Sönnichsen, F.D., 1997. A diminished role for hydrogen bonds in antifreeze protein binding to ice. Biochemistry 36, 14652-14660.

Chen, L., DeVries, A.L., Cheng, C.H.C., 1997a. Evolution of antifreeze glycoprotein gene from a trypsinogen gene in Antarctic notothenioid fish. Proceedings of the National Academy of Sciences, USA 94, 3811-3816.

Chen, L., DeVries, A.L., Cheng, C.H.C., 1997b. Convergent evolution of antifreeze glycoproteins in Antarctic notothenioid fish and Arctic cod. Proceedings of the National Academy of Sciences, USA 94, 3817-3822.

Cheng, C-H.C., DeVries, A.L., 1991. The role of antifreeze glycopeptides and peptides in freeze avoidance of cold-water fish. In: Life under Extreme Conditions, Prisco, G.D. (Ed.), Springer-Verlag, Berlin pp.1-14.

Cheng, C-H.C., 1998. Evolution of the diverse antifreeze proteins. Current Opinion in Genetics and Development 8, 715-720.

Cotton, R.T., St. George, R.A., 1929. The meal worms. Technical Bulletin of the United States Department of Agriculture 95, 1-37.

Danks, H.V. 1978. Modes of seasonal adaptation in the insects. I. Winter survival. Canadian Entomologist 110, 1167-1205.

Danks, H.V., 1981. Arctic Arthropods. A review of the systematics and ecology with particular reference to the North American fauna. Entomological Society of Canada, Ottawa, ON, 608 p.

Danks, H.V., 1987. Insect Dormancy: An Ecological Perspective. Biological Survey of Canada, Ottawa, ON, 439 p.

Danks, H.V., Kukal, O. and Ring, R.A. 1994. Insect cold-hardiness: insights from the Arctic. Arctic 47, 391-404.

Davies, P.L., Fletcher, G.L., Hew, C.L. 1989. Fish antifreeze protein genes and their use in transgenic studies. Oxford Surveys on Eukaryotic Genes 6, 85-109.

Davies, P.L., Hew, C.L. 1990. Biochemistry of fish antifreeze proteins. FASEB Journal 4, 2460-2468.

Davies, P.L., Sykes, B.D., 1997. Antifreeze proteins. Current Opinions in Structural Biology 7, 828-834.

Denlinger, D.L., Joplin, K.H., Chen, C-P., Lee, R.E. Jr., 1991. Cold shock and heat shock. In: Lee, R.E., Denlinger, D.L. (Eds.) Insects at Low Temperature. Chapman and Hall, New York, NY, pp. 131-148.

Doucet, D., Tyshenko, M.G., Kuiper, M.J., Graether, S.P., Sykes, B.D., Daugulis, A.J., Davies, P.L., Walker, V.K. submitted. Structure-function relationships in spruce budworm antifreeze protein revealed by isoform diversity. European Journal of Biochemistry.

Duman, J.D., Xu, L., Neven, L.G., Tursman, D., Wu, D.W., 1991. Hemolymph proteins involved in insect subzero-temperature tolerance: ice nucleators and antifreeze proteins. In: Lee, R.E., Denlinger, D.L. (Eds.) Insects at Low Temperature. Chapman and Hall, New York, NY, pp. 94-127.

Duman, J.G., Li, N., Verleye, D., Goetz, F.W., Wu, D. W., Andorfer, C.A., Benjamin, T., Parmalee, D.C. 1998. Molecular characterization and sequencing of antifreeze proteins from larvae of the beetle *Dendroides canadensis*. Journal of Comparative Physiology B 168, 225-232.

Ewart, K.V., Lin, Q., Hew, C.L., 1999. Structure, function and evolution of antifreeze proteins. Cellular and Molecular Life Science 55, 271-283.

Gauthier, S.Y., Kay, C.M., Sykes, B.D., Walker, V.K., Davies, P.L., 1998. Disulfide bond mapping and structural characterization of spruce budworm antifreeze protein. European Journal of Biochemistry 258, 445-453.

Graether, S.P., Kuiper, M.J., Gagné, S.M., Walker, V.K., Jia, Z., Sykes, B.D., Davies, P.L., 2000. β-helix structure and ice-binding properties of a hyperactive insect antifreeze protein. Nature 406, 325-328.

Graham, L.A., Liou, Y-C., Walker, V.K., Davies, P.L. 1997. Hyperactive antifreeze protein from beetles. Nature 388, 727-728.

Graham, L.A., Walker, V.K., Davies, P.L. submitted. Developmental and environmental regulation of antifreeze proteins in the mealworm beetle *Tenebrio molitor*. European Journal of Biochemistry.

Grimstone, A.V., Mullinger, A.M., Ramsay, J.A., 1968. Further studies on the rectal complex of the mealworm *Tenebrio molitor*, L. Philisophical Transactions B 253, 343-382.

Han, E-N., Bauce, E., 1993. Physiological changes and cold hardiness of spruce budworm larvae, *Choristoneura fumiferana*, during pre-diapause and diapause development under laboratory conditions. Canadian Entomologist 125, 1043-1053.

Han, E-N., Bauce, E., 1995. Glycerol synthesis by diapausing larvae in response to the timing of low temperature exposure, and implications for overwintering survival of the spruce budworm, *Choristoneura fumiferana*. Journal of Insect Physiology 41, 981-985.

Hansen, T.N., Baust, J.G. 1988. Differential scanning calorimetric analysis of antifreeze protein activity in the common mealworm, *Tenebrio molitor*. Biochimica et Biophysica Acta 957, 217-221.

Haymet, A.D., Ward, L.G., Harding, M.M., Knight, C.A., 1998. Valine substituted winter flounder 'antifreeze': preservation of ice growth hysteresis. FEBS Letters 430, 301-306.

Hew, C.L., Kao, M.H., So, Y-P., Lim, K-P., 1983. Presence of cystine-containing antifreeze proteins in the spruce budworm, *Choristoneura fumiferana*. Canadian Journal of Zoology 61, 2324-2328.

Hew, C.L., Wang, N.C., Joshi, S., Fletcher, G.L., Scott, G.K., Hayes, P.H., Buettner, B., Davies, P.L., 1988. Multiple genes provide the basis for antifreeze protein diversity and dosage in the ocean pout, *Macrozoarces americanus*. Journal of Biological Chemistry 263, 12049-12055.

Horwarth, K,L., Duman, J.G. 1983. Photoperiodic and thermal regulation of antifreeze protein levels in the beetle *Dendroides canadensis*. Journal of Insect Physiology 29, 907-917.

Jia, Z., DeLuca, C.I., Chao, H., Davies, P.L. (1996) Structural basis for the binding of a globular antifreeze protein to ice. Nature 384, 285-288.

Joanisse, D.R., Storey, K.B., 1996. Fatty acid content and enzymes of fatty acid metabolism in overwintering cold-hardy gall insects. Physiological Zoology 69, 1977-1095.

Knight, C.A., Cheng, C.C., DeVries, A.L., 1991. Adsoption of alpha-helical antifreeze peptides on specific ice crystal surface planes. Biophysical Journal 59, 409-418.

Knight, C.A., Duman, J.G., 1986. Inhibition of recrystallization of ice by insect thermal hysteresis proteins: a possible cryoprotective role. Cryobiology 23, 256-262.

Knight, C.A., Hallett, J., DeVries, A.L., 1984. Fish antifreeze protein and the freezing and recrystallisation of ice. Nature 308, 295-296.

Lee, R.E., 1991. Principles of insect low temperature tolerance. In: Lee, R.E., Denlinger, D.L. (Eds.) Insects at Low Temperature. Chapman and Hall, New York, NY, pp. 17-46.

Li, N. Kendrick, B., Manning, M. Carpenter, J., Duman J.G., 1998a. Secondary structure of antifreeze proteins from overwintering larvae of the beetle *Dendroides canadensis*. Archives of Biochemistry and Biophysics 360, 25-33.

Li, N., Chibber, B., Castellino, F.J., Duman, J.G., 1998b. Mapping of the disulfide bridges in antifreeze proteins from overwintering larvae of the beetle *Dendroides canadensis*. Biochemistry 37, 6343-6350.

Liou, Y-C., Thibault, P., Walker, V.K., Davies, P.L., Graham, L.A., 1999. A complex family of highly heterogeneous and internally repetitive hyperactive antifreeze proteins from the beetle *Tenebrio molitor*. Biochemistry 38, 11415-11424.

Liou, Y-C., Daley, M.E., Graham, L.A., Kay, C.M., Walker, V.K., Sykes, B.D. and Davies, P.L., 2000a. Folding and structural characterization of highly disulfide-bonded beetle antifreeze protein produced in bacteria. Protein Expression and Purification 19, 148-157.

Liou, Y-C., Tocilj A., Davies, P.L., Jia, Z., 2000. Mimicry of ice structure by surface hydroxyls and water of a β-helix antifreeze protein. Nature, 406, 322-324.

Ramsay, J.A., 1964. The rectal complex of the mealworm *Tenebrio molitor* L. Philosophical Transactions B 248, 279-314.

Sakagami, S.F., Tanno, K., Tsutsui, H., Honma, K., 1985. The role of cocoons in overwintering of the soybean pod borer *Leguminovora glycinivorella*. Journal of the Kansas Entomological Society 58, 240-247.

Sönnichsen, F.D., Sykes, B.D., Davies, P.L., 1995. Comparative modeling of the three-dimensional structure of Type II antifreeze protein. Protein Science 4, 460-471.

Storey, K.B., Storey, J.M. 1983. Biochemistry of freeze tolerance in terrestrial insects. Trends in Biochemical Science 8, 242-245.

Storey, K.B., Storey, J.M. 1986. Winter survival of the gall fly larva, *Eurosta solidaginis*: profiles of fuel reserves and cryoprotectants in a natural population. Journal of Insect Physiology 32, 549-556.

Storey, K.B., Storey, J.M., 1991. Biochemistry of cryoprotectants. In: Lee, R.E., Denlinger, D.L. (Eds.) Insects at Low Temperature. Chapman and Hall, New York, NY, pp. 64-93.

Tyshenko, M.G., Doucet, D., Davies, P.L., Walker, V.K., 1997. The antifreeze potential of the spruce budworm thermal hysteresis protein. Nature Biotechnology 15, 887-890.

Tyshenko, M.G. and Walker, V.K., 1997. Towards a reconciliation of the introns early or late views: triosephosphate isomerase genes from insects. Biochimica et Biophysica Acta 1353, 131-136.

Walker, V.K., Rancourt, D.E., Duncker, B.P. 1995. The transfer of fish antifreeze genes to *Drosophila*: a model for the generation of transgenic beneficial insects. Proceedings of the Entomological Society of Ontario 126, 3-13.

Insect Timing: Circadian Rhythmicity to Seasonality
D.L. Denlinger, J. Giebultowicz and D.S. Saunders (Editors)
© 2001 Elsevier Science B.V.   All rights reserved.

Using ice-nucleating bacteria to reduce winter survival of Colorado potato beetles: development of a novel strategy for biological control

R. E. Lee, Jr.[a], L. A. Castrillo[a,*], M. R. Lee[b], J. A. Wyman[c], and J. P. Costanzo[a]

[a]Department of Zoology, Miami University, Oxford, OH 45056, USA

[b]Department of Microbiology, Miami University, Oxford, OH 45056, USA

[c]Department of Entomology, University of Wisconsin, Madison, WI 53706, USA

A major factor in the overwintering survival of insect pests is their ability to seasonally enhance their cold tolerance by increasing their capacity to supercool and, thus, avoid the lethal effects of internal ice formation. It has now been established that the supercooling capacity of a variety of insects can be significantly reduced by ingestion or surface application of ice-nucleating active (INA) microorganisms. Our recent studies use the Colorado potato beetle, *Leptinotarsa decemlineata,* (Coleoptera: Chrysomelidae) as a model system to address basic questions regarding the seasonal regulation of cold tolerance and insect-microbial interactions. These studies provide a foundation for novel approaches in biological control by manipulating insect cold-hardiness and overwintering survival using INA microorganisms.

1. INTRODUCTION

Most insects are not able to survive internal ice formation. Thus, a key factor in their winter survival is the regulation of the temperature at which they freeze. This temperature is termed the supercooling point (SCP) or the temperature of crystallization (Lee, 1991). As an insect is cooled below the melting point of its body fluids, freezing usually is not immediate. Instead, insects typically supercool many degrees below 0°C before ice nucleation occurs.

A seasonal pattern in the supercooling capacity is common among insects. For many species summer SCP values are often between -4 and -8°C, but gradually decrease in the autumn to -15°C or lower. Many freeze-intolerant species increase their cold tolerance by synthesizing antifreeze proteins and/or

---

*Current address: USDA-ARS, US Plant Soil and Nutrition Lab., Tower Rd., Ithaca, NY 14853

214

by accumulating large amounts of glycerol and other low molecular weight polyols and sugars (see review by Lee et al., 1996).

Various mechanisms play a role in regulating supercooling capacity (Zachariassen, 1992; Duman et al., 1995; Lee and Costanzo, 1998). For many overwintering insects, seasonal increases in supercooling capacity are correlated with increases in glycerol production (Salt, 1968; Somme, 1982). The supercooling limit is determined by the presence of ice catalysts, termed ice-nucleating agents. These compounds function as seeds for ice crystal growth in supercooled liquids. A number of authors have suggested that food material or dust within the gut may function as ice-nucleating agents (see review by Cannon and Block, 1988). Accordingly, cessation of feeding and emptying of the gut are often associated with increases in supercooling capacity. Inoculative freezing, in which contact with external ice initiates freezing of the body fluids, may result in little or essentially no supercooling of the body fluids prior to ice formation. Obviously, this markedly reduces cold tolerance for freezing-intolerant species, including most insect pests.

2. ICE-NUCLEATING ACTIVE MICROORGANISMS

An important implication of these relationships for biological pest control is that any agent that limits the supercooling capacity of a freeze-intolerant insect will increase the likelihood of injury or death following exposure to subzero temperatures. In the 1970s a unique class of biological nucleators, ice-nucleating active (INA) bacteria, was discovered (Maki et al., 1974; Lindow et al., 1978). These bacteria are remarkable for their ability to catalyze ice nucleation at temperatures as high as -1 to -2°C. Ice-nucleating activity is conferred by the presence of *ina* genes (also called *ice* genes), which code for ice-nucleating proteins localized on the bacterium's outer membrane (Warren, 1995). Ice-nucleating proteins have a 16-amino acid sequence repeat and may aggregate to function as templates for the formation of small ice crystal seeds termed "ice nuclei" (Yankofsky et al., 1981; Wolber and Warren, 1989). Ice nuclei activity has been classified by the range of temperatures in which they initiate freezing: type 1 are active between -2 to -5°C, type 2 are active between -5 to -7°C, and type 3 between -7 to -10°C (Yankofsky et al., 1981).

2.1. INA bacteria in insects and other animals

Although most of the reported ice-nucleating bacteria are epiphytic (Gurian-Sherman and Lindow, 1993), some strains have been isolated from the gut of frogs (Lee et al., 1995) and insects (Kaneko, 1991; Lee et al., 1991; Takahashi et al., 1995; Olsen and Duman, 1997). It has been proposed that these microorganisms enhance the survival of freeze-tolerant organisms in winter by triggering freezing at relatively high sub-zero temperatures, a strategy that decreases the chance of osmotic shock and intracellular freezing (Lee, 1991). In freeze-intolerant insects, however, the presence of these ice-nucleating bacteria in the gut generally reduces cold hardiness and increases the likelihood of mortality at subzero temperatures.

2.2. Potential of INA bacteria as biological control agents

The potential use of ice-nucleating bacteria as biological control agents against insect pests first became apparent with the demonstration of the ability of these microorganisms to elevate the SCP of the lady beetle *Hippodamia convergens* (Strong-Gunderson et al., 1990). For freeze-intolerant insect pests whose SCP can be elevated by exposure to INA bacteria, and which are exposed to sufficiently low environmental temperatures to initiate freezing of their body fluids, these microorganisms offer an alternative means of control (Lee et al., 1998).

Several studies have shown that ice-nucleating bacteria may be used to decrease the cold tolerance of a variety of insects. Target insects include storage pests (Fields, 1992, 1993; Fields et al., 1995; Lee et al., 1992), the Russian wheat aphid, *Diuraphis noxia* (Armstrong et al., 1998), the pear psylla, *Cacopsylla pyricola* (Lee et al., 1999), the mulberry pyralid, *Glyphodes pyloalis* (Watanabe et al., 2000), and the Colorado potato beetle, *L. decemlineata* (Lee et al., 1994; Costanzo et al., 1998; Castrillo et al., 2000a, b).

3. USING INA BACTERIA AS BIOLOGICAL CONTROL AGENTS AGAINST COLORADO POTATO BEETLES

We have used the Colorado potato beetle, *L. decemlineata*, as our primary model for these studies because it appears to be an exceptionally well-suited candidate for applications of INA bacteria. This beetle is the most serious defoliating pest of potatoes, *Solanum tuberosum* L., in North America (Hare, 1990). Adults overwinter after burrowing shallowly into the soil in late summer or early autumn (Ushatinskaya, 1978). When they emerge from dormancy they can significantly reduce yields by defoliating the early growth stages of potato plants (Shields and Wyman, 1984). This pest is notorious for rapidly developing resistance to a wide range of pesticides, including synthetic pyrethroids (Casagrande, 1987). Consequently, alternative methods that are compatible with other pest management strategies are urgently needed for controlling this pest.

In our initial study, we determined that the Colorado potato beetle is a freeze-intolerant species that dies when it freezes (Lee et al., 1994). However, overwintering beetles survive to temperatures immediately above their supercooling point, indicating that death is due to the onset of internal ice formation, and not low temperature *per se*. This result also indicates that in this species the SCP may be used as a measure of the lower lethal temperature.

Considering the relatively high SCP of overwintering adults (-7 to -9°C), it is clear that this species lacks exceptional cold tolerance (Lee et al., 1994). Furthermore, because overwintering beetles burrow shallowly (7-14 cm) in certain soils, they may be exposed to subzero temperatures. The elevation of beetle SCPs to as little as 2 to 4 °C could be of major significance in decreasing the proportion of beetles surviving the winter. We demonstrated that exposure of Colorado potato beetles to INA *Pseudomonas syringae* significantly increased SCPs from -7.6 $\pm$ 0.2 °C to -3.7 $\pm$ 0.1°C (Lee et al., 1994).

3.1. Selection of INA bacterial strains for use in biological control

Although killed preparations of INA *P. syringae* sprayed onto CPB adults resulted in elevated SCPs, the effect persisted for only seven days (Lee et al., 1994). Consequently, we considered the feasibility of colonizing the gut of overwintering beetles with INA bacteria to achieve longer lasting effects, since temperatures low enough to kill beetles will occur long after application of INA bacteria. To explore this possibility, we first conducted a survey of the types and prevalence of bacterial flora in the digestive tract of Colorado potato beetles. Our results show a diverse flora that was present in actively feeding beetles in summer, as well as in overwintering adults (see Table 1). The relative abundance of bacterial gut flora was lower in overwintering adults than in actively feeding ones, suggesting that gut evacuation prior to winter, which eliminates food contents, also eliminates some of the gut flora.

Table 1

Bacterial flora isolated from the gut of summer and winter populations of Colorado potato beetles. (M.R. Lee et al., unpublished data)

Bacterial species	Relative abundance	
	Summer	Winter
Alcaligenes sp.		+
Acinetobacter spp.		+
Citrobacter freundii	+++	
Enterobacter agglomerans	++++	+
E. cloacae	++++	
E. taylorae	++++	+
Enterobacter spp.	++++	+
Flavobacterium odoratum	++++	+
Klebsiella oxytoca		+
K. pneumonia	++	
Pseudomonas aeruginosa	+++	
P. fluorescens		+
P. maltophilia	+	
P. paucimobilis	++	
P. stutzeri		+
Xanthomonas maltophilia		+

Ice-nucleating strains that are of the same species as the bacteria retained during the voiding process and which thrive in gut conditions of overwintering beetles are likely candidates for use against Colorado potato beetles.

The feasibility of colonizing the gut of overwintering adults with INA bacteria was confirmed when it was shown that ingestion of living ice-nucleating strains of *P. fluorescens* and *P. putida* persisted for 10 weeks after initial exposure (Costanzo et al., 1998). To simulate natural overwintering conditions, field-collected beetles in autumn were fed INA bacteria before they burrowed into the soil. Beetles were assayed for SCP and gut flora 1.5 hours after feeding, at the conclusion of a 2-week diapause induction regimen, and in mid-winter after the beetles had been in diapause for 2.5 months. Ingestion of INA bacteria caused an immediate increase in the beetle's SCP. Most noteworthy, however, is the fact that *P. fluorescens* and *P. putida* caused not only an initial SCP elevation, but that the SCP remained elevated for 2.5 months. Also of significance is the fact that the SCP remained elevated even after the beetles extensively purged their gut contents in preparation for burrowing in the soil and overwintering in reproductive diapause. Thus, this result demonstrated that natural defense mechanisms in the gut against freezing could be overcome.

Additional laboratory experiments identified two other strains of *P. fluorescens*, frog-derived F26-4C and insect-derived 88-335, able to persist and maintain activity in the beetle gut for 2 to 12 weeks after exposure (Fig. 1). Moreover, positive correlation between elevated SCP in treated beetles and the presence of INA bacteria was confirmed by use of Polymerase chain reaction (PCR) technique (Castrillo et al., 2000a). This method provided molecular evidence for the presence or absence of ice-nucleating bacteria fed to beetles. Using primers specific for the *ina* gene in the different *Pseudomonas* spp. tested, persistence of the two *P. fluorescens* strains was confirmed by the presence of their ice-nucleating gene, *inaW*, in beetles up to 12 weeks after initial exposure (Fig. 2). A band of approximately 4.5 Kb, corresponding to the *inaW* gene, was detected in treated beetles exhibiting elevated SCPs.

Consequently, *P. fluorescens* strains F26-4C and 88-335 were selected for further studies as potential biological control agents for the following reasons:

1. high levels of ice-nucleating activity with some cells active at -2°C,

2. initiate freezing in beetles at temperatures as high as -2.6°C,

3. natural isolates from the gut of frogs and insects (i.e., non-genetically engineered),

4. persist in the gut of overwintering beetles,

5. relatively easy to culture and grow at low temperatures (4°C), and

6. detectable with cultural and molecular (PCR) methods.

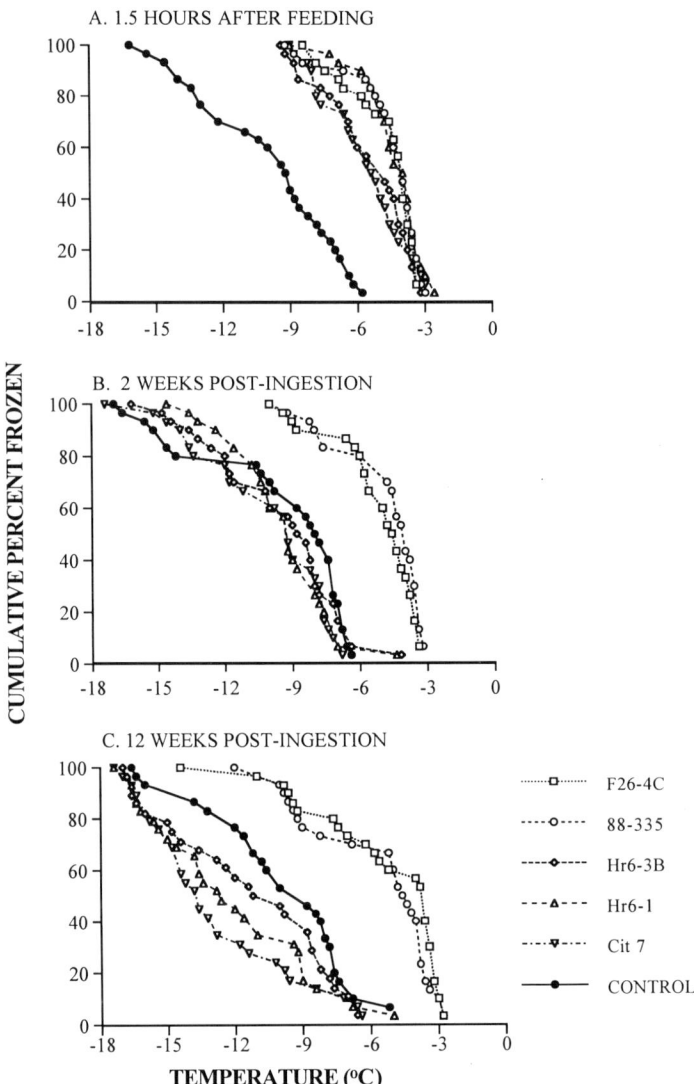

Figure 1. Cumulative freezing profiles based on individual supercooling points of Colorado potato beetles fed potato slices coated with ice-nucleating active *Pseudomonas* spp. Supercooling point values of beetles treated with each bacterial strain, along with control beetles, were measured 1.5 hours (A), 2 weeks (B), and 12 weeks (C) after ingestion. (From Castrillo et al., 2000a.)

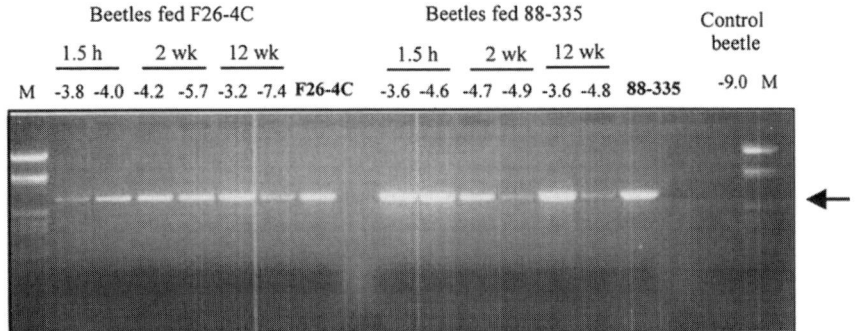

Figure 2. Detection of ice-nucleating active *Pseudomonas fluorescens* strains F26-4C and 88-335 in the gut of Colorado potato beetles at different sampling times after feeding: 1.5 hours, 2 weeks, and 12 weeks. Arrow indicates bands corresponding to the *inaW* gene. (From Castrillo et al., 2000a.)

3.2. Evaluation of bacterial efficacy and persistence in the field

Our field studies during the past two winters support the contention that it is possible to establish INA bacteria in the gut of overwintering adults (Castrillo et al., 2000b; unpublished data). In early autumn, beetles were fed slices of potato tubers coated with INA *P. fluorescens* F26-4C and released in the test arenas in the field. In spring of the following year, before beetles emerged from the ground, the soil was excavated and beetles were recovered. After overwintering for 7 months in the soil, treated beetles still had elevated SCPs, (-4.2 ± 0.1°C) compared to control beetles (-6.4 ± 0.1°C) (Fig. 3, Castrillo et al., unpublished data). Furthermore, SCP values of the recovered treated beetles were comparable to those observed 1.5 hours after exposure to INA bacteria (-4.4 ± 0.2°C) indicating that gut conditions in overwintering adults were favorable for expression of ice-nucleating activity. These data also provide evidence that laboratory results may accurately reflect bacterial activity in the field and that strains persisting for 2 weeks post-ingestion in diapausing beetles are likely to persist through the winter.

Even though the supercooling capacity of overwintering beetles was compromised by the presence of INA bacteria in their gut, overwintering survival of treated beetles with *P. fluorescens* F26-4C (31.8%) remained statistically indistinguishable from control beetles (42.8%) (Castrillo et al., unpublished data). Our field studies show that depth of burrowing determines the temperature overwintering beetles experience, which in turn determines the effect of INA bacteria on beetle survival. Given the unseasonably mild winters during which our field studies were conducted, soil temperatures did not drop low enough to affect a significant proportion of the population. Most of the recovered beetles were in the upper 15 cm of the soil strata and only beetles closer to the soil surface were subject to extreme temperature fluctuations and

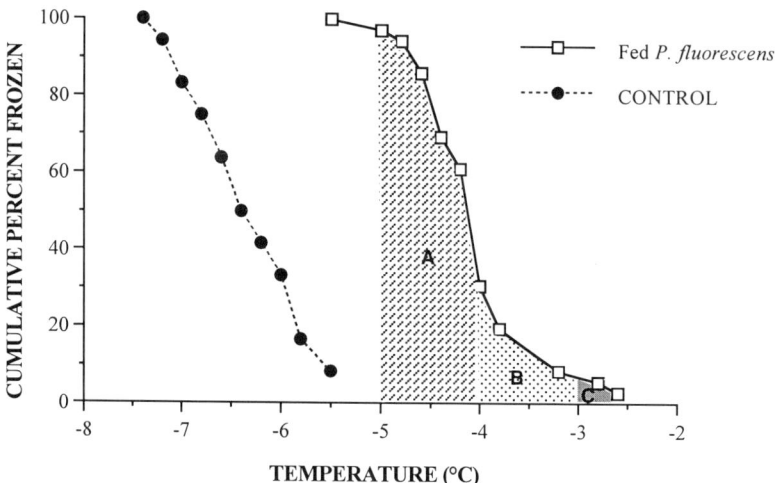

Figure 3. Cumulative freezing profile based on supercooling points of individual Colorado potato beetles fed *Pseudomonas fluorescens* F26-4C and measured after overwintering in the field for 7 months. Hatched areas in treated beetles indicate theoretical mortality at subzero temperatures: A) >95 to 50% mortality at -5 to -4°C; (B) ~50 to 10% at -4 to -3; and (C) <10% at above -3°C. (L.A. Castrillo, R.E. Lee, J.A. Wyman, M.R. Lee, and S.T. Rutherford, unpublished data.)

severe cold. Nevertheless, the relatively shallow depth of burrowing at which most beetles were found suggest that most overwintering beetles could be subject to lethally low temperatures during normal winters. For example, when temperatures in the upper 15 cm of the soil strata drop below -5.0°C, the presence of ice-nucleating *P. fluorescens* in the gut of overwintering adults could initiate internal freezing and, thus, cause notable mortality in the population. Given the range of SCPs of beetles recovered in spring (Fig. 3), 50 to 95% mortality would occur in overwintering adults exposed to soil temperatures as low as -4 to -5°C. In contrast, milder winters with warmer temperatures will reduce the likelihood of mortality due to ice-nucleating bacteria.

The impact of ice-nucleating bacteria on survival of overwintering beetles is likely affected by a number of physical factors (i.e., soil type and soil moisture) that affect depth of burrowing. Heavy soils, which tend to have high moisture levels, limit depth of burrowing and also increase the beetle's sensitivity to cold by increasing their susceptibility to inoculative freezing (Minder, 1966; Kung et al., 1992; Costanzo et al., 1997). Thus, in moist soils where beetles burrow shallowly, the effect of ice-nucleating bacteria on overwintering beetles would be enhanced during a severe winter.

3.3. Enhancement of bacterial activity

Our previous studies indicated that the degree of SCP elevation in treated beetles was affected not only by the number of bacterial cells retained in the gut, but also by the variability in the ice-nucleating activity of individual cells (Castrillo et al., 2000 a, b). Consequently, we conducted experiments to maximize ice-nucleating activity in *P. fluorescens* F26-4C by enhancing the expression of type 1 cells. Although these cells are most desirable in increasing beetle SCP, they generally make up only a small fraction of the bacterial population. In addition to our goal of maximizing bacterial activity, it was also our objective to develop a liquid medium for mass production for future field applications. Previously, bacterial culture in our laboratory was limited to a solid medium (nutrient agar plates with 2.5% glycerol), known to enhance expression of bacterial ice-nucleating activity (Lindow et al., 1982).

We adapted culture conditions used previously to enhance ice-nucleating activity in epiphytic strains (Nemecek-Marshall et al., 1993; Fall and Fall, 1998), and found that *P. fluorescens* F26-4C grown in liquid media could be induced to increase expression of type 1 ice nuclei by shifting the growth medium from 23°C to either 4 or 15°C (Castrillo et al., unpublished data).

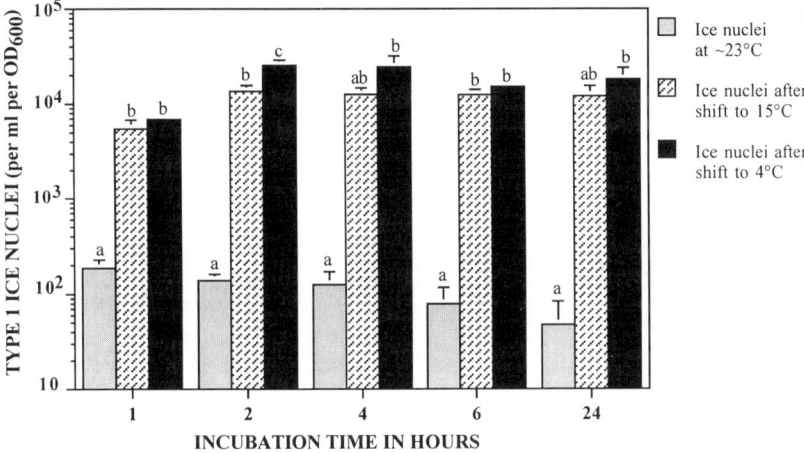

Figure 4. Effect of induction time and temperature on expression of type 1 ice nuclei in *Pseudomonas fluorescens* F26-4C grown in nutrient broth at ~23°C. Bacteria were grown to the stationary phase and then transferred to 4 or 15°C, or maintained at 23°C. Bars with the same letter are not significantly different. (L.A. Castrillo, S.T. Rutherford, R.E. Lee, and M.R. Lee, unpublished data.)

Table 2
Efficacy of ice-nucleating active *Pseudomonas fluorescens* with induced type 1 ice nuclei against overwintering Colorado potato beetles. (L.A. Castrillo, S.T. Rutherford, R.E. Lee, and M.R. Lee, unpublished data.)

Bacterial Culture Conditions			Beetle SCP (°C)*	
Medium	Phase	Induction†	1.5 Hour	2 Weeks
Nutrient agar w/ 2.5% glycerol	Solid	No	$-4.2 \pm 0.2a$	$-4.5 \pm 0.3a$
Nutrient agar w/ 2.5% glycerol	Solid	Yes	$-3.8 \pm 0.3a$	$-4.6 \pm 0.6a$
Nutrient broth w/ 2.5% glycerol	Liquid	Yes	$-4.1 \pm 0.3a$	$-4.3 \pm 0.1a$
L broth with limiting N and P, and with 5% dextrose	Liquid	Yes	$-4.3 \pm 0.1a$	$-4.1 \pm 0.3a$
CONTROL			$-6.9 \pm 0.9b$	$-7.5 \pm 0.3b$

*Mean values $\pm$ SEM ($N = 30$) within a column followed by the same letter are not significantly different at $\alpha = 0.05$ (Fisher protected LSD test).
†Type 1 ice nuclei expression in *P. fluorescens* was induced at 4°C for 2 hours.

Ice-nucleating activity can be enhanced during the exponential phase and into the stationary phase of growth, with greater levels of enhancement during the exponential phase. During the stationary phase optimal induction was achieved after shifting the bacteria to 4°C for 2 hours (Fig. 4). Growth in a defined medium with 5% dextrose or galactose and with limited levels of nitrogen or nitrogen and phosphorus, coupled with a temperature shift to 4°C resulted in the greatest induction of type 1 ice nuclei (Castrillo et al., unpublished data)

Bacterial cells grown in liquid media with induced type 1 ice nuclei were observed to be comparable to cells grown on solid medium in their efficacy against overwintering beetles (Castrillo et al., unpublished data). Beetle SCPs were significantly elevated 1.5 hours after feeding and even after 2 weeks post-ingestion (Table 2). Because *P. fluorescens* F26-4C will be growing under conditions prevailing in the gut of overwintering beetles in the field, induction of bacterial ice-nucleation activity prior to field application may not be necessary. Elevated SCPs of treated beetles were maintained for at least two weeks indicating that the conditions within the beetle gut permit bacterial growth and expression of ice-nucleating activity. Therefore, the low temperature shift that induces bacterial ice-nucleating activity will be provided naturally as soil temperatures decrease during winter.

4. FUTURE STUDIES

4.1. Trap-crop application strategy

One of the challenges that must be met if INA microorganisms are to be used for pest control is to effectively deliver INA microorganisms to the target insect. We envision the following scenario in using INA bacteria for the biological control of Colorado potato beetles. In late summer and early fall, shortly before entry into diapause, a high proportion of the adults disperse from the crop to seek overwintering sites in protected areas, often in close proximity to the crop (Milner et al., 1992; Weber and Ferro, 1993). During dispersal, adults typically congregate in large numbers on surviving pockets of crop foliage where feeding continues until vines are defoliated. Since the vines in virtually all fall potato fields are artificially killed 2-3 weeks prior to harvest, adult beetles migrate in large numbers to areas of fields where vine desiccants are not used (Wyman et al., 1994).

We propose to use unkilled strips of potato vines on field edges as trap crops where INA bacteria can be delivered. Trap crops will be sprayed with suspensions of INA bacteria that will be ingested by the beetles as they feed on the leaves. Shortly after feeding, beetles burrow into the soil to overwinter. Retention of the ingested INA bacteria will reduce cold tolerance, and, thus, increase mortality when soil temperatures decrease to subzero temperatures in midwinter.

4.2. Environmental persistence

Before any large-scale applications of ice-nucleating *P. fluorescens* are conducted, studies on environmental contamination and possible non-target effects need to be considered. How long do INA bacteria persist in the environment? Possible contamination of potato tubers harvested several days after bacterial application and/or possible soil accumulation leading to contamination of crops and tubers the following year are issues of concern that need to be addressed. Although *P. fluorescens* F26-4C is a gut-derived strain and may not survive outside of its overwintering host, sampling of the soil in the trap crop area and in adjacent sites, along with harvested crops should determine its persistence in an environment outside of the insect gut.

4.3. Non-target effects on beneficial insects

Because INA bacteria will be applied to potato foliage where beetles will actively feed, possible contamination of other insects, specifically parasitoids and predators, is a major concern. Do other insects become contaminated with INA bacteria applied in the field? If so, do they acquire sufficient numbers of bacteria to increase their SCPs? Are any such effects retained during the following weeks or into the winter? Although our previous experience (Lee et al., 1996) with the regulation of supercooling and ice nucleation in insects suggests that casual contact with ice-nucleating agents is unlikely to affect the cold tolerance of non-target insect species found in potato fields, further studies are needed.

The use of ice-nucleating bacteria as biological control agents is based on their potential for reducing cold hardiness in freeze-intolerant insects. In the Colorado potato beetle, a pest that has expanded its habitat range from Mexico to southern Canada (Boiteau and Coleman, 1986) through its burrowing behavior and ability to supercool, the use of these bacteria may provide an additional means of control in regions with cold winter conditions. The feasibility of this control method is further enhanced with the identification of INA strains that are efficacious in elevating beetle SCPs and that persist in the gut of overwintering adults, and with the development of growth media for bacterial production.

ACKNOWLEDGMENTS

This research was supported by grant No. 96-35302-3419 from the Cooperative State Research Education and Extension Service, United States Department of Agriculture and grant No. IBN-9728573 from the National Science Foundation.

REFERENCES

Armstrong, J.S., Lee, R.E., Peairs, F.B., 1998. Application of ice-nucleating active bacteria decreases the supercooling capacity of the Russian wheat aphid (Homoptera: Aphididae). In: Quisenberry, S.S., Peairs, F.B. (Eds.), Response Model for an Introduced Pest-the Russian Wheat Aphid. Entomological Society of America, Maryland, pp. 248-257.

Boiteau, G., Coleman, W., 1996. Cold tolerance in the Colorado potato beetle, *Leptinotarsa decemlineata* (Say) (Coleoptera: Chrysomelidae). Canadian Entomologists 128, 1087-1099.

Cannon, R.J.C., Block, W., 1988. Cold tolerance of microarthropods. Biological Review 63, 23-77.

Casagrande, R.A., 1987. The Colorado potato beetle: 125 years of mismanagement. Bulletin of the Entomological Society of America 33, 142-150.

Castrillo, L.A., Lee, R.E., Lee, M.R., Rutherford, S.T., 2000a. Identification of ice-nucleating active *Pseudomonas fluorescens* strains for biological control of overwintering Colorado potato beetles. Journal of Economic Entomology 93, 226-233.

Castrillo, L.A., Lee, R.E., Lee, M.R., Wyman, J.A., 2000b. Long-term retention of ice-nucleating *Pseudomonas fluorescens* by overwintering Colorado potato beetles. Cryo-Letters 21, 5-12.

Costanzo, J.P., Moore, J.B., Lee, R.E., Kaufman, P.E., Wyman, J.A., 1997. Influence of soil hydric parameters on the winter cold hardiness of a burrowing beetle, *Leptinotarsa decemlineata* (Say). Journal of Comparative Physiology B 167, 169-176.

Costanzo, J.P., Humphreys, T.L., Lee, R.E., Moore, J.B., Lee M.R., Wyman, J.A., 1998. Long-term reduction of cold hardiness following ingestion of ice-nucleating bacteria in the Colorado potato beetle, *Leptinotarsa decemlineata*. Journal of Insect Physiology 44, 1173-1180.

Duman, J.G., Olsen, T.M., Yeung, K.L., Jerva, F., 1995. The roles of ice nucleators in cold tolerant invertebrates. In: Lee, R.E, Warren, G.J., Gusta, L.V. (Eds.), Biological Ice Nucleation and its Applications. APS Press, St. Paul, pp. 201-220.

Fall, A., Fall., R., 1998. High level expression of ice nuclei in *Erwinia herbicola* is induced by phosphate starvation and low temperature. Current Microbiology 36, 370-376.

Fields, P.G., 1992. The control of stored-product insects and mites with extreme temperatures. Journal of Stored Products Research 28, 89-118.

Fields, P.G. 1993., Reduction of cold tolerance of stored-product insects by ice-nucleating-active bacteria. Environmental Entomology 22, 470-476.

Fields, P., Pouleur, S., Richard, C., 1995. The effect of high temperature storage on the capacity of an ice-nucleating-active bacterium and fungus to reduce insect cold-tolerance. The Canadian Entomologist 127, 33-40.

Gurian-Sherman, D., Lindow, S. E., 1993. Bacterial ice nucleation: significance and molecular basis. FASEB Journal 7, 1338-1343.

Hare, J.D., 1990. Ecology and management of the Colorado potato beetle. Annual Review of Entomology 35, 81-100.

Kaneko, J.T., Yoshida, T., Owada, T., Kita, K., Tanno, T., 1991. *Erwinia herbicola*: Ice nucleation active bacteria isolated from diamondback moth, *Plutella xylostella* L. pupae. Japanese Journal of Applied Entomology and Zoology 35, 247-251.

Kung, K-J.S., Milner, M., Wyman, J.A., Feldman, J., Nordheim, E., 1992. Survival of Colorado potato beetle (Coleoptera: Chrysomelidae) after exposure to subzero thermal shocks during diapause. Journal of Economic Entomology 85, 1695-1700.

Lee, M.R, Lee, R.E., Strong-Gunderson, J.M., Minges, S.R., 1995. Isolation of ice-nucleating bacteria from the freeze-tolerant frog *Rana sylvatica*. Cryobiology 32, 358-365.

Lee, R.E., 1991. Principles of insect low temperature tolerance. In: Lee, R.E. , Denlinger, D.L. (Eds.), Insects at Low Temperature. Chapman and Hall, New York. pp. 17-46.

Lee, R.E. and J.P. Costanzo. 1998. Biological ice nucleation and ice distribution in cold-hardy ectothermic animals. Annual Review of Physiology 60, 55-72.

Lee, R.E., Costanzo, J.P., Kaufman, P.E., Lee, M.R., Wyman, J.A., 1994. Ice-nucleating active bacteria reduce the cold-hardiness of the freeze-intolerant

Colorado potato beetle (Coleoptera: Chrysomelidae). Journal of Economic Entomology 87, 377-381.

Lee, R.E., Costanzo, J.P., Lee, M.R., 1998. Reducing the cold-hardiness of insect pests using ice-nucleating active microbes. In:. Hallman, G.J., Denlinger, D.L. (Eds.), Temperature Sensitivity in Insects and Application in Integrated Pest Management. Westview Press, Colorado, pp. 97-124.

Lee, R.E., Costanzo, J.P., Mugnano, J.A., 1996. Regulation of supercooling and ice nucleation in insects. European Journal of Entomology 93, 405-418.

Lee, R.E., Litzgus, J.D., Mugnano, J.A., Lee, M.R., Horton, D. R., Dunley, J., 1999. Reduction of supercooling capacity and cold-hardiness of the winterform of pear psylla *Cacopsylla pyricola* (Homoptera: Psyllidae) using ice-nucleating microorganisms. The Canadian Entomologist 131, 715-723.

Lee, R.E., Strong-Gunderson, J.M., Lee, M.R., Davidson, E.C., 1992. Ice-nucleating bacteria decrease the cold-hardiness of stored grain insects. Journal of Economic Entomology 58, 371-374.

Lee, R.E., Strong-Gunderson, J.M., Lee, M.R., Grove, K.S., Riga, T.J., 1991. Isolation of ice-nucleating active bacteria from insects. Journal of Experimental Zoology 257, 124-127.

Lindow, S.E., Arny, D.C., Upper, C.D., 1978. Distribution of ice nucleation-active bacteria on plants in nature. Applied and Environmental Microbiology 36, 831-838.

Lindow, S. E., Hirano, S. S., Barchet, W. R., Arny, D. C., Upper C. D. 1982. Relationship between ice-nucleation frequency of bacteria and frost injury. Plant Physiology 70, 1090-1093.

Maki, L.R., Galyan, E.L., Chang-Chien, M-M., Caldwell, D.R., 1974. Ice nucleation induced by *Pseudomonas syringae*. Applied Microbiology 28, 456-459.

Milner, M., Kung, K-J.S., Wyman, J.A., Feldman, J., Nordheim, E., 1992. Enhancing overwintering mortality of Colorado potato beetle (Coleoptera: Chrysomelidae) by manipulating the temperature of its habitat. Journal of Economic Entomology 85,1701-1708.

Minder, I.F., 1966. Hibernation conditions and survival rate of Colorado potato beetle in different types of soils. In: Arnold, K.V. (Ed.), Ecology and Physiology of Diapause in the Colorado Potato Beetles. Indian National Scientific Documentation Center, New Delhi. pp. 28-58.

Nemecek-Marshall, M., LaDuca, R., Fall, R., 1993. High level expression of ice nuclei in a *Pseudomonas syringae* strain is induced by nutrient limitation at low temperature. Journal of Bacteriology 175, 4062-4070

Olsen, T.M., Duman, J. G., 1997. Maintenance of the supercooled state in the gut of overwintering pyrochroid beetle larvae, *Dendroides canadensis*: role of ice nucleators and antifreeze proteins. Journal of Comparative Physiology B 167, 114-122.

Salt, R.W., 1968. Location and quantitative aspects of ice nucleators in insects. Canadian Journal of Zoology 46, 329-333.

Shields, E.J., Wyman, J.A., 1984. Effect of defoliation at specific growth stages of potato yields. Journal of Economic Entomology 7, 1194-199.

Sømme, L., 1982. Supercooling and winter survival in terrestrial arthropods. Comparative Biochemistry and Physiology 73A, 519-543.

Strong-Gunderson, J.M., Lee, R.E., Lee, M.R., Riga, T.J., 1990. Ingestion of ice-nucleating active bacteria increases the supercooling point of the lady beetle *Hippodamia convergens*. Journal of Insect Physiology 36, 153-157.

Takahashi, K., Watanabe, K., Sato, M., 1995. Survival and characteristics of ice nucleation-active bacteria on mulberry trees (*Morus* spp) and in mulberry pyralid (*Glyphodes pyloalis*). Annals of the Phytopathology Society of Japan 61, 439-443.

Ushatinskaya, R.S., 1978. Seasonal migration of adult *Leptinotarsa decemlineata* (Insecta, Coleoptera) in different types of soil and physiological variations of individuals in hibernating populations. Pedobiology 18, 120-126.

Warren, G. J. 1995., Identification and analysis of ina genes and proteins. In: Lee, R.E, Warren, G.J. and Gusta, L.V. (Eds.), Biological Ice Nucleation and its Application. APS Press, St. Paul, pp. 85-99.

Watanabe, K., Abe, K., Sato, M., 2000. Biological control of an insect pest by gut colonizing *Enterobacter cloacae* transformed with ice-nucleating gene. Journal of Applied Microbiology 88, 90-97.

Weber, D.C., Ferro, D.N., 1993. Distribution of overwintering Colorado potato beetle in and near Massachusetts potato fields. Entomologia Experimentalis et Applicata 66, 191-196.

Wolber, P., Warren, G., 1989. Bacterial ice-nucleation proteins. Trends in Biochemical Science 14, 179-182.

Wyman, J.A., Feldman, J., Kung, S.K., 1994. Cultural control of Colorado potato beetle: off-crop management. In: Zehnder, G.W., Powelson, M.L., Jansson, R.K., Raman, K.V. (Eds.), Advances in Potato Pest Biology and Management. APS Press, St. Paul. pp. 376-385.

Yankofsky, S.A., Levin Z., Bertold, T., Sandlerman, N., 1981. Some basic characteristics of bacterial freezing nuclei. Journal of Applied Meterology 20, 1013-1019.

Zachariassen, K.E., 1992. Ice nucleating agents in cold-hardy insects. In: Somero, G.N., C.B. Osmond, and C.L. Bolis (Eds.), Water and Life. Springer-Verlag, Berlin, pp. 261-281.

Species Index

Subject Index